Health Informatics and Technological Solutions for Coronavirus (COVID-19)

Health Informatics and Technological Solutions for Coronavirus (COVID-19)

Edited by
Suman Lata Tripathi, Kanav Dhir, Deepika Ghai,
and Shashikant Patil

CRC Press
Taylor & Francis Group
Boca Raton London New York

CRC Press is an imprint of the
Taylor & Francis Group, an **informa** business

First edition published 2021
by CRC Press
6000 Broken Sound Parkway NW, Suite 300, Boca Raton, FL 33487-2742

and by CRC Press
2 Park Square, Milton Park, Abingdon, Oxon, OX14 4RN

First edition published by CRC Press 2021

CRC Press is an imprint of Taylor & Francis Group, LLC

ISBN: 9780367704179 (hbk)
ISBN: 9780367751210 (pbk)
ISBN: 9781003161066 (ebk)

Typeset in Times
by codeMantra

Contents

Preface

Technology and Science has enthused forward, too. We have perceived an added intricate depiction of COVID-19 and new scientific patterns; the foremost data from serum trials; first results from randomized meticulous drug studies; cheering publications on monoclonal neutralizing antibodies and serological indication about the number of people who have come into interaction with SARS-CoV-2. Regrettably, we have also witnessed the initial science scandal with fake data published. And we face new tasks like long-term effects of COVID-19 and a syndrome in children. For moderately some time, prevention will endure to be the prime pillar of epidemic control. In forthcoming waves of the SARS-CoV-2 epidemic, we will emphasize on the circumstances under which SARS-CoV-2 is communicated: swarming, closed (and noisy) places and spaces. However sickbays are not noisy; they are jam-packed and pad-locked, and the combat against the new coronavirus will be decided at the very epicenter of our healthcare system. Over the subsequent months and maybe years, one of all of our top primacies will be to give all healthcare workforces and patients perfect personal protective equipment.

We should be certain that the present state needs a new type of reference book. Civilization is antagonizing a mysterious and intimidating disease which is often unadorned and fatal. Healthcare systems are speechless. There is no demonstrated treatment and serums will not be available soon. Such a state of affairs has not occurred since the flu epidemic in 1918.

We have confidence in a clear head is vital in eras of over-information, with lots of methodical articles and manuscripts published *every day*, newscast about hundreds of studies being prearranged or previously on the mode and social media amalgamation hard data with stories and fake news. The deadly work of airing the technical works and the technical data has to be done—repeatedly and continually, like a Swiss watch.

The book is addressed primarily to the Health and Biomedical Professionals and Biomedical Engineers who are in the field of Engineering, Biology, Medicine and Pharmaceutical Sciences but it is also intended to provide guidance to the researchers, clinicians, physicians, hospitals, policy-makers, instrument manufactures and vendors as well as students, academicians and universities too. As far as possible, the information is presented in a form readily accessible to interested engineering professionals, researchers and multidisciplinary fields and medical professionals; other sources of supplementary materials are presented and vital information is specified as concisely as possible.

The book is envisioned to give direction on the full range of significant health issues associated with COVID-19 pandemic disease. The roles of the health occupation, the pharmaceutical industry and diagnosis methods and doctors' role in treating patients and patients' role in avoiding COVID-19 problems are documented. The endorsements address the well-being risks associated with different types of treatments spreading analysis, risk analysis as well as diagnosing and treatment methods. Although there is a suggestion that more educated people take suitable preclusion measures or receive proper treatment, recent cases deserve certain attention because they are at a higher risk of suffering certain health problems.

In this edition, chapters and case studies have been added. The wide-reaching spreading of the foremost communicable disease is shown in swotted, more detailed and precise stuffs. The core communicable diseases that pose latent health intimidations for people have been illustrated in a lucid manners as well as corresponding preventive measures and information on several factors that may have adverse effects on the health of common and well-being.

CHAPTER ORGANIZATION

This book is organized into 23 chapters in total.

Chapter 1 describes the analysis of post COVID-19 sentiments clamouring for "vocal for local" including the reduce imports dependence and embark upon import substitution

Chapter 2 presents AI-based model that has the ability to perform tasks related to speech recognition, facial recognition, and translation of various languages, decision-making and improved risk stratification.

Chapter 3 summarizes recent developments in work on COVID-19 epidemiology, pathogenesis and treatment, and the review of scientific and technological advancements in the fight against the outbreak of modern coronavirus.

Chapter 4 gives the understanding about the genetics of coronaviruses that includes the scientific advancement and research to combat the novel pandemic COVID-19.

Chapter 5 describes presumable strategies to combat the COVID-19 deleterious effects and also the role of artificial intelligence and nanotechnology toward analysis for the possible solutions.

Chapter 6 focuses on the early stage identification of COVID-19 in a prescribed group and the identified patients are handled in the different methods so as to avoid the major spread among the region of any country.

Chapter 7 deals with the different strategies and preventive measures to cope up with such a pandemic situation.

Chapter 8 describes the strategies for prevention of corona virus disease including the detailed description of different identified corona viruses.

Chapter 9 attempts to explore the impact of coronavirus in our day to day lives with regard to works, public places, use of transportation, social gatherings, etc.

Chapter 10 explains the position that IoT, machine intelligence, deep learning, Smart Sensors to see the possibilities in corona virus detection and analysis.

Chapter 11 provides the details about the steps to prevent corona viruses and the analysis of corona infection cases in Tamil Nadu up to June 23, 2020.

Chapter 12 deals with impact of COVID-19 on consumer and also the latest developments of Digital Marketing in Indonesia.

Chapter 13 elaborates the virus-related curve of symptomatic disease peaks around May including the pathogenic features, symptoms, protections, treatment statics.

Chapter 14 provides a case study for the prediction of COVID-19 in India using machine learning models.

Chapter 15 deliberates on the utilities of food habits in the Indian context in contrast to the foreign land people.

Chapter 16 gives background about corona virus and elaborates a prototype system developed for the detection of corona virus affected human body.

Chapter 17 provides information about accessing the correct data regarding symptoms of disease, factors responsible for occurrence and outbreak of the disease, and corrective measures any individual can adopt appropriate steps to avoid and combat with the disease.

Chapter 18 explores progress of COVID-19 epidemic in India including safety and corrective measures to control the virus effect.

Chapter 19 proposes and describes a neutrosophic-based edge detection model for COVID-19 CT images.

Chapter 20 is related to an application to isolate all suspected and confirmed cases based on the questionnaire of publicly available information that is conducted for a better understanding of person symptoms.

Chapter 21 concludes the integration of biomedical depictions that has manifested to be functional for the advancement of the clinical dependence of usage of biomedical imaging for analysis and diagnosis.

Chapter 22 narrates the various applications of artificial intelligence (AI) based tools to detect, diagnose, monitor, and assist treatment of the novel corona virus disease.

Chapter 23 documented and analyzed the practices, technologies, and innovations in the Philippines' response to the COVID-19 pandemic from January to May 2020.

Chapter 12 deals with impact of COVID-19 on [illegible] and 13, the impact of [illegible] on Digital Marketing to Millennials.

Chapter 14 elaborates the structural [illegible] aspects [illegible] including the pathogenic features, mutations, pathogenesis, treatment, [illegible]

Chapter 15 provides a case study for the prediction of COVID-19 [illegible] in [illegible] learning techniques.

Chapter 16 [illegible] the statistics of [illegible] the [illegible] people.

Chapter 16 [illegible] and [illegible] developed for the [illegible] in human body.

Chapter 17 provides information about [illegible] of disease, factors responsible for [illegible] and outbreak of the disease and [illegible] with the disease.

Chapter 18 explores [illegible] of COVID-19 [illegible] safety and [illegible] effective [illegible] of the virus spread.

Chapter 19 proposes and describes [illegible] based edge detection model for COVID-19 [illegible] images.

Chapter 20 is related to [illegible] and confirmed cases [illegible] publicly available [illegible] using [illegible]

Chapter 21 [illegible] the [illegible] of [illegible]

Chapter 2[illegible] [illegible]

Chapter 2[illegible] [illegible] COVID-19 [illegible]

Editors

Dr. Suman Lata Tripathi has completed her Ph.D. in the area of microelectronics and VLSI from MNNIT, Allahabad. She did her M.Tech in Electronics Engineering from UP Technical University, Lucknow and B.Tech in Electrical Engineering from Purvanchal University, Jaunpur. She is associated with Lovely Professional University as a Professor with more than seventeen years of experience in academics. She has published more than 45 research papers in refereed journals and conferences. She has organized several workshops, summer internships, and expert lectures for students. She has worked as a session chair, conference steering committee member, editorial board member, and reviewer in international/national IEEE Journal and conferences. She has been nominated for the **Research Excellence Award** in 2019 at Lovely professional university. She had received the best paper at IEEE ICICS-2018. She has published an edited book titled "Recent advancement in Electronic Device, Circuit and Materials" by **Nova science publishers**. She has also submitted edited books, "Advanced VLSI Design and Testability Issues" and "Electronic Devices and Circuit Design Challenges for IoT application" in **CRC Taylor & Francis and Apple Academic Press**. She is also associated as an editor of book Series on "Green Energy: Fundamentals, Concepts, and Applications," to be published by **Scrivener Publishing, Wiley**.

Her area of expertise includes microelectronics device modeling and characterization, low power VLSI circuit design, VLSI design of testing, and advance FET design for IoT, Embedded System Design and biomedical applications, etc.

Dr. Kanav Dhir has completed his Ph.D. in the area of Inorganic and Medicinal Chemistry from Punjab Engineering College (Deemed to be University), Chandigarh. He did his M.Sc. in Chemistry from DAV College, Chandigarh and B.Sc. from A.S. College, Khanna. He is associated with DAV College, Chandigarh as an Assistant Professor with more than ten years of experience in academics. He has published more than 19 research papers in refereed journals and conferences and two book chapters. He has actively participated and conducted workshops, seminar, and summer training programs. He has worked as a session chair, conference steering committee member, editorial board member, and reviewer in international/national Journal and conferences. His area of expertise includes Inorganic Chemistry and Bioinorganic Chemistry. Currently, he is guiding M. Sc. and Ph.D. scholars in the area of Bioinorganic and Medical Chemistry.

Dr. Deepika Ghai has completed her Ph.D. in the area of Signal and Image Processing from Punjab Engineering College (Deemed to be University), Chandigarh. She did her M.Tech in VLSI Design & CAD from Thapar University, Patiala and B.Tech in Electronics and Communications Engineering from Rayat Institute of Engineering and Technology, Ropar. She is associated with Lovely Professional University as an Assistant Professor with more than eight years of experience in academics. She has published more than 25 research papers/chapters/proceedings in refereed journals/books/conferences. She has worked as a session chair, conference steering committee member, editorial board member, and reviewer in international/national IEEE Journal and conferences. Her area of expertise includes Signal and Image Processing and VLSI Signal Processing. She has actively participated and conducted in several workshops, seminars and real-time projects. Currently, she is guiding M. Tech. and Ph.D. scholars in the area of signal and image processing.

Prof. Shashikant Patil is a Senior ACM Member, Senior Member IEEE as well as Fellow Member of Institution of Engineers, India (FIE); Fellow Member of IETE India (FIETE); and Fellow Member of IRED(UK) FIRED. He is having over 20 years of teaching experience and a recipient of Best Researcher Award 2014 of SVKMs NMIMS Shirpur. In addition to this, he had been nominated as IEEE Day 2014 Section Ambassador for region 10. Initially, he had been selected as a Member Executive Committee on the IEEE CSI India Council as well as Core Team Member of Social Media and Online Content Management Team, Visibility Committee of IEEE SPS Society at the International Level. He has also volunteered as a Regional Lead Ambassador for Region 10 for IEEE Day 2015 and IEEE Day 2016 event. He is a member of IEEE RFID Technical Council and SIG Member of IoT. His research interests are Signal Processing and Imaging. He is also serving as an Associate Editor on the IEEE RFID Journal; IEEE SDN Journal; IEEE RFID Steering Committee and EiC for IEEE C-RFID Newsletter. He has also rendered his services as a Member of Jury for the IEEE SIGHT (AIYEHUM) 2016 Competition at Asia Pacific level for R10. Recently he shouldered the responsibility of a Student Travel Grant Chair for the IEEE DySPAN 2017 USA Conference. As a Technical Program

Committee Member (TPC) he has served more than 250 International Conferences of IEEE and other organizations. He is having more than 51 publications at national and international level in various conferences and Journals. He is also IEEE Designated Reviewer and TPC Member for ICASSP; MMSP; ICME and more than 300 Conferences; Reviewer for Elsevier Springer and Taylor and Francis and more than 50 Journals; Editor-in-Chief IEEE C-RFID Newsletter; Editorial Member IEEE SDN; IEEE Day Lead Ambassador Region 10 for 2015 and 2016; IEEE SPS SIG IoT Member; IEEE SPS Visibility Committee Member; EDGE Program Reviewer; IEEE CSI All India Council Secretary and many more. Recently, he is designated as Chartered Engineer by Institution of Engineers, India (CEng.), for his contributions to the Education and Professional Field.

Committee Member, TPC. He has served more than 250 International Conferences of IEEE and other organizations. He is having more than 51 publications at national and international level in various conferences and Journals. He is also IEEE Designated Speaker and TPC Member for ICASSP, MWSCAS, ICME and more than 200 Conferences; Reviewer for Elsevier, Springer and Taylor and Francis, and more than 50 Journals; Editor-in-Chief IEEE CRED Newsletter; Editorial Member IEEE SSCS; IEEE SPS Lead Ambassador, Region 10 for 2015 and 2016; IEEE SPS SC Member, IEEE SPS [illegible] Committee Member; IEEE [illegible] Forum, [illegible] All India Council Secretary; and many more. Recently, he is delighted as Chartered Engineer by Institution of Engineers, India (Chapter) for his contribution to the Education and Profession IEEE.

Contributors

Harshit Agarwal
Department of Physics, Institute of Science
Banaras Hindu University
Varanasi, India

Shayan Ahmed
Department of Biosciences, Faculty of Natural Sciences
Jamia Millia Islamia
New Delhi

Ajay Kumar H.
Department of Computer Science and Engineering
Mar Ephraem College of Engineering and Technology
Elavuvilai, India

C. Andy Jason
Research Scholar, ECE
Sreyas Institute of Engineering & Technology
Hyderabad, India

Krishan Arora
Assistant Professor and Head, School of Electronics and Electrical Engineering
Lovely Professional University
Phagwara, India

M. N. Arumugham
Department of Chemistry
Thiruvalluvar University
Vellore, India

Shakti Bhushan
Department of Rasa Shastra, Faculty of Ayurveda, Institute of Medical Sciences
Banaras Hindu University
Varanasi, India

G. Boopathi Raja
Department of Electronics and Communication Engineering
Velalar College of Engineering and Technology
Erode, India

Miriam Caryl Carada
Institute for Governance and Rural Development
University of the Philippines
Los Baños, Laguna, Philippines

Ginbert Permejo Cuaton
Research Postgraduate Student
The Hong Kong University of Science and Technology
Clear Water Bay, Hong Kong

N.S. Damle
Department of Electronics Engineering
Shri Ramdeobaba College of Engineering & Management
Gittikhadan, Nagpur, India

V. Dhinakaran
Centre for Applied Research
Chennai Institute of Technology
Chennai, India

A. Dinesh Karthik
Department of Biochemistry
Shanmuga Industries Arts and Science College
Tiruvannamalai, India

Avinash Dubey
Department of Mechanical Engineering
GLA University
Mathura, India

Souvik Ganguli
Department of Electrical and Instrumentation Engineering
Thapar Institute of Engineering and Technology
Patiala, India

V. Gowrishankar
Department of Electronics and Communication Engineering
Velalar College of Engineering and Technology
Erode, India

Neha Gupta
Department of Biosciences
Faculty of Natural Sciences, Jamia Millia Islamia
New Delhi, India

Balazs Gulyas
Lee Kong Chian School of Medicine
Nanyang Technological University
Nanyang, Singapore

Nupur Gupta
Department of Physics
Guru Nanak Dev University
Amritsar, India

B.M. Hardas
Department of Electronics Engineering
Shri Ramdeobaba College of Engineering & Management
Gittikhadan, Nagpur, India

Saumyadip Hazra
Department of Electrical and Instrumentation Engineering
Thapar Institute of Engineering and Technology
Patiala, India

Tushar H. Jaware
R C Patel Institute of Technology
Shirpur, India

L. R. Jonisha Miriam
Department of Computer Science and Engineering
Mar Ephraem College of Engineering and Technology
Elavuvilai, India

Rahul Koshti
Department of Electronics & Telecommunication Engineering
SVKMs NMIMS, MPSTME, MPTP
Shirpur, India

Abhimanyu Kumar
Department of Electrical & Instrumentation Engineering
Thapar Institute of Engineering and Technology
Patiala, India

Ashwini Kumar
Ramanujan College
University of Delhi
New Delhi, India

Gaurav Kumar
Department of Mechanical Engineering
Vidya College of Engineering
Meerut, India

Sandeep Kumar
Professor, ECE
Sreyas Institute of Engineering & Technology
Hyderabad, India

S. N. Kumar
Department of Electrical and Electronics Engineering
Amal Jyothi College of Engineering
Kanjirapally, India

A Lenin Fred
Department of Computer Science and Engineering
Mar Ephraem College of Engineering and Technology
Elavuvilai, India

Vandana Meena
Department of Rasa Shastra, Faculty of Ayurveda, Institute of Medical Sciences
Banaras Hindu University
Varanasi, India

Amit Mishra
E & TC Department
SIEM Sandip Foundation
Nashik, India

Smita Nirkhi
GHRIET
Nagpur, India

Mahak Nischal
Department of Physics, Amity Institute of Applied Sciences
Amity University
Noida, India

S. Nithya
Department of Electrical and Electronics Engineering
SRMIST
Chennai, India

Parasuraman Padmanabhan
Lee Kong Chian School of Medicine
Nanyang Technological University
Nanyang, Singapore

Dipak P. Patil
E & TC Department
SIEM Sandip Foundation
Nashik, India

Pooja
Department of Instrumentation and Control Engineering
DR BR Ambedkar National Institute of Technology
Jalandhar, India

Pooja Pant
Department of Physics, Institute of Science
Banaras Hindu University
Varanasi, India

M. Parimaladevi
Department of Electronics and Communication Engineering
Velalar College of Engineering and Technology
Erode, India

Pooja Pathak
Department of Mathematics
GLA University
Mathura, India

Poongothai A.
PG and Research Department of Biochemistry
Sacred Heart College (Autonomous)
Tirupattur, India

Seprianti Eka Putri
Department Management Faculty Economics
Business University of Bengkulu
Bengkulu, Indonesia

S. Saiteja
Research Scholar, ECE
Sreyas Institute of Engineering & Technology
Hyderabad, India

T. Sathya
Department of Electronics and Communication Engineering
Velalar College of Engineering and Technology
Erode, India

Aditi Saxena
Department of Physics, Amity Institute of Applied Sciences
Amity University
Noida, India

Meenakashi Sharma
Department of Computer Science Engineering
Global Institute of Management and Emerging Technologies

Yash Srivastava
GLA University
Mathura, India

Satyendra Pratap Singh
Department of Physics, Amity Institute of Applied Sciences
Amity University
Noida, India

A. Sivasankari
PG and Research Department of Computer Science
Shanmuga Industries Arts and Science College
Tiruvannamalai, India

Shalini Soman
Department of Physics, Amity Institute of Applied Sciences
Amity University
Noida, India

V. Shalini
Department of Chemistry
Thiruvalluvar University
Vellore, India

R. Subasri
Department of Biochemistry
Shanmuga Industries Arts and Science College
Tiruvannamalai, India

M. Sudha
Department of Biochemistry
Shanmuga Industries Arts and Science College
Tiruvannamalai, India

Surbhi
Ramanujan College
University of Delhi
New Delhi, India

R. Surendran
Centre for Applied Research
Chennai Institute of Technology
Chennai, India

Manish Tongo
Economics and Management
Priyadarshini College of Engineering
Nagpur, India

Suman Lata Tripathi
School of Electronics and Engineering
Lovely Professional University
Phagwara, India

M. Varsha Shree
Centre for Applied Research
Chennai Institute of Technology
Chennai, India

Karan Veer
Department of Instrumentation and Control Engineering
DR BR Ambedkar National Institute of Technology
Jalandhar, India

K. Vijayakumar
Chief Librarian, Department of Library
Shanmuga Industries Arts and Science College
Tiruvannamalai, India

Sahil Virk
Department of Testing and Customer Support
Mentor Graphics Pvt. Ltd.
Noida, India

Pravin Wararkar
Department of Electronics & Telecommunication Engineering
SVKMs NMIMS, MPSTME, MPTP
Shirpur, India

Nirupuma Yadav
Ramanujan College
University of Delhi
New Delhi, India

1 Corona Virus Disease (COVID-19) Pandemic
Globalization to Localization

Manish Tongo
Priyadarshini College of Engineering

Smita Nirkhi
GHRIET

CONTENTS

1.1 OVERVIEW OF GLOBALIZATION

Integration across geographical boundaries has been an insatiable quest of human being for different altruistic purposes. The connect between the nations existed informally for the purposes of trade and cooperation. The trade has always been the most instrumental and effective vehicle for such integration among the different countries across the geographical boundaries.

The geographical and economic history is full of evidences which convince us of the fact that an appetite for globalization existed since the time immemorial. Craving for multimodal exchanges has fostered the globalization over the period of time. Through exchange of products, ideas, overtly or covertly, we find that other aspects of human life like philosophy, religion, language, and arts have started exerting influence and impression over others [1].

The footprints of globalization could be traced backed to the 15th and 16th century. In the 19th century, it toed along with technology exchange, transport and sophisticated infrastructure [2].

1.2 ETYMOLOGY AND USAGE

The term globalization refers to the autonomous surfacing of social and economic systems which aims for international cooperation. Social, economic and management scientist have given different expressions to integration of material, human and intelligence across the nations. It is a fallacy to understand that the globalization is an international phenomenon of speeding and widening the connections across the geographical boundaries. Globalization within localization and localization within globalization have later become the approach styles for many proponent of globalization. In 2000, the International Monetary Fund (IMF) gave vivid expressions to different building blocks of globalization [3].

Today on the threshold of the 21st century, we realize the world is effectively revolving around these four fundamental blocks of globalization as expressed by IMF [4].

1.3 MODERN GLOBALIZATION

Modern globalization was an era of transportation network, corridors for trade facilitation and intentional and deliberate attempts to reduce the cost of logistics and transaction cost of conducting the business between the two countries. Importantly the modern globalization has essentially evolved and shaped by imperialism engineered by the nation of shopkeepers. Imperialism in Africa and Asia is testimony to this fact.

The wave and the rise of globalization received a setback during the First World War. It was proved that political animosity and undeclared economic embargo can never go hand in hand. For healthy and undisrupted globalization, we need healthy cooperation among the nations of the world [5].

The Breton Woods conference was a step in a right direction which provided meaning and order to the concept of globalization. It was a political steering for economic cooperation which hinged around the philosophy of comparative advantage. Modern technology and modern science with the state of the infrastructure in developed countries of the world along with patronage from international agencies like World Bank and IMF gave much needed boost to the wave of globalization.

Initially, the General Agreement on Tariffs and Trade (GATT) accelerated the process of removing the gray areas by dismantling the trade restrictions so as to provide level-playing field to all participants. Excessive protectionism and globalization are the anomalies they cannot go hand in hand. Hence the birth of World Trade Organization took place to facilitate orderly free trade with reasonable degree of protectionism to safeguard the interest of less developed countries of the world [6].

The World Trade Organization laid out the carpet in the form of a formal institution to facilitate international trade. However, the Doha round was a stigma for globalization which receded the global partnership to a great extent though transitory.

A final nail was put in the coffin of GATT in the Doha round. With acute discordance and lack of consensus, the nations started exploring on their own that culminated into bilateral trading partnership on one to one basis. The purple patch of GATT got effaced from the international mindsets.

1.4 GATT: AN ENGINE OF GLOBALIZATION

However, it is pertinent to glance through the achievement canvass of GATT. It was indeed regarded as an engine of growth for global economies. The aim of GATT was to achieve "substantial reduction of tariffs and other trade barriers and the elimination of preferences, on a reciprocal and mutually advantageous basis." GATT took her birth from the ambers of International Trade Organization which was an international platform for negotiations for the governments of different nations.

The currency period of GATT was from the year 1958 till 1993, when it was replaced by the World Trade Organization in 1995. It is not that with eclipse of GATT all the relics are forgotten, some of the clauses and understanding of GATT have found their honorable entry into World Trade Organization (WTO). Some of the traces of GATT can still be found under WTO which is kept on the anvil from time to time for bringing about apt and time-tested changes [7].

As stated earlier, the World Trade Organization took her birth from the ambers of failure of Uruguay round. The success story of WTO has been phenomenal in ending the disputes and creating parity of justice in global economics and commerce. WTO has been massively successful in reducing the tariff walls and expanding the international trade. The dispute settlement and redress mechanism of WTO played an important role to bring the nations together on a common table for creating a win-win situation. The success story of WTO lies in the fact that many nations are joining the bandwagon of globalization today and WTO is holding the fort quite successfully.

The conflict of interest and resolving them amicably has been the founding principle of WTO. The nations eying for experts markets and nations who were fighting tooth and nail to protect the domestic agriculture market was a regular scene of discordance in so far as WTO is concerned.

Agriculture was a special subject which received special cognizance in both the arrangements. Agricultural sector was a bone of contention. The club of some countries called as "cairn group" wanted agricultural products to be brought under WTO for free trade. This group was not ready to sign WTO agreement in the absence of free trade of agriculture products. Hence over the years, it paved the way for inclusion of agricultural products into the ambit of WTO.

1.4.1 Economic Globalization

Economic globalization is a deep integration through value chain of trade in products and services as well as setting up of facilities across the world for creation of wealth. For fruitful economic cooperation, reduction in tariff walls and creation of congenial atmosphere for international trade have always been the hallmark of international policy.

China is perhaps the classic example who started exploiting the trade liberalization since 1980. Success story of China lies in optimum exploitation of foreign direct investment (FDI) as a route to invite foreign capital. China as a country has surpassed many nations in so far as market openness is concerned [8].

As a result, the stock of wealth in China is massive which is finding exit route through currency devaluation. China is known for global manufacturing hub and dumping king of the world. It all happens because China exploited comparative advantage of low-cost labor economy to the fullest. India is a low-cost labor country yet the growth rate of China and India makes a huge difference [9].

Economic globalization of India which took off in 1991 has started yielding the desired results in the second decade onward. It is said that due to active globalization of India, 300 million people could escape the extreme poverty in India. Indian products and services found the attractive international markets and it is today an attractive destination for business outsourcing from the developed economies.

1.5 POLITICAL GLOBALIZATION

With economic interest, the nations over the long run tend to develop geopolitical interest as well. The hunt of virgin markets for exports and competitive markets for cheap imports makes the case fit for competitive spirit in international trade [10]. The trade value ultimately determines the quality of nation's political relations with other nations. The relations get spoiled in the events of trade and economic embargos which often take place due to pursuance of nuclear programs, etc.

1.6 CULTURAL GLOBALIZATION

Naturally, the free movement of goods, services and manpower has direct as well as indirect bearing on the culture of the nation. Many hence argued and continue to fight against the philosophy of globalization because they feel that behind cultural pollution globalization is the main culprit [11]. The term globalization bound to alter to some extent the cultural practices. Certain aspects of the culture may get lost if degree of globalization is more or they may assume another form. Americanization and westernization are the common terms which are being used to express the cultural pollution being driven recklessly by globalization.

1.7 GLOBALIZATION AND GLOBAL WORK FORCE

In the context of globalization, the popular term coined as global village. The work force of the global village is soaring day by day giving birth to colloquial debate issue brain drain or brain gain. The labor is finding an attractive destination for appropriate reward. This way the factors of production are excited for freedom of choice. The movement of work force across the globe is the result of active globalization. The pool of international worker is soaring across the globe. By the year 2019, it already crossed the 5 billion mark [12].

Globalization has opened the flood gates for international intellectual capital to be parked in attractive destination. It has reduced the technological gap, investment gap, in the formation of capital and provided good deal of benchmarking values for every country. There is a competitive war between low-cost labor countries verses high-cost labor countries. The jobs of later are being siphoned off to low-cost labor countries.

1.8 CRITICISM OF GLOBALIZATION

It is true that the path of the globalization for any nation was not festooned. There has been acrid opposition to the concept of globalization from the viewpoints of environmental damage, financial terrorism, cultural pollution, crony capitalism, brain drain, dumping tactics, and geopolitical tensions to name a few.

Of late, the most negative consequence of globalization was manifested in utter abhorrence for migrant workers in America, Britain, and in some Asian countries to some extent. Similarly, the outsource model of business has irked the workers and politicians especially the tea party members in America. Talks were afoot to make immigration policy of America more stringent and pro-locals of this sort of views are dampening the free spirit of factors of production.

1.9 ANTIGLOBALIZATION MOVEMENT

Globalization did not fare well for many hearts and souls. The perception value of the globalization is bound to be different from people to people [13,14]. Naturally some political organizations started raising antiglobalization voices stating abuse of environment and cultural pollutions to name a few [15].

1.9.1 Opposition to Capital Market Integration: Anticapitalist Movements

Free play of capital, few rich and much poor, financial maneuverability, round tripping of capital, tax avoidance were some of the instrument al factors responsible for developing anger for financial globalization. Financial abuses and wild financial parking without any regard to the social equality were causing havoc in the minds of the people.

In light of the economic gap between rich and poor countries, movement adherents claim "free trade" without measures in place to protect the under-capitalized will contribute only to strengthening the power of industrialized nations.

1.10 JUSTIFIABLE WORLD ORDER ARGUMENT

The fair play protagonists and advocates of level-playing field continually seek the fair game rule of globalization. Such activists do not prefer any lopsided rules of the game as well as they oppose the harmful tactics of shifting the goal post according to the convenience of developed nations. In a rush of things, such people are being labeled as antiglobalists by the main stream media but there is a stampede of claims and counter-claims and such activists claim that they are in favor of fair globalization [16].

1.10.1 Anticonsumerism

Some groups of activists argue that growing consumerism in some societies does not augur well for the humanity. An epidemic of consumerism is spreading all over the world through the agent of globalization force. It in the process killing the spiritual power of renunciation and making people addicted to instant appeasement which often leads to a mad race for garnering the additional income which is never sufficient [17]. This kind of sabotage is unearthing plethora of problems in a chain reaction to the citizens of the globe.

1.10.2 Antiglobal Governance

Since World War 1 and World War 11, the global citizens had lost their faith in common umbrella of government. They started viewing the intentional exchange of labor with acute skepticism. Many economic institutions and world organizations were the strong notaries of global governance which was opposed tooth and nail by many [18].

1.10.3 Environmentalist Opposition

Perhaps the most hue and cries we hear in the context of globalization are about perennial environmental damages. Interestingly the proponents of globalization are today the advocates of planet protection. Efforts are on the anvil to mitigate the carbon footprints from the planet through collective responsibilities and accountability. Human life and ecological balance are getting disturbed because of extreme pro-wealth attitude of the world. Tsunami effects in the recent times are alarming signals to the mankind that enough is enough [19]. Various antiglobalization movements which took place in Berlin, Seattle, Japan, Washington have gone on record for acute animosity toward globalization.

1.10.4 Globalization of Indian Economy

Closed economy model which was pursued for a very long time did more harm than good to Indian economy. On the front of economic growth, economic development and human development index, the largest democracy of the world with largest population based lagged far behind.

Closed economy model presented India with multiplicity of problems like huge technological gap, yawning investment gap, poor industrialization, poverty and poor standard of living.

The ascribed concept of Hindu growth rate of 3°c per annum at maximum was like a biggest mental road block. Much water had already flown under the bridge economic situation of India was getting from bad to worse. External economic shocks, adverse balance of payment situation, depreciation in the value of rupee compelled Indian policy makers to get back to the drawing board again.

In the year 1991, government of India took a revolutionary decision to bid a good bye kiss to closed economy model and to adopt market-oriented open economy model. The first decade of globalization of Indian economy was cautious, of vacillations and skeptical.

The waves of internal liberalization echoed and reverberated in international arena as well. A nation develops an appetite for creating niche area or positioning in international market. Year 1991 heralded new beginning for Indian economy. However, many economists believe that the actual vestige of liberalization and globalization of Indian economy could be seen in 1980s. Later in the second half of 1980s, the late Prime Minister of India Mr. Rajiv Gandhi introduced what many call to it as creeping globalization the move that was not bold and confident enough. The official announcement came in the year 1991 which meant business for everybody in letter and spirit.

By now government of India nailed her color to the mast. The purple patch of liberalization and globalization started to be appearing on the horizon during the first decade of globalization which was a period of teething problem for Indian economy. Initially government of India initiated creeping measures of liberalization and globalization making its intention more than clear that open economy model shall prevail in India.

1.10.5 Global Corona Virus Disease of 2019: Policy Gyration on Globalization

The COVID-19 pandemic of late has triggered the sense of dire urgency for self-reliance among the nations. India as a nation has adopted open economic model in the year 1991. This model is popularly known as LPG model (Liberalization, Privatization and Globalization).

The root cause of shunning closed economy model and embracing open economy model was to undo the plethora of disadvantages surfaced due to closed economy model such as investment gap, technological gap, poor industrialization, massive unemployment and concomitant poverty. Today India is into the third decade of successful globalization era. India has got benefited handsomely and could mitigate the gaps in technology, investment, industrialization to name a few. During the era of globalization, India moved from localization to globalization. In other words it started producing products and services which address the needs of not only local markets of India but also that of overseas markets [20].

Post-COVID-19 sentiments are clamoring for "vocal for local." It essentially means reduce imports dependence and embark upon import substitution. International trade is an engine of economic growth for global economies. Exports led economic growth model calls for maximum exports and less imports thereby to strengthen country's foreign exchange reserve and reduce its import bills service obligations.

COVID-19 is responsible for shrinking international trade all over the world. Many economies have registered negative growth rate with no signs of economic revival in near future. Exports have dwindled drastically with no scope for imports. It has thrown in disarray the entire value chain of all the sectors of an economy [21].

In this gloomy scenario a domestic consumption story of India has emerged as a light at the end of the dark tunnel. Indian economy needs to flex its every nerve to cater to its own consumption needs. This would eventually insulate Indian economy from external shocks like COVID-19. This would help Indian economy achieve near full employment status. All said and done moving from globalization to localization calls for exploitation of underutilized resources. It entails massive public and private capital expenditure. It envisages huge training cost to the human resource with near 100 percentage literacy level. It calls for massive amount of production of intermediary goods which shall further promote manufacturing with little or no dependence on foreign countries for imports of specialized sophisticated heavy machineries.

What is otherwise feasible for surplus economies of the world may not be feasible for deficit economies like India at least for initial period? Moving from globalization to localization involves a long gestation period. All said and done it also calls for both eyes to be set on quality norms to avoid distress in consumer psychology. Although we aim at catering to domestic markets our quality standard should no way be less than that of international standard. This quality benchmark shall result into profitable exports basket for the country which will earn India her precious foreign exchange reserve. This foreign exchange reserve can be utilized for selective imports so that India can strengthen her rupee in international markets.

Furthermore India will have to expand its production possibility frontier. Public Provident Fund (PPF) is an index that shows nation's total caliber in producing goods and services domestically. Tapping untapped potential reserves of India is a key factor here. Expansion in production possibility frontier shall not take place unless the huge training cost is involved in turning manpower into human resource. Access to primary, secondary, higher and professional education is a key to unlock the potential of India.

We are living in a digital age. India as a nation has to successfully paddle the education 4.0 and industry 4.0. The nation needs to focus on speed, skill and scale in all its sectors. India as a nation needs to live by a common dream. The small nation on the earth like Singapore for decades lived for fulfillment of her dream. India needs to have such concrete dream on nation's radar [22].

The COVID-19 has to some extent exposed India's health index which suggest much needs to be done. Economic growth development of the nation and its self-reliance can hardly be achieved without drastically improving nation's health parameters. In so far as health parameter of India is concerned out of pocket expenses are more and per capita density of the bed is quite less. While per capita investment on health infrastructure is less the dream of self-reliance shall remain a distant dream [23].

Indian health industry contributes at 4% to India's GDP. The health industry of India is growing at a constant computer annual growth rate (CAGR) of 14°/c. India is lagging far behind in various health indicators. The performance of Indian health sector is even abysmal when it comes to low and medium income countries. The bed density in India is lagging, out of pocket expenses are more, the incidences of epidemics are more, and mortality rate is high [24].

As of now for government of India the priority sectors of government spending were other than health industry. The social security measures are not guaranteed to all in India. Incidences of out of pocket expenses are more. Preventive index in Indian society is low and in recent times it is increasing. The private equity capital and funding needs to be attracted in India's health industry. India has enough potential to become a medical tourism country. Yoga and spiritual practices have their genesis in India. Indian can capitalize on her pristine knowledge in Yoga and Wellness industry to attract foreign capital and foreign industry.

According to World Health Organization (WHO) report India is on the threshold of becoming a global capital for diabetic disease and mental health-related disorders. Dire urgency staring straight in the face of India to take urgent measures to infuse sufficient capital into the health industry of India to compensate for the gaps accumulated thus far. Thus we can infer that the impact of globalization on India's health industry is desired to be taken up the value chain.

1.10.6 Inference

India as a nation in the year adopted the policy shift of globalization. In a planned way LPG of Indian economy carried out. The economic reforms and institutional reforms were carried out in all sectors of Indian economy. India adopted exports led model of economic growth. The first decade of globalization of Indian economy witnessed certain teething troubles. However from second decade onward the revolutionary decision of policy shift started showing some positive results for Indian economy. These positive results can be summarized as reduction in technology gap, reduction in investment gap, reduction in poverty, and unemployment, industrialization got accelerated and the brand image of India got elevated in the eyes of developed countries of the world.

In the year 2019 the whole world came under the grip of an invisible enemy which brought all the nations to the knees. A novel Corona virus disease brought the economic wheels of the whole world to the grinding halt. The governments were compelled to announce fiscal and monetary measures to tide over the economic crisis.

With vaccine to treat Corona patients seems to take long period of FDA approval the social and economic shut down seems inevitable and new normal practices are being thought of. Corona pandemic being started to be considered as a necessary evil of globalization. Hence a silent and sometimes whispering is making the rounds of moving from globalization to localization.

The honorable Prime Minister of India appealed to the fellow countryman for being vocal for local. All said and done this does not in any way indicates that India should take a lasting divorce from globalization. Instead it does mean that India needs to reduce its reliance on overseas markets for lubricating its own economy. The basic thought is that India should remain insulated from external shocks like Corona Virus and other shocks of economic and social nature.

Achieving self-reliance is a herculean task, it calls for massive spade work to reorient Indian economy toward internalization. Although herculean it is achievable with the statesmanship form central and state leadership with a concrete dream constantly radiating not only on the National radar of India but also in the hearts, minds and souls of 1.3 billion populace of India which is a demographic dividend of India.

REFERENCES

1. Al-Rodhan, Nayef R.F. and Gérard Stoudmann. (2006, 19 June). "Definitions of Globalization: A Comprehensive Overview and a Proposed Definition." GCSP Occasional Papers, Geneva: Geneva Centre for Security Policy (GCSP).
2. Albrow, Martin and Elizabeth King (eds.). (1990). *Globalization, Knowledge and Society*. London: Sage. p. 8. "all those processes by which the peoples of the world are incorporated into a single world society." ISBN 9780803983243.
3. Stever, H. Guyford. (1972). "Science, Systems, and Society." *Journal of Cybernetics*, 2(3): 1–3. doi: 10.1080/01969727208542909.
4. Frank, Andre Gunder. (1998). *ReOrient: Global Economy in the Asian Age*. Berkeley: University of California Press. ISBN 9780520214743.
5. O'Rourke, Kevin H. and Jeffrey G. Williamson. (2000). "When Did Globalization Begin?" NBER Working Paper No. 7632.
6. "Globalization." Oxford English Dictionary Online. September 2009. Retrieved 5 November 2010.
7. The Battle of Armageddon. October 1897. pp. 365–370. Retrieved 31 July 2010, Pastor-russell.com.
8. Hopkins, Anthony G. (ed.). (2004). *Globalization in World History*. London: Norton. pp. 4–8. ISBN 9780393979428.
9. Robertson, Roland. (1992). *Globalization: Social Theory and Global Culture* (Reprint. ed.). London: Sage. ISBN 0803981872.
10. Giddens, Anthony. (1991). *The Consequences of Modernity*. Cambridge: Polity Press. p. 64. ISBN 9780745609232.
11. Held, David, et al. (1999). *Global Transformations*. Cambridge: Polity Press. ISBN 9780745614984.
12. Larsson, Thomas. (2001). *The Race to the Top: The Real Story of Globalization*. Washington, DC: Cato Institute. p. 9. ISBN 9781930865150.
13. Friedman, Thomas L. (2008) "The Dell Theory of Conflict Prevention." *Emerging: A Reader*. Ed. Barclay Barrios. Boston: Bedford, St. Martins. p. 49.
14. International Monetary Fund. (2000). "Globalization: Threats or Opportunity." 12th April, 2000. IMF Publications.
15. Fotopoulos, Takis. (2001, July). "Globalization, the Reformist Left and the Anti-Globalization 'Movement.'" *Democracy & Nature: The International Journal of Inclusive Democracy*, 7(2). http://www.democracynature.org/vol7/takis_globalisation.htm
16. Lee, Adela C.Y. "Ancient Silk Road Travellers." Silkroad Foundation: Silk-road.com. Retrieved 31 July 2020.
17. Lewis, David and Karl Moore. (2009). "The Origins of Globalization: A Canadian Perspective." *Ivey Business Journal*, (May/June).

18. Hobson, John M. (2004). *The Eastern Origins of Western Civilisation*. Cambridge: Cambridge University Press. ISBN 0521547245.
19. Weatherford, Jack. (2004). *Genghis Khan and the Making of the Modern World*. New York: Three Rivers Press. ISBN 0609809644.
20. Tracy, James D. (1993). *The Rise of Merchant Empires: Long-Distance Trade in the Early Modern World, 1350–1750*. Cambridge: Cambridge University Press. p. 405. ISBN 0521457351. https://www.cambridge.org/core/journals/business-history-review/article/abs/rise-of-merchant-empires-longdistance-trade-in-the-early-modern-world-edited-by-james-d-tracy-new-york-cambridge-university-press-1990-xviii-442-pp-charts-tables-notes-and-index-4750/1D42E009B88B3CDF05772ABC476F0E2B. Retrieved 28 November 2010.
21. Leong, Ho Khai and Khai Leong Ho. (2009). *Connecting and Distancing: Southeast Asia and China*. Singapore: Institute of Southeast Asian Studies. p. 11. ISBN 9812308563. https://books.google.co.in/books/about/Connecting_and_Distancing.html?id=EwnzBiM0LmAC&redir_esc=y. Retrieved 28 November 2010.
22. Roubini, Nouriel. (2009, 15 January). "A Global Breakdown of the Recession in 2009". Forbes. https://www.forbes.com/2009/01/14/global-recession-2009-oped-cx_nr_0115roubini.html?sh=61c615f0185f.
23. Faiola, Anthony. (2009). "A Global Retreat As Economies Dry Up." *The Washington Post*. http://www.washingtonpost.com/wp- dyn/content/article/2009/03/04/AR2009030404221.html. Retrieved 5 March 2020.
24. Gjelten, Tom. (2009). "Economic Crisis Poses Threat to Global Stability." *NPR*. https://www.npr.org/templates/story/story.php?storyId=100781975. Retrieved 18 February 2009.

2 Enrich to Rich—an Indigenous Model to Combat COVID-19

B.M. Hardas and N.S. Damle
Shri Ramdeobaba College of Engineering & Management

CONTENTS

2.1 INTRODUCTION

In the first principles of infectious diseases, the worst bugs are by breaking the cycle of human contact or keeping contact in a sterile state. If you think you have started an infection, you need to visit a doctor or a nearby hospital. Keep yourself away from others until the diagnosis or symptoms are resolved. Traveling to the affected areas or contacting people in those areas should be done only after government consultation or investigation. A number of test kits have been made so far to detect corona virus infection. It also has expensive and inexpensive test kits, based on blood samples from the spit, to tell if the person is infected with the corona. In the coming time, it can be ascertained whether the person in the house is infected with a corona, depending on the speed of breathing of the person. Scientists at the University of Massachusetts Institute of Technology have designed a similar device.

The sensor installed in this wireless device will monitor the movement and breathing of people in the house and will tell if anyone has a corona infection. Scientists have named this wireless device as "Emerald." It is a kind of laboratory named

Computer Science and Artificial Intelligence Laboratory. It has been developed by an MIT (Massachusetts Institute of Technology) Professor Dina Katabi and her team. It can be installed on any wall in the house. After connecting it to wi-fi it starts working. According to the researchers, it would be better to stay at home for the members of the household who have been reported as corona positive. If anyone in the house is suffering from corona then this device will alert the rest of the members in the house. That is, it is also for general use in the house, so that one should be careful to avoid infection. The device works on the basis of artificial intelligence (AI). This device monitors the physical activities of the people in the household. Wireless signals emanating from the device have been examining the activities of the person, i.e., the way of walking, his sleeping habits and most importantly, the speed or rate of breathing.

MIT has suggested that with this device, when you find that your breathing rate is not normal, it is changing, and then you should understand that you need medical advice. This device tells how much your breathing is changing. The rate of breathing has already increased or is not reduced since a person with corona infection has trouble breathing, medical advice should be sought on changes in respiratory rate. According to researcher William McCrery of Dr. Qatabi's team, the device is easy to use. You just have to apply it once. It observes the activities of the people of the house itself. It does not require every person to wear or hold it, nor does it require any command. Since the elderly are at greater risk of corona infection, this tool will prove more useful to monitor their movements. Currently the trial of this equipment is going on in Boston. It will be available in the market on success. With this help, medical data can be collected.

Indian Ayurveda is an eco-friendly science that is very useful even in the ancient and modern age of the world. There are three fundamental doshas: Vata, Pitta and Kapha, and good Ayurvedic philosophy is about balancing the body, mind and spirit. There is no doubt that Ayurveda, based on the three-fold nature of cough, Vaat and Pitta, will be the focal point of human existence in the coming millennium. This chapter presents a critique of the corona epidemic by combining Indian Ayurveda with AI.

2.1.1 Corona Virus (Novel Corona Virus/2019-NCOV)

There is no cure yet. Its recent outbreak has caused concern to all over the world, as well as created an atmosphere of fear all around. This fear is especially prevalent in Wuhan, China, where the virus was first detected. The main reason for the panic among people is that the symptoms of this corona virus are similar to the common flu, which makes it difficult to distinguish or distinguish between the two. Because corona viruses belong to the group of pathogens, they also affect mammals, birds and humans. Most viruses belonging to this group are not dangerous. They affect the human respiratory system and cause diseases such as the common flu, Severe Acute Respiratory Syndrome (SARS) and Middle-East Respiratory Syndrome (MERS). There is nothing completely ignorant about the corona virus in medical science (Figure 2.1) [1].

In fact, it is more common than you might imagine. Viruses that affect humans include NL63, 229E, HKU1 and OC43 and usually cause upper respiratory tract infections (mild to moderate) and are as contagious as colds. The World Health Organization has officially named the disease caused by the virus as "COVID-19." Scientists researching the corona virus have described the disease as "severe acute

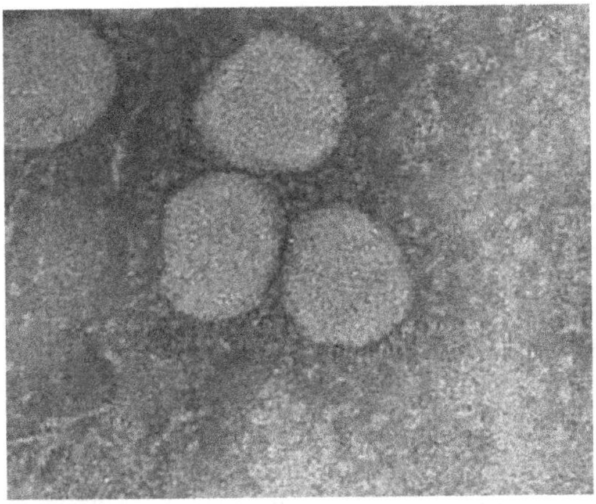

FIGURE 2.1 Corona virus from SARS isolated in FRhK-4 cells. Thin section electron micrograph and negative stained virus. (Department of Microbiology, The University of Hong Kong and the Government Virus Unit, Department of Health, Hong Kong SAR China.)

respiratory syndrome corona virus 2" or "SARS-COV-2." The first patient associated with this outbreak was found in the last week of December 2019. On January 8, 2020, it was confirmed that the epidemic was caused by the corona virus. As of March 8, 2020, the pandemic has infected 1,05,586 people worldwide, with 4,027 deaths. The contagious disease, which originated in the Chinese province of Hubei, has spread across the borders of 100 countries. In India, the first case of this corona was reported in Kerala. SARS-Co-V2 is the third contagious virus to be found in the last 20 years. In 2003, SARS-CoV caused the pneumonia, the most common form of SARS, to spread worldwide. Ten years later, MERS-CoV (Middle East Respiratory Syndrome) virus appeared in the Middle East. The virus has infected 2,494 people and caused about 900 deaths. The SARS co-op spread more rapidly but the number of deaths from it was relatively low: 8000 infected patients and 800 deaths. MERS-CoV has been reported in animals. This is called zoonosis. The same is true of "Sars Covey 2." Zoonosis is transmitted from animals to humans. The diseases MERS COV and SARS COV were transmitted to humans by camel and otter cats (civet cats).

2.1.2 Genetic Analysis

According to the genetic analysis of SARS-CoV-2, very close similarities between these viruses have been found in the virus. It is called corona virus. The mediating animal between bats and humans is the scaly cat. The scaly cat is known for its strange-looking scales and its habit of wrapping itself when its life is in danger, but they are smuggled into China for the scales and their meat. It mediates viral transmission from animals to humans. A virus has to reach human cells. It tries to establish a relationship with the proteins on the viral cell. Each cell contains some trace proteins. These proteins are called receptor proteins, e.g., the HIV virus only associates with CD4 lymphocytes. Once in contact with the scar protein, the virus enters the cell.

The virus divides into cells. Thus the host cells become infected. If this does not happen, the virus cannot enter the cell. The corona virus has found a very simple way to get into the cells of the shelter. The glycoprotein spikes on the surface of the virus (spikes with spikes on the soles of the feet for walking on ice) are constantly peeking out of the surface. These are called glycoproteins. This is because this protein is accompanied by a sugar molecule. The glycoprotein that comes out makes the virus look like a crown of thorns. This is why the virus is called corona virus. Spike protein is attached to the actual shelter cell. This is called "S1 unit." The structure of the "S1 unit" is so varied that the "S1 unit" is attached to many mammalian cells. Scientists have discovered exactly which endothelial cell protein is associated with the S1 unit of the SARS CoV2 virus. This pair is like the key to a seven-sided lock. Sticking to the surface cells of the human respiratory tract is a very minor matter in terms of corona virus. The SARS CO-V2 virus infects the protein in the receptor cell called ACE2 and causes further disease. Another worrying virus is influenza. All influenza viruses originate from waterfowl such as ducks, geese, terns, gulls and the like. Reaching humans from birds is the closest way to them. Because birds and humans are warm-blooded, their body temperatures are similar. In many cases, the parasite's survival in the new shelter is a dead end. Bird flu has spread from bird to human but has not usually spread from one human to another, with the exception of the 2009 swine flu outbreak of H1N1. The virus spread so rapidly in humans that the outbreak of swine flu had to be declared ubiquitous. In 1918, the bird flu outbreak was global [2].

In a study based on DNA analysis, the closest virus to SARS-Co-V2 is in bats. It is spreading the disease in the human respiratory tract by replacing the original bird flu virus from a viral mixture of pigs and bats. Corona virus and influenza diseases have become universal (pandemic) due to the method of jumping into its shelter. In biological evolution, organisms with low immunity are more likely to develop viral and bacterial diseases. Interbreeding and Captive breeding in the same community are important causes of impaired immunity. Bird flu has spread to Chinese ducks and spread to humans. Raising millions of ducks for eggs and meat in a closed duck rearing center was a lucrative business—closed breeding of the same type of chicken. When the swine flu came from domestic pigs, the pigs ate pet droppings, which changed the bird virus to swine flu virus. The origin of the corona virus is yet to be determined.

2.1.3 Effect on Immune System

Those with a weakened immune system are more likely to be affected. Therefore, senior citizens, pregnant women, children as well as those with heart disease or cancer are more likely to be infected with the corona virus. The symptoms of the corona virus are similar to those of the common flu, as mentioned earlier. Here is a list of signs and symptoms that you need to be aware of so that they can be remedied (Figure 2.2).

- Persistent runny nose
- Coughing pain
- Not feeling well
- Breathing difficulty
- Lung swelling/pneumonia

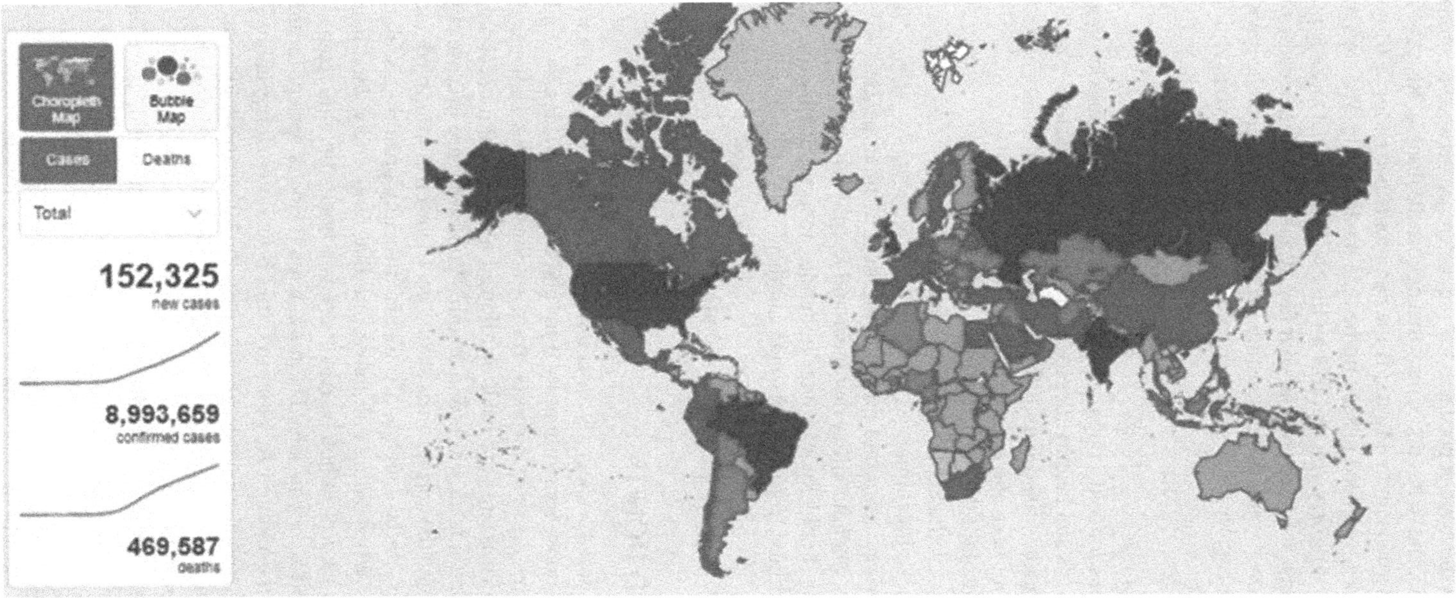

FIGURE 2.2 WHO COVID-19 dashboard. (Globally, as of 3:58 pm CEST, 23 June 2020, there have been 8,993,659 confirmed cases of COVID-19, including 469,587 deaths, reported to WHO.)

It is difficult for the average person to tell the difference between a corona infection and the common flu, and so it is not so easy to diagnose the disease. Also, the incubation period of corona virus is up to 14 days; if the symptoms persist for 7–10 days and the discomfort increases, you need to seek immediate medical help. We are sure that this news will continue to worry you, but it is an illness. There are also some preventative measures to prevent it and keep you and your family safe. The biggest cause of infection with this virus is low immunity. So boosting the immune system is the obvious way to keep the disease at bay. Here are some ways you can take care of yourself [3].

1. Stay tuned
2. Get plenty of rest
3. Take medicine for cold, cough, sore throat and fever on the advice of a doctor
4. Eat a nutritious and healthy diet for a strong immune system
5. Personal hygiene: dispose used tissues properly and disinfect items that do not touch your nose and eyes frequently
6. Avoid crowds until the end of spring
7. Avoid contact with people who have a cold or cough
8. Avoid contact with animals
9. Avoid eating raw meat and do not drink milk without heating it properly. Wear gloves when handling raw meat and keep it separate to prevent reinfection. You should know some guidelines to prevent this disease from being transmitted from one person to another.

General guidelines for protecting against corona virus (Figure 2.3).

2.2 DO'S AND DON'TS

When traveling abroad, especially in countries where there is an outbreak of the virus, you should boost your immunity a lot and keep medicines for colds, coughs and fevers close by. At most airports, train stations and bus stations, passengers infected with the virus are sent back or allowed after a health check-up, so avoid traveling if you have a fever, chills and cough. People who are in close contact with patients infected with the corona virus should get themselves examined immediately. If you have a fever, chills, cough and you have difficulty breathing, seek immediate medical attention. When eating out, make sure that food safety, food handling and hygiene are taken care of in hotels. Also here is a bonus tip, this recipe will boost your immunity and keep you safe. The masala tea recipe is perfect for tea flavor as well as nutrients like Vitamin A, Vitamin C and B vitamins. It has antioxidant, antifungal and antiinflammatory properties that help the human body to repair and boost the immune system. Note: This action is not recommended for those who have breast cancer, uterine cancer or any disease caused by excessive estrogen.

While the corona virus has literally spread all over the world, as well as the lack of any effective or preventive modern medicine against it, an atmosphere of fear has been created in all societies. The ever-increasing number of victims is shocking. Research is now underway to find a cure for the virus. But, it will be worthwhile to see if there is any thought in Ayurveda regarding such ailments.

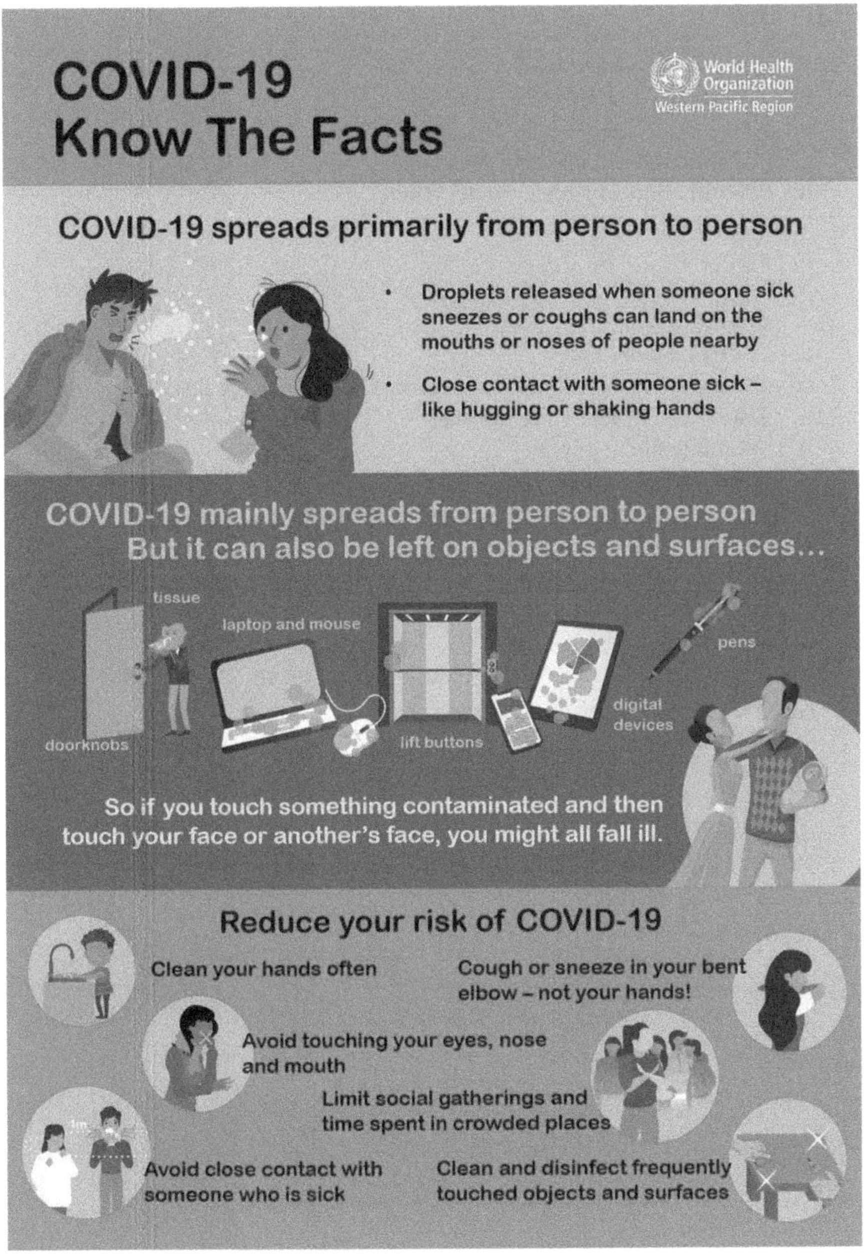

FIGURE 2.3 COVID-19 know the facts. (World Health Organization.)

2.2.1 Quiz

1. How big is the corona virus?
   a. 1 billionths of a meter in diameter
   b. 8 billionths of a meter in diameter
   c. 80 billionths of a meter in diameter
   d. 800 billionths of a meter in diameter
2. Which of these is NOT listed by the WHO as a symptom of corona virus?
   a. Fever
   b. Dry cough
   c. Blurred vision
   d. Nasal congestion
3. How long does the novel corona virus survive outside the body?
   a. A week in the air and on surface
   b. Several hours to days
   c. Up to a two and half weeks
4. Which of the following viruses causes COVID-19?
   a. SARS-CoV-2
   b. COVID-19 virus
   c. Wuhan corona virus
   d. MERS virus
5. The virus that causes COVID-19 was believed to have originated in which species?
   a. Bat
   b. Pig
   c. Bird
   d. Horse
6. How is a diagnosis of COVID-19 most commonly confirmed?
   a. RT-PCR Testing
   b. Chest CT scan
   c. Viral culture testing
   d. Bacterial culture testing

Answers:

1	2	3	4	5	6
c	c	b	a	a	a

2.3 ARTIFICIAL INTELLIGENCE

AI has also been instrumental in the all-out war against the corona virus. Till yesterday, AI was being said to rob human work and snatch employment. That is what is happening to us now. A start-up in Mumbai has started using AI based X-ray technology to detect corona virus infection. The company developed this method to check tuberculosis infection (TB) with a chest X-ray which has become suitable for

COVID-19 screening. It not only tracks the spread of corona virus in the lungs and determines the proportion of the affected part, but also automatically analyzes tuberculosis and lung dysfunction and its degeneration and cardiovascular disorders. Blue Dart, an AI-based Canadian company, warned in the last week of December 2019 about the spread of the corona virus and fears of rapid infecting people. Later, millions of AI-based cameras in China used corona to identify infected people. It is now playing a key role in drug and vaccine development to eradicate the corona virus. The world will recover from the Corona crisis. After this, whatever new events take place in the whole world will prove to be a revolutionary change. The world market will leave no stone unturned in turning this crisis into new opportunities. The result will be that the world will probably change completely.

In the second half of the 20th century, the two major fields of science that the world has done are information technology and biotechnology. Thanks to these two, the 21st century began with fleet changes. The charismatic landscape that technology is creating for the betterment of human beings by using new technologies will rapidly increase their acceptance rate. Under the pretext of privacy and security threats, the areas in which these technological experiments have so far been prevented may soon be able to overcome these obstacles and raise its flag. With these two technologies, AI has made a huge difference. Intelligence enables and motivates man to act in his environment by sensing appropriate and far-sighted results. When we take the risk of assigning this responsibility to a technician, after thorough scientific scrutiny, we ensure that the said device learns, thinks, plans, speaks and processes according to our circumstances and makes decisions in that direction in an orderly manner. Going forward has started working. AI works on this principle. In the development of which previous human experience, research, statistics, knowledge and technology have coordinated and done important yoga. On the basis of them, today it has emerged as an ally of man. The development of civilization is the result of the intellectual development of man. In the same evolutionary order, now machines have also been endowed with intellect by human beings, which is called AI.

2.3.1 Role of Information Technology

Information technology has worked like icing on the cake in this direction. That is why successful experiments of driving unmanned cars, buses, computers and even factories are taking place in this direction. Of course, the dangers and the consequences of this are not yet visible. But it is certain that AI will emerge as an ally of the time to come as an ally of man. Its expansion and development can no longer be avoided. AI can put modern man at risk. This is the speculation of the future. But from what we have just seen during the Corona Crisis, it is clear that it has emerged as a shield in the time of the greatest danger. When a virus has challenged the entire human existence. Meanwhile, China monitored the smart phones of its people and saw its good use [4]. Used technology to identify faces from smartphone screens. Even got to know people's body temperature. Taking the location and keeping the infected people directly under the observation of specialists. Not only this, he even found out through the phone which person has met where and from whom. All of them were contacted and kept in solitude and treated. All this was made possible by AI. AI is a capability that can perform risky tasks automatically with full efficiency. Even without direct human intervention. The corona crisis in the world is a direct proof of this. AI is successfully intervening beyond human intellect. Saudi Arabia's intelligent robot Sophia is a successful proof of AI in front of the world. Which was also granted full citizenship by Saudi Arabia on 25 October 2017. This is the first instance of a machine getting citizenship of a country like a citizen, which is giving us a glimpse of the strong future of AI. Not only that, whenever Sofia makes a mistake, she will be prosecuted under the law of Saudi Arabia.

AI is being used in the IT sector. Its worldwide market is growing at a rate of about 63%. Automated cars, robots, Go computers and digital assistants are all marvels of AI. In the foreseeable future, disease, poverty and growing inequality on a large scale on earth can be reduced only with the help of AI. The use of AI can prove to be a revolutionary change in the health sector. From cancer treatment to bacterial diseases in which there is a high risk of human infection. AI is the greatest and most amazing event in the history of world civilization [5].

It is going to hand over the power of AI to human beings. The only way to avoid this risk is to find a modern human being. Otherwise, the danger of losing everything will continue to hover over human civilization with it. AI can play a big role in the near future in increasing the capacity in the field of health, improving education and bringing quality, good transparent governance for the citizens. Recognizing the importance of AI, India is determined to make steady progress in that direction with a positive outlook [6].

2.4 ADMINISTRATIVE MEASURES

NITI Aayog has constituted a committee and made special provisions for this in its budget. Under which the government is working on a special plan to enhance research, training, human resource and skill development in the fields of AI, Robotics, Digital Manufacturing, Big Data Intelligence, Real Time Data and Quantum Communication for which a seven-point strategy has been decided. In the future, the monitoring of technology will increase on the world society which cannot be avoided for long. Because both the market and the government seem to stand on this side. There is so much fighting in the world regarding data. The intention behind it is that we should know everything about everyone and move forward by formulating our policy and strategy accordingly. A click of the internet reveals everything from our location to our body temperature. Our blood pressure and heart rate can be monitored only by the screen of our mobile phone. In the future, AI will be able to mark the moments of man's laughing and crying and understand his emotions from anywhere. It can also be considered a threat in the name of privacy and the future of modern man. Under which our life will get new speed and progress. However, a technologically advanced future is the destiny of our future generations, it is certain. Its dangers and benefits will continue to explain our human survival and existence. It will depend a lot on the development of new ethics whether AI will stand with us or against us. AI has come in handy in the recent threat to humanity that has deepened with the Corona Crisis.

2.5 AYURVEDA

Ayurveda and Yoga, the ancient medical sciences of India, are popular abroad today under the glamorous name of Ayurveda and Yoga. Ayu means life and Veda means science. Ayurveda, which knows the science of life or body, has existed since time immemorial. Specialization Abroad Ayurveda is used in the United States as a complementary and alternative medicine. It is difficult to trace the origin of Ayurveda but it is said that Ayurveda was mentioned in Rig-Veda 6000 years ago. *Lord Dhanvantari* is considered to be the father of Ayurveda. According to the mythology, he got this knowledge from Brahma himself. According to Ayurveda, our body is occupied by the five great beings, earth, (land), (you) water, air, light (fire) and sky (universe). According to this, there are three natures of the body, namely, vata, pitta and kapha. Similarly, there are seven metals like juice, blood, mass, fat, bone, marrow and Venus. These seven metals are influenced by the above three natures. Ayurveda is very useful for balancing body, mind and soul with these nature and saptadhatu. Ayurveda says that in order to maintain the balance of overall physical and mental health, one should follow one's daily routine and menstrual cycle without using Ayurveda only when one falls ill.

According to the Ayurvedic treatment method, eight main departments were developed—External & Internal medicine, Surgery, Otolaryngology (ENT), Ophthalmology, Toxicology, Pediatrics and Aphrodisiacs and Psychiatry, the science of rejuvenation. Information on each of these subjects is written in the Charak Samhita and the Sushruta Samhita. Before 1500 BC, the two main divisions of Ayurveda were Atreya and Dhanvantari. Those who adopted the Atreya method used to cure the patients with medicines and Dhanvantari Vaidya used to perform surgeries on the patients [7].

Charak sages wrote the Charak Samhita before 1000 BC. Based on the references written by Atreya, he wrote medicines for many ailments that are still used today. The Sushruta Samhita was written by the Sushruta sage of the Dhanvantari division. It covers cosmetic surgery, caesarean section, hand, foot, brain, nose, ear and eye treatments. After 500 AD, Vagbhata wrote Ashtanga Hriday Samhita which includes both Atreya and Dhanvantari sections. The Charak Samhita is in the original Sanskrit language and contains 8,400 verses. In the past, oral knowledge was given to each other. So the code is written this way. Charaka wrote internal and external therapies for the body, mind and soul according to the Atreya method. Therefore, he emphasized on making a diagnosis before treatment. So much so that it was suggested to consider the time, time, regional structure, birth of a child, etc., to diagnose a disease. Charaka describes the onset of a disease in eight stages. Charaka also suggests considering the time and method of collecting herbs [3,8].

2.5.1 Ayurvedic Medicines and Remedial Measures

Over the last thousands of years, many Ayurvedic medicines like ginger, turmeric, black pepper, cinnamon, garlic, amla, honey, jaggery, basil, neem, etc., have cured many ailments and maintained good health. No one in the whole world would mind that all these drugs are so safe. Corona has forced doctors all over the world, giant hospitals, specialists and super specialists who claim to be experts in the field, giant medical companies, state-of-the-art technology, ICCUs, ventilators, all so-called state-of-the-art medical labs to kneel down.

In the 1880s, Louis Pasteur reversed his earlier theory and formulated a new one: Not the Seed, Soil Is Important Cause of Infective Diseases. This means that any infectious disease is not caused by a virus or a bacterium, but by the weakness of a particular organ and the disease it causes. Bacteria or viruses are constantly changing their appearance. Anxiety is the cause of Ras Kshay (Rasa-Loss) according to Ayurveda. When the first metal Rasa decays, all subsequent metals decay. Here are some simple steps you can take today. Be it body emotion or mind emotion.

1. **Milk:** Babies who drink milk are often happier. Milk is vital, enlightening, life-giving and beneficial in mental disorders. Cow's milk is the best. Buffalo's milk is sleepy. Milk, ghee and sugar are the best chemicals. Patients with diabetes should not do this. A healthy person should take one cup of milk, one teaspoon of sugar and one teaspoon of ghee. Drink one cup of milk in the morning and evening.
2. **Ghee:** Ghee is a chemical. Intellectual and intellect enhancer. It is life-enhancing. Ghee, however, should preferably be taken from cows. The meal should contain ghee. Add a tablespoon of ghee to the hot rice-chapati. Ghee is important in stabilization.
3. **Padabhyanga (foot massage):** Massage both feet with ghee or oil every night before going to bed. Acharya Sushruta says "Nidrakaro Dehasukhchakshushya: Shramasuptikrit." Regular footwork improves sleep and reduces fatigue. Stability comes.
4. **Pranayama and Shavasana:** Pranayama is important along with yoga. The best solution is to break your fast and focus on your breathing. No one would possibly say I do not believe in breathing. So pranayama is great for both believers and atheists. Since the body carries oxygen, it needs to be regulated.
5. **Medications:** Take 1/2 teaspoon of Ashwagandha powder or 1 (500 mg) of Ashwagandha before going to bed. Ashwagandha is a sedative and anti-tuberculosis. After 2-2 teaspoons of Saraswatarisht, take half a cup of plain water. Before going to bed at night, apply two drops of Panchendriya Vardhan Oil in both nostrils.
6. There are common remedies for healthy people not to get upset. Those who have severe stress, depression and stress should consult a doctor in time. The potential loss of wealth is enough to cause the loss of life. It is said that no one has found the shirt of happiness. Nothing is bigger than your life. Looking at zero does not solve the problem. Strengthen the mind. The questions get easier!!

2.6 ENRICH TO RICH

Experts say that in the changing technology scenario, India now needs courses and professional training programs in AI. Pearson Professional Programs' Vice President Varun Dhamija said that in the digital era and the rapidly changing industrial landscape, AI is affecting many industries, changing job roles. In almost every industry, the world is looking toward AI and it is being considered as the next major technological change in the industrial and smart phone revolution. According to the survey, 60% of Indians believe that the world is now moving toward a model where people participate in education for life. More and more professional, novice young and mid-level employees have now realized that their employees need some new teaching and formal training in AI and other fields. In view of this change, we will not only see the demand for short duration or vocational education but will see the full curriculum based on AI. In particular, AI refers to the simulation of the human brain and logic that specifically resolves and plans things according to the surrounding landscape. AI is one of the most innovative developments in technology, and according to experts, AI will form the basis of our future generations. AI is understood to provide various benefits in various domains such as healthcare, banks, information technology, businesses, and more. AI has a notable application in the field of healthcare, in which machines will be operated on people for various check-ups without human intervention.

With the use of complex algorithms, AI can provide doctors with data-driven solutions that in turn help them make better decisions about the health of the patient. Professional performance has a great function of AI in which chat bots respond to queries promptly and provide 24/7 support to customers. This ultimately saves the time and effort required to carry out a successful business. Nowadays, you use Google Maps, Uber Cabs and many other applications that automatically detect your questions and answers to your questions in no time—all thanks to AI. The maps usually instruct you to take a different route as the heavy traffic congestion in which you are going to walk is all possible with the help of AI machines. AI has greatly influenced various other fields related to manufacturing, education, finance, gaming, arts, government, and more. AI has the ability to perform tasks related to speech recognition, facial recognition, translation of various languages, decision-making, etc. Therefore, it is the future of technology as it can solve highly complex problems and gives you results without time. AI is currently being used as an effective tool in various measures. It may be more effective for a fragmented and colossal country like India to use AI, including its ancient Ayurvedic therapies, in combating corona or even biological weapons. Indian ancient literature is an inexhaustible source of knowledge and science. Ayurveda is the crown jewel of medicine.

It is imperative to coordinate ancient Ayurveda and contemporary AI by understanding the opportunity without considering a catastrophic epidemic like COVID as a disaster. At a time when humanity is at cross-roads, it is only natural that the whole world should look at India with optimism. The positive message will be that Ayurvedic physicians, scientists, researchers and practitioners will come together and keep trying. The corona virus has plagued over 7.5 billion people worldwide. In this endless ocean of worries about the present and worries about the future, all human beings are helpless. But for the last four or six months, the whole world has struggled for survival. Today, it is time for all human beings to forget their grievances and work together. Many issues like environmental imbalance and industrialization were discussed a lot, but what happened? Basically, the global thinking in the Indian way of life will be the inspiration for the world to come. For that, Indians should put the Indian way of life into practice and present it to the world. The world now wants India's ideology, but are those givers in India? This is the real question. There is no reason to oppose any technology. But the addition of modern technology to the ancient Indian health system will in no doubt be a revolution.

2.7 CONCLUSION

Enrich to rich—An Indigenous Model to Combat COVID-19; it is the request to the scholars and researchers to come together and provide this ideological secret to all manhood. AI has the ability to perform tasks related to speech recognition, facial recognition, and translation of various languages, decision-making, and improved risk stratification. AI could enhance researchers' understanding of the virus, leading them to better target and treat COVID-19. Investigators can use AI tools to understand the relationship between COVID-19 lung pathology and immune inflammation. It is imperative to coordinate ancient Ayurveda and contemporary AI by understanding the opportunity without considering a catastrophic epidemic like COVID as a disaster, and in addition at a time when humanity is at cross-roads, it is only natural that the whole world should look at India with optimism. The positive message will be that Ayurvedic physicians, scientists, researchers and practitioners will come together and keep trying [3,7,8].

REFERENCES

1. Wang C, Horby PW, Hayden FG, Gao GF (February 2020). "A novel corona virus outbreak of global health concern". *Lancet.* 395 (10223): 470–473. doi: 10.1016/S0140-6736(20)30185-9. PMID 31986257.
2. Hui DS, I Azhar E, Madani TA, Ntoumi F, Kock R, Dar O, et al. (February 2020). "The continuing 2019-nCoV epidemic threat of novel corona viruses to global health—The latest 2019 novel corona virus outbreak in Wuhan, China". *International Journal of Infectious Diseases.* 91: 264–266. doi: 10.1016/j.ijid.2020.01.009. PMC 7128332. PMID 31953166.
3. Lau EH, Hsiung CA, Cowling BJ, Chen CH, Ho LM, Tsang T, et al. (March 2010). "A comparative epidemiologic analysis of SARS in Hong Kong, Beijing and Taiwan". *BMC Infectious Diseases.* 10: 50. doi: 10.1186/1471-2334-10-50. PMC 2846944. PMID 20205928.
4. Report of the WHO-China joint mission on corona virus disease 2019 (COVID-19)" (PDF). World Health Organization. February 2020.
5. COVID-19 dashboard by the Center for Systems Science and Engineering (CSSE) at Johns Hopkins University (JHU). ArcGIS. Johns Hopkins University.
6. Vinay C, et al. "Comprehensive review of the COVID-19 pandemic and the role of IoT, drones, AI, Blockchain, and 5G in managing its impact". Special section on deep learning algorithms for internet of medical things, digital object identifier. doi: 10.1109/ACCESS.2020.2992341
7. Charaka Samhita Vd. Harish Chandra Kushwaha. New Delhi: Choukhamba Prakshan. ISBN: 9788176371490.
8. KR Bhavana, S (October-December 2014) Medical geography in Charaka Samhita. *'AYU' – An International Quarterly Journal of Research in Ayurveda.* 35(4): 371–377. doi: 10.4103/0974–8520.158984.

3 An Overview of Evolution, Transmission, Detection and Diagnosis for the Way Out of Corona Virus

V Dhinakaran and M Varsha Shree
Chennai Institute of Technology

Suman Lata Tripathi
Lovely Professional University

Meenakashi Sharma
Global Institute of Management and
Emerging Technologies

CONTENTS

3.1 INTRODUCTION

The COVID-19 pandemic will exacerbate the underlying mental health problems and contribute to further cases in children and adolescents due to the unusual mix of public health epidemics, social exclusion and economic contraction. Economic downturns are associated with elevated problems of youth emotional well-being and may be affected by how economic downturns affect individual employment, mental well-being for parents and child abuse. Coronaviruses are a common family of viruses which cause disease in humans and animals alike. COVID-19 coronavirus, a new form, was first discovered in Wuhan, China. Nevertheless, the epidemic has circulated rapidly across much of the international community and has triggered a pandemic in recent times, according to the International Health Organization (WHO). In addition, the COVID-19 is now watched by every nation in the world. There are several methods for coronavirus diagnosis including surgical examination of photographs taken from chest computerized tomography (CT) scans and blood test results. Documented condition COVID-19 manifests as fatigue, weariness and dry cough. More specifically, several techniques such as medical detection kits may be employed to detect the virus' initial results. Thus, these devices bear massive costs that require months of installation and use. Thus this article suggests a new detection method for COVID-19 utilizing built-in mobile sensors [1]. Because most radiologists already have smartphones for various daily uses, the proposal provides a low-cost solution. Not only this, but average users will even be using the software on their smartphones for virus identification purposes. Smartphones are now powerful, with current computer-rich cups, memory space and a wide range of sensors including cameras, microphones, temperature sensors, inertial sensors, proximity sensors, light sensors, moisture sensors and wireless chipsets as well as sensors. The device developed by Artificial Intelligence (AI) reads the mobile sensor signal measurements to predict the severity grade of pneumonia and to predict the outcome of the illness. In fact, 35% of young people who provided any mental health support from 2012 to 2015 have sought their mental well-being therapy from school settings. The closing of schools will be especially averse to the behavioral health services supported by the public (Figure 3.1) [2].

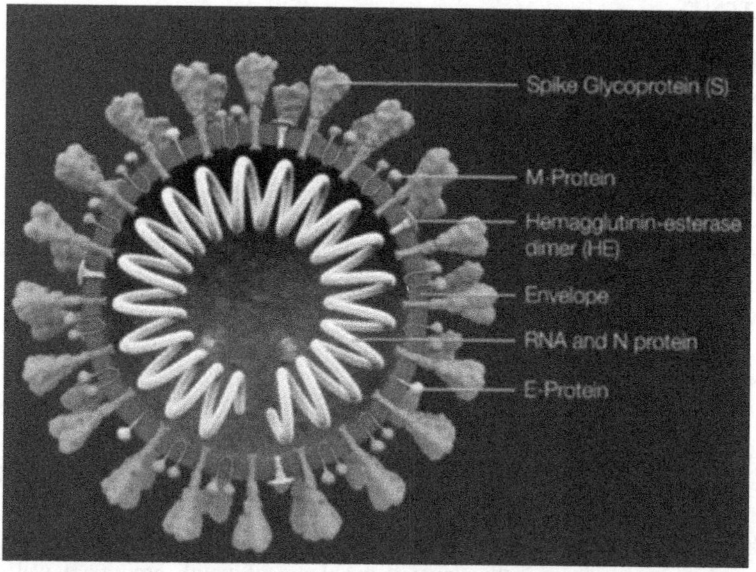

FIGURE 3.1 Structure of corona virus [3].

It is therefore necessary to note that school delays can be even more detrimental to mental health services for certain young people. Adolescents in racial and ethnic minority backgrounds and low family wealth eligible with public health services were more likely to try out school-only mental health therapy. The family resources and existing clinical relationships may be lacking for these students to gain quick access to alternative community-based services. Policy-makers will seek solutions to tackle short-term changes to children's mental health programs across COVID-19, while simultaneously setting foundations for strengthening children's mental healthcare in the long run. The United States Department of Health and Human Services has approved a narrow waiver in reaction to COVID-19 to relax the requirements of the Health Insurance Portability and Accountability Act (HIPAA). This means resources that were not HIPAA-compliant (e.g., FaceTime Apple) are now accessible for clinicians to analyze and manage [4]. This temporary change can be instructive when it comes to whether, if the relaxed rules improve access to care, HIPAA regulations can be rebalanced over the longer term. Other privacy restrictions could be reconsidered, even for a temporary period. For example, the United States Federal Student Rights and Privacy Act prohibits the release of details from education reports. This may be a deterrent to organized treatment for adolescents who are already expected to seek services beyond school but whose physicians are unsure of the appropriate details that could be accessible to the hospital. As well as regulating privacy, clinicians should be aware that other privacy considerations may hinder the use of mental health services on the telephone [5]. And physical distancing recommendations in places where, aside from parents and family, most young people may not be eligible to access mobile mental wellness services in a private environment. For low-income households with limited living spaces, this challenge may be especially important.

3.2 COVID PANDEMIC: LOCALIZATION TO GLOBALIZATION

The study attempts to determine the size of the effect COVID-19 has on the global economy. Since its shrouded in uncertainty, however, about 35% of the respondents were unable to determine the situation. About 33% of respondents expect a major downturn in the global economy, of which 49% were in the services market. Overall, only 17% feel the slowdown continues until the lockdown is in effect, post they expect to quickly recover. Roughly 84% of market leaders anticipate a significant realignment from global supply chains as a consequence of the pandemic, although digital/tech penetration will grow faster. Nearly 77% of the respondents, however, are optimistic that the market impact would be short-term because they anticipate the emergence of new business models. Companies on lockdown took on a hit with India like never before [6]. Through this survey, a report from the Bengaluru-based B2B market analysis and corporate consultancy firm Feedback Consultancy attempted to determine the effect of COVID-19 on companies across sectors and the economy. The survey noted that remote work (29%), digital/tech adoption (23%), employee health scheme concentration (17%), and deglobalization (18%) would be the main resets in industry in the postpandemic environment. Around 46% of businesses currently plan to invest more and more on employee engagement, while 66% of leaders agree the role is gaining more prominence than ever. "Around 146 business leaders took part in this web-based survey in April 2020, which was carried out across various sectors, including Automobiles, Agriculture, Banking, Financial Services and Insurance, Chemicals, Energy, Pharmaceuticals, Healthcare, Education, Construction, IT and Services, to name a few," a research showed. The effect of the pandemic on the volatile world environment will function as a catalyst for globalization, and 69% believe it is more important than ever. Currently 59% of businesses expect to work from home to become a standard in a postpandemic setting. Though business leaders were largely hopeful they were dismissive of the government's initiatives. One out of three business leaders believe Indian recovery is better than the global economy. UNESCAP has published the annual COVID-19 economic and social report, which is an enormous risk that would be a greater blow to the world economy than the global financial crisis, powered mainly by a blow to market. That includes a demand and supply blow that is already evolving. It is now apparent that many economies would also shrink developing countries, as would those heavily dependent on tourism and commodity trade in the Asia Pacific region. Commodity rates have been down to their lowest over the past 10 years [7]. But for India, owing to low oil and commodity prices, there is a small silver lining as we are net importers and, even because the government does not require a complete passage of lower global prices, it implies that there is some fiscal space through commodity price reductions. Nevertheless, the disturbance in the workplace is extremely bad, particularly in Ministry of Micro, Small & Medium Enterprises (MSMEs), which are the backbone of manufacturing, commerce and services. This is a really big blow to the global economy and a lot of things will shift once we come out of it. Social science analysis suggests that social patterns change as a consequence of a crisis if three conditions are met: First, the activities in question have to be identified as the cause or at least the aggravation of the crisis. Second, alternatives to the behavioral modes defined above must be practical and not too expensive. Third, if the activities involved before the recession were still deteriorating, a recession is particularly apt to bring about transition. For starters, the Second World War contributed to a significant increase in decolonization, not least because colonialism had already passed past its height. Under this context, it must be debated whether economic globalization, too, will become a survivor of COVID 19 [8]. There is one clear reason for that. The virus as such is not a product of globalization but the absence of localization or undue reliance on it. In each case, the onset of the pandemic occurred locally and then spread epidemically to the region. The various, more or less compact circles on the global contagion diagram reflect this, and each of these stands for a regional disease.

3.3 GENETIC TRANSMISSION STAGES

The latest 2019 coronavirus disease (COVID-19) is an infectiously developing disease that was first identified in Wuhan City, China, and was infected with a new coronavirus called Extreme Acute Coronavirus Syndrome 2 (SARS-CoV-2). The term move refers to the transmission of microorganisms through direct contact, droplets or indirect communication from one infectious individual to another noncontaminated material, such as surface contamination. SARS-CoV-2 is a β-coronavirus in

Coronavirus (CoV) and is a nonsegmented positive sensory RNA virus, collectively classified into four genera. All α and β-CoV may get contaminated with mammals while both α- and β-CoV are more specific to the birds [9]. Six small pathogenic CoVs with mild respiratory symptoms are traditionally referred to as prone human viruses, similar to the common cold, caused by α-CoVs HCoV-229E, HCoV-NL63, β-CoVs HCoV-HKU1 and HCoV-OC43. The other two β-CoV, SARS-CoV and MERS-CoV, were known to cause serious and potentially fatal infections of the respiratory tract. The SARS-CoV-2 genome sequence was observed to be 96.2% similar to that of RaTG13 bat CoV but it is 79.5% similar to the SARS-CoV. Based on viral sequence findings and evolutionary theory, Bat has been believed as a natural host with virus origin, with the SARS-CoV-2 moving via uncertain middle hosts from bats to human infections. A single-stranded coronavirus with positive RNA is SARS-CoV-2. Two thirds of viral RNA codes are 16 nonstructural proteins which are mainly in the first open reader frame (ORF 1a/b) [10]. There are four basic protein structures left over from the virus genome codes, including glycoprotein (S) spicy protein, protein (E), protein (M) matrix, and nucleocapsids (N) protein and other compound proteins. A crucial step in the entry of SARS-CoV-2 glycoprotein into cell host receptors is the enzyme-conversion angiotensin 2 (ACE2). There is also no information about the possible SARS-CoV-2 endocytosis molecules that allowed membrane invagination. Lot of protein, including viruses, can induce pathogenesis. Infection resistance and disease incidence may also be affected by the host conditions (lower table). SARS-CoV-2 is toxic and tends to grow to severe circumstances for older adults and people with serious disabilities. RBD, the domain of receptor binding; HR1, repeats heptad 1. Human SARS-CoV-2 is mainly spread to people, including family and associates who have observed patients or incubation carriers near. The propagating efficiency of SARS-CoV-2 has proved high and has recorded higher reproductive numbers of H1N1 virus in 2009 [11]. There have been several discussions on exposure routes, particularly on the function of aerosol transmission, that have contributed to this high transmissibility. The infected can release SARS-CoV-2, which contains aerosols and droplets, by coughing or snoring as with a respiratory infective disease (Figure 3.2).

3.4 CORONAVIRUS REPLICATION AND PATHOGENESIS

Until first, the viruses were divided into different genera based on serology, but now they are distinguished by phylogenetic clusters. Both viruses have a good understanding of nonsegmented RNA viruses in the Nidovirales family. We still have very small genomes of RNA virus and some of the most widely recognized viruses consisting of up to 33.5 kilobases (kb) [13]. As the cryo-electron tomography and cryo-electron microscopy in latest studies have shown, the coronavirus virus is spherical with an average diameter of 125 nm. Coronavirus is most commonly found in club-formed spike projections originating from the center of the virus. Such peaks are the distinctive nature of the virus and give the look of a solar corona, which induces the term coronavirus. Within one virus genome, nucleocapsids are detected. Coronaviruses have symmetrical nuclear capsids that are rare in RNA viruses with positive sense but are much more common in RNA viruses with negative meaning. The next step

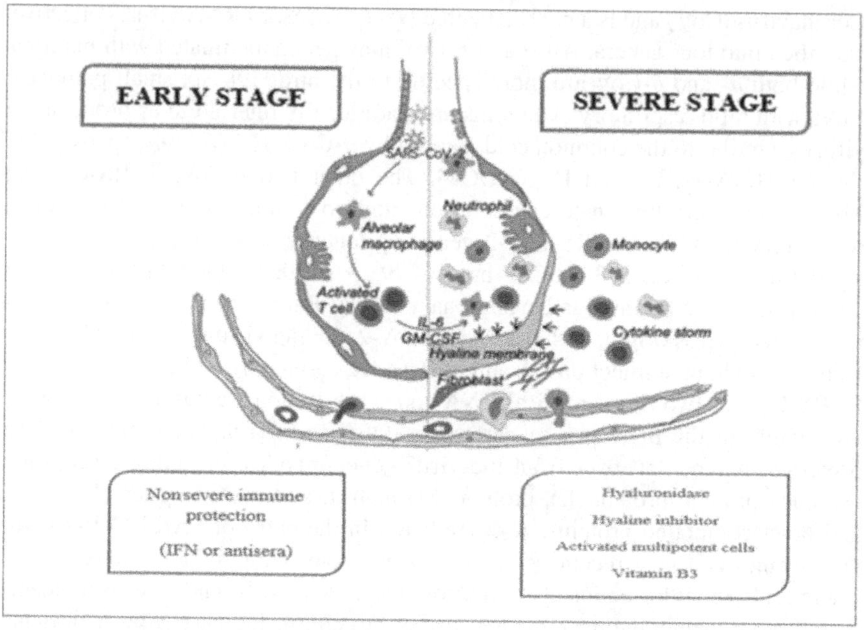

FIGURE 3.2 Stages of COVID 19 [12].

in the life cycle of the coronavirus is to translate the gene into genomic RNA replicas. The replicas gene codes two essential oRF, repla and replb, generating two coterminals, ppla and pplab, polyproteins. The virus uses a smooth sequence and a pseudoknot of the RNA (induction of a ribosomal frameshifting from the read-frame repla into the replb ORF) to transfer all polyproteins [14]. In certain cases, the ribosome relaxes the structure of the pseudoknot and starts to repeat the stop codon. The pseudoknot sometimes slows the ribosomes and allows them to stop on the slanting axis. The reading frame can be modified by pulling down a nucleotide, a-1 frameshift. The replb translation leads to the pplab phase thereby. In vitro experiments show that ribosomal frame shifting is as frequent as 25%, although virus infection has not been determined. Although it is not clear that these viruses use frameshifting for proteins, either the exact replb and repla proteins ratios are tracked, or the processing of replb products is postponed until the repla products have produced an adequate environment for RNA replication [15].

3.5 DETECTION AND DIAGNOSIS

The virus blood culture and a high-production analysis of the genome as a whole would certainly classify SARS-CoV-2. However, because of hardware reliance and high cost, the use of high-performance sequencing technology in clinical diagnostics is minimal. The RT-qPCR method in the respiratory and blood secretions is therefore the most general, successful and easy for pathogenic diseases. Many firms in China have released rapidly RT-qPCR research kits for therapeutic evaluation since SARS-CoV-2 outbreak [16]. If all results are positive, a laboratory-verified infection can be identified. In respiratory testing, samples from two SARS-CoV-2-infected patients had to be checked as accurate, all the negative control samples had been reported with zero. A further analysis showed that 91.7% (11/12) use of RT-qPCR (nonsample SYBR-based fluorescence signal) for the positive SARS-CoV-2 outcome of self-collected saliva suggests that salivary is an important noninvasive tool for patients who need to diagnose, track and treat SARS-CoV-2. SARS-CoV identification by RT-qPCR can only be 50%–79% prone, according to the procedure and sample size, as well as according to the number of clinical specimens collected [17]. Therefore, the risk of RT-qPCR diagnosis for SARS-CoV-2 infection would be improved. There are numerous other disadvantages to RT-qPCR, including certain biological safety dangers caused by patient samples preparation and evaluation, lengthy nuclear acid detection periods and longtime studies. Although RT-qPCR is accurate for diagnosing COVID-19, its falsified downside is unavoidable because it becomes more flexible because of the drastic consequences, and many health professionals are of the opinion that CT scans are one of the diagnostic tools. Among relatively healthy individuals, a successful and negative SARS-CoV-2 infection screening can be included in daily RT-qPCR testing and chest CT screening with RT-qPCR (Figure 3.3) [18].

For early detection and disease evaluations of patients with SARS-CoV-2, high-resolution chest CT is particularly important. CT scans for COVID-19 are highly clinical, especially in the high-prevalence SARS-CoV-2 region, according to some reports. CT scans may have risks, such as the pathological hysteresis of CT images and other viral pneumonia. The success of the tests relies on a variety of factors, including the time from the initiation of the outbreak, the virus presence in the specimen, the quality and treatment of specimens, and the exact composition of reagents in the research kits. The findings of the studies include centered on experience of other respiratory disorders, including influenza, for antigen-dependent RDTs influenza, where patients with flu virus, as demonstrated by COVID-19, are similarly prone to respiratory samples from 34% to 80%. The regulation of population-based

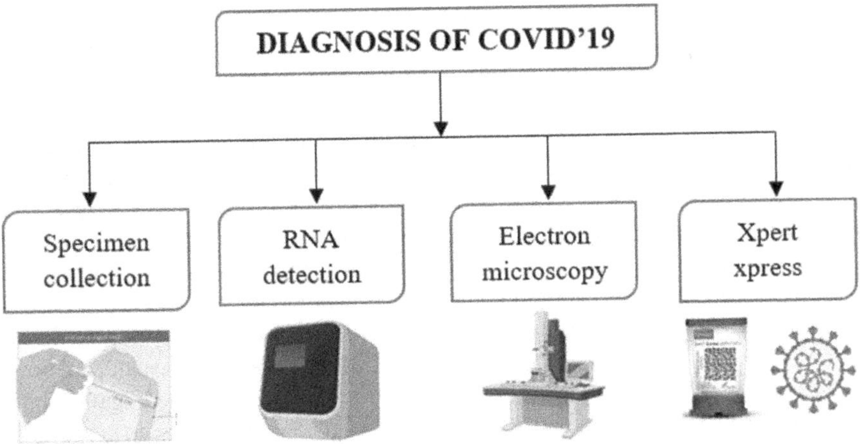

FIGURE 3.3 Detection and diagnosis of COVID-19 [19].

COVID-19 antibodies would be crucial for the growth and enhancement of our awareness of infection in unrecognized people through successful case recognition and survival. Practices, group attack rates and risk of infection [20]. Nonetheless, these assessments are of no benefit for clinical diagnosis because acute illness or procedures required to determine recovery cannot be accurately detected or administered. In cases with a negative molecular test, but strong epidemiologic relationship to COVID-19 infection, some clinicians have made presumptive diagnosis by these antibody response tests of the new COVID-19 disease has demonstrated elevated levels of anticourses and combined blood samples (acute and recoveries).

3.6 STRATEGIES FOR PREVENTION

COVID-19 is a serious, acute coronavirus 2 (SARS-CoV-2) infection. Fast patient diagnosis and treatment became the founding pillar in curbing the accelerated spread of communicable disease. Nonetheless, care should be practiced in detecting asymptomatic viral carriers instantly. Enhanced environmental protection and sanitation regulation are officially approved at community level in order to shield the public from contaminants. Burn-out was induced in most hospitals in the province of Hubei, particularly in Wuhan City, through excessive workloads and a lack of personal protective equipment for physicians. The lack of respiratory devices (i.e., supply of oxygen, ventilators) exacerbated this, which prevented the mechanical ventilation of certain hypoxemic patients. COVID-19 patients also needed to quickly protect themselves from certain feverish circumstances. The Chinese Government has urgently recommended that publics should not travel to Wuhan City and has ordered all local governments to monitor the quantities of breathing treatment, microbiology and infectious disease specialists in the network of respiratory drugs, and the China Disease Prevention and Control Centre to track the amount of cases in real time. Approximately 80% of patients have been reported to recover without significant symptoms from the disease. However, one in six COVID-19 patients may become critically sick and have respiratory problems [21]. In more severe cases, illness may contribute to acute pneumonia and other diseases that can only be handled in facilities of greater standard (District Hospitals and above). In a few cases, that can also cause death. COVID-19 spreads primarily by droplets created by coughing or sneezing an individual infected with COVID-19. It can happen in two ways: by being in direct contact with COVID-19 patients (within 1 m of the infected person), one can get the infection, especially if they do not cover their faces when they cough or sneeze. Indirect contact: The droplets linger several days on surfaces and clothing. And the illness can be spread by touching some other infected object or cloth and then rubbing the mouth, nose or eyes. Using at least 20 seconds of moist soap and rinse, then rub your face. The lather's gamut of wrists and fingernails beneath the palms. This can also be used as antibacterial agents and as antivirals. If you cannot wash your hands well, use the sanitizer. Many times a day, particularly after handling things like your phone or laptop, get your hands rewashed again. Should not cover your face or head with your lips, nose and ears. Stop your fingernails just scraping. It can give SARS-CoV-2 the opportunity to travel from your mouth into your bloodstream. Likewise, do not touch any men. Touching your skin can move

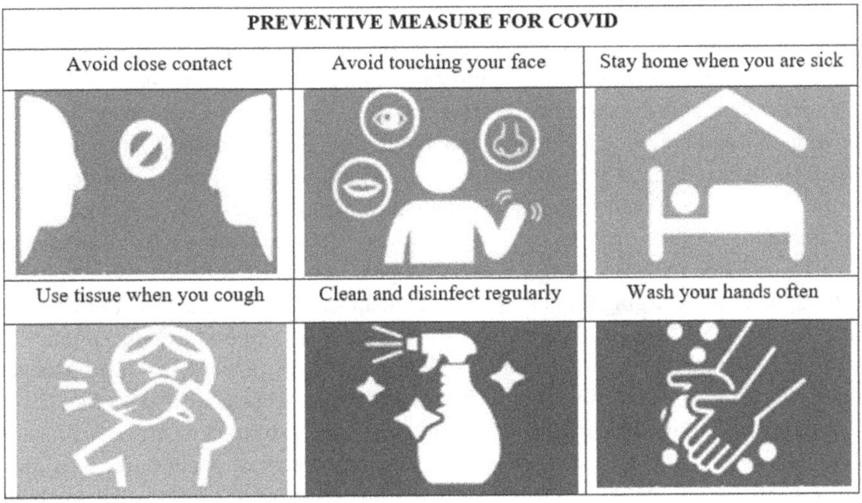

FIGURE 3.4 Prevention of COVID [23].

SARS-CoV-2 from person to person. It is also essential not to indulge in utensils and straws. Teach children to learn their own cup, straw and other recycled items. The nose and the mouth display high amounts of SARS-CoV-2 [22]. This means air droplet will transfer on to others when you are coughing, sneezing or chatting. It can collapse and remain in it for up to 3 days on hard surfaces. Using the tissue to keep your hands as warm or sneeze your elbow as possible. Wash your mouth regardless, whether you are hacking or sneezing. Use solutions for white vinegar or hydrogen peroxide between the disinfecting surfaces for general cleaning. Being in a party or getting to know each other ensures that you can interact more closely with others. That involves avoiding all public service sites, as you could be too close to another party to sit or stand. Not clustered either in beaches or in parks. The virus could spread to foods, tables, utensils and cups. Many users on the platform may also be temporarily airborne. The latest coronavirus tends to spread farther into the nose than through the lungs and other areas of the body, a 2020 study said (Figure 3.4).

3.7 AI FOR COVID

AI is being used as a weapon to help combat the pandemic virus that has infected the entire planet since the early 2020s. Press and the scientific community echo the high hopes that the coronavirus may be confronted by data science and AI, and that science can still fill in the blanks. Three experts on the issue of monitoring devices on cell phones were put together in the battle against COVID-19. The theories and the reality of these programs were discussed by Michael Veale, lecturer at University College London, created the anonymized authentication protocol DP-3T. This technique is by definition autonomous, with activities that the epidemiologists perform as useful [24]. The software will allow them to process mass data while restricting the burden of gathered (centralized) data at one location—which could be misused or cross-referenced with other records. Michael Veale provided an overview of how a contact tracing or proximity tracing device operates. If someone is ill (voluntarily) in the client files, the program recomposes a number of individuals he/she has been in touch with during the last few days at a gap of 2–3 m. Then, by data analysis, these individuals would provide a completely anonymous message, with options for action. Based on his experience at the task force of the Italian Government, Dino Pedreschi claims that the use of data and models will help counter the outbreak, particularly after containment. There are two potential approaches: either pressuring citizens to engage in this automated monitoring, which would appear to be inconsistent with democratic criteria, or actually urging them to do so, which appears to be the case with European propositions [25]. Michael Veale‘s DP-3T protocol fully supports these “firemen's” work, by allowing them to trace more contacts. It is actually a way to extend a person's memory that has been testing positively to know what his or her contacts have been over the last few days. The use of mobile phones can help to track communications in a completely private and accessible way. It also automates an epidemiologist's work because people will get an email telling them that they are in contact with someone who has been diagnosed with the virus and that they will apply for the authority to be voluntarily checked. Those completely decentralized approaches need to be thought out, with data shared with none and based on the infected people's voluntary approach to contribute to public health. Because the pandemic spreads across the globe, revolutionary AI innovations have collected in many specific forms. Location-based messaging was a pivotal tool in South Korea's fight to reduce disease transmission. In China, Alibaba unveiled an AI algorithm claiming to be able to diagnose suspicious cases with 96% accuracy within 20 seconds (almost 45 times faster than human detection). Autonomous automobiles were easily placed to use in situations that could have been too risky for humans. Robots delivered food, medicine, and equipment to patients in hospitals or quarantined families in the Hubei and Guangdong provinces of China, many of them had lost home due to the virus [26]. Computer scientists in California are focusing on apps that will track the safety of elderly people in their homes remotely and include warnings as they are sick with COVID-19 or other diseases.

3.8 SMART SENSORS FOR EARLY STAGE DETECTION

Researchers at the Swiss Federal Materials Physics Laboratories (EMPA), ETH Zurich and the University Hospital of Zurich are examining whether an optical device tested on SARS may be used to reliably classify COVID-19 in the air in real time. The sensor incorporates optical and thermal impact to detect viruses in a safe and effective way. The sensor is based on tiny gold constructions, the so-called gold nanoislands, on a glass substratum, according to a press release by EMPA. The artificially generated DNA receptors are grafted onto the nanoislands to suit different RNA virus sequences [27]. The coronavirus is a so-called RNA virus: its genome is not a double strand of DNA, like it is in living species, but a single strand of RNA; sensor surface receptors are therefore the complementary sequences to the special sequences of RNA viruses that can accurately recognize the virus. Localized surface plasmon resonance is the technique of detection, an optical phenomenon that occurs in metallic nanostructures. They modulate the incident light in a wide range of wavelengths when excited and generate a near-field plasmonic around the nanostructure. Inside the plasmonic excited near-field the local refractive index differs, as the molecules bind to oxygen. To calculate this shift, an optical sensor located at the back of the specimen may be used to decide whether the sample includes the respective RNA strands. Through contrast, most laboratories use a molecular method named polymerase chain reverse transcription reaction (RT-PCR) to classify viruses in respiratory infections. This program is well established and can recognize only limited numbers of viruses so it can be time-consuming and vulnerable to error at the same time. Further research work is required to determine the concentration of the corona virus in the air until the sensor is equipped. But researchers say the principle could be applied to other viruses once it is ready and help to detect and stop epidemics early on. Therefore, only those RNA strands that exactly fit the DNA receptor on the sensor need to be rendered. This is why a second impact on the sensor falls into action with the photothermal plasmonic effects. If the same nanostructure is excited by a laser with a certain wavelength on the sensor, it will emit localized heat. Nevertheless, the sensor is currently not available to calculate the concentration of corona virus in the soil, e.g., at the main train station in Zürich [28]. A system that draws in the air, for example, concentrates the aerosols in it and releases the RNA from the viruses, which still requires a number of developmental steps. "It also requires an analysis of the development," says Wang. But once the sensor is enabled the theory can be extended to other viruses and aid in early detection and prevention of epidemics. "Vayyar Imaging," the world pioneer in 4D imaging sensor technology, and Meditemi (JV of Ogawa and Medisana), a revolutionary in robotic healthcare solutions, have partnered to develop a smarter medical robot that provides a range of remote monitoring capabilities including early detection of COVID-19 symptoms, and safety decline and fall alerts [29]. The solution equips Meditemi's health robots with Vayyar's 4D intelligent RF sensors to easily, stably and reliably track and control COVID-19 symptoms in the early stages. A person approaching the robot within 1 m of it can conduct a fast, touchless scan to determine their heart rate (BPM), respiratory rate (RPM), waveform and temperature. The care robot can regain public trust by delivering rapid, automated screening across all public spaces, including building exits, warehouses, public transit, airports, trade shows and border crossings, with zero personnel intervention or sanitation requirements. This will also require surveillance of people in their residences, alerting health care professionals or declining health conditions [30].

3.9 CONCLUSION

As of 5 March 2020, the COVID-19 outbreak in China was rapidly sweeping and spreading to 85 countries, abundant territories and areas outside China. Scientists have taken measures to classify the present coronavirus and are working extensively on therapies and vaccines for viruses. This research review outlined SARS-CoV-2's existing awareness as follows:

- First, on morbidity and mortality, SARS-CoV-2-induced evolving pneumonia, COVID-19, shows high infectivity, but less virulence compared to SARS and MERS.

- The evolution of corona virus is clearly described in the introduction section and the methodologies to detection and diagnosis are explained in the following sections.
- The influence of smart sensors and the role of AI in treating the corona virus are clearly examined and delivered.
- Further, unravel the mystery of the molecular mechanism for viral entry and replication that provides the basis for future research into the development of targeted antiviral drugs and vaccines.

REFERENCES

1. COVID, CDC, and Response Team. "Severe outcomes among patients with coronavirus disease 2019 (COVID-19)—United States, February 12–March 16, 2020." *Morbidity and Mortality Weekly Report* 69, no. 12 (2020): 343–346.
2. Lodigiani, Corrado, Giacomo Iapichino, Luca Carenzo, Maurizio Cecconi, Paola Ferrazzi, Tim Sebastian, Nils Kucher et al. "Venous and arterial thromboembolic complications in COVID-19 patients admitted to an academic hospital in Milan, Italy." *Thrombosis Research* 191 (2020): 9–14.
3. *Structure of corona virus*. https://www.dnasoftware.com/wp-content/uploads/Screenshot-174-e1580840718372.png
4. Fauci, Anthony S., H. Clifford Lane, and Robert R. Redfield. "Covid-19—Navigating the uncharted." *The New England Journal of Medicine* 382, no. 13 (2020): 1268–1269.
5. Mehta, Puja, Daniel F. McAuley, Michael Brown, Emilie Sanchez, Rachel S. Tattersall, Jessica J. Manson, and HLH across Speciality Collaboration. "COVID-19: Consider cytokine storm syndromes and immunosuppression." *Lancet (London, England)* 395, no. 10229 (2020): 1033.
6. World Health Organization. "Coronavirus disease 2019 (COVID-19): Situation report, 72." (2020).
7. Hollander, Judd E. and Brendan G. Carr. "Virtually perfect? Telemedicine for COVID-19." *New England Journal of Medicine* 382, no. 18 (2020): 1679–1681.
8. Bai, Yan, Lingsheng Yao, Tao Wei, Fei Tian, Dong-Yan Jin, Lijuan Chen, and Meiyun Wang. "Presumed asymptomatic carrier transmission of COVID-19." *JAMA* 323, no. 14 (2020): 1406–1407.
9. Helmy, Yosra A., Mohamed Fawzy, Ahmed Elaswad, Ahmed Sobieh, Scott P. Kenney, and Awad A. Shehata. "The COVID-19 pandemic: A comprehensive review of taxonomy, genetics, epidemiology, diagnosis, treatment, and control." *Journal of Clinical Medicine* 9, no. 4 (2020): 1225.
10. Wang, Huihui, Xuemei Li, Tao Li, Shubing Zhang, Lianzi Wang, Xian Wu, and Jiaqing Liu. "The genetic sequence, origin, and diagnosis of SARS-CoV-2." *European Journal of Clinical Microbiology & Infectious Diseases* 39 (2020): 1629–1635.
11. Kolifarhood, Goodarz, Mohammad Aghaali, Hossein Mozafar Saadati, Niloufar Taherpour, Sajjad Rahimi, Neda Izadi, and Seyed Saeed Hashemi Nazari. "Epidemiological and clinical aspects of Covid-19; a narrative review." *Archives of Academic Emergency Medicine* 8, no. 1 (2020): e41.
12. *Stages of COVID 19*. https://media.springernature.com/full/springer-static/image/art%3A10.1038%2Fs41418-020-0530-3/MediaObjects/41418_2020_530_Fig1_HTML.png
13. Chen, Yu, Qianyun Liu, and Deyin Guo. "Emerging coronaviruses: Genome structure, replication, and pathogenesis." *Journal of Medical Virology* 92, no. 4 (2020): 418–423.
14. Liu, Jia, Xin Zheng, Qiaoxia Tong, Wei Li, Baoju Wang, Kathrin Sutter, Mirko Trilling, Mengji Lu, Ulf Dittmer, and Dongliang Yang. "Overlapping and discrete aspects of the pathology and pathogenesis of the emerging human pathogenic coronaviruses SARS-CoV, MERS-CoV, and 2019-nCoV." *Journal of Medical Virology* 92, no. 5 (2020): 491–494.

15. Iqbal, H. M., Kenya D. Romero-Castillo, Muhammad Bilal, and Roberto Parra-Saldivar. "The emergence of novel-coronavirus and its replication cycle: An overview." *Journal of Pure and Applied Microbiology* 14, no. 1 (2020): 13–16.
16. Khan, Asif Iqbal, Junaid Latief Shah, and Mohammad Mudasir Bhat. "Coronet: A deep neural network for detection and diagnosis of COVID-19 from chest x-ray images." *Computer Methods and Programs in Biomedicine* 105581 (2020): 105581–105589.
17. Hao, Wendong and Manxiang Li. "Clinical diagnostic value of CT imaging in COVID-19 with multiple negative RT-PCR testing." *Travel Medicine and Infectious Disease* 34 (2020):101627.
18. Sabino-Silva, Robinson, Ana Carolina Gomes Jardim, and Walter L. Siqueira. "Coronavirus COVID-19 impacts to dentistry and potential salivary diagnosis." *Clinical Oral Investigations* 24, no. 4 (2020): 1619–1621.
19. https://microbenotes.com/wp-content/uploads/2020/04/Sample-collection-and-Diagnosis-of-COVID-19.jpeg
20. Nguyen, Trieu, Dang Duong Bang, and Anders Wolff. "2019 novel coronavirus disease (COVID-19): Paving the road for rapid detection and point-of-care diagnostics." *Micromachines* 11, no. 3 (2020): 306.
21. Mamun, Mohammed A. and Mark D. Griffiths. "First COVID-19 suicide case in Bangladesh due to fear of COVID-19 and xenophobia: Possible suicide prevention strategies." *Asian Journal of Psychiatry* 51 (2020): 102073.
22. Watkins, John. "Preventing a covid-19 pandemic." (2020).
23. *Prevention of COVID.* https://www.gillettechildrens.org/images/made/assets/blog/COVID-19-Updated-Mar12-KHM-Graphic-1500px-_1500_1110_70.jpg
24. Vaishya, Raju, Mohd Javaid, Ibrahim Haleem Khan, and Abid Haleem. "Artificial Intelligence (AI) applications for COVID-19 pandemic." *Diabetes & Metabolic Syndrome: Clinical Research & Reviews* 14, no. 4 (2020): 337–339.
25. Yang, Zifeng, Zhiqi Zeng, Ke Wang, Sook-San Wong, Wenhua Liang, Mark Zanin, Peng Liu et al. "Modified SEIR and AI prediction of the epidemics trend of COVID-19 in China under public health interventions." *Journal of Thoracic Disease* 12, no. 3 (2020): 165.
26. Gozes, Ophir, Maayan Frid-Adar, Hayit Greenspan, Patrick D. Browning, Huangqi Zhang, Wenbin Ji, Adam Bernheim, and Eliot Siegel. "Rapid ai development cycle for the coronavirus (covid-19) pandemic: Initial results for automated detection & patient monitoring using deep learning CT image analysis." arXiv preprint arXiv:2003.05037 (2020).
27. Allam, Zaheer and David S. Jones. "On the coronavirus (COVID-19) outbreak and the smart city network: Universal data sharing standards coupled with artificial intelligence (AI) to benefit urban health monitoring and management." *In Healthcare* 8, no. 1 (2020): 46. Multidisciplinary Digital Publishing Institute.
28. Mohammed, M. N., Halim Syamsudin, S. Al-Zubaidi, Ramli R. AKS, and E. Yusuf. "Novel COVID-19 detection and diagnosis system using IOT based smart helmet." *International Journal of Psychosocial Rehabilitation* 24, no. 7 (2020): 1–7.
29. Mujawar, Mubarak A., Hardik Gohel, Sheetal Kaushik Bhardwaj, Sesha Srinivasan, Nicolerta Hickman, and Ajeet Kaushik. "Aspects of nano-enabling biosensing systems for intelligent healthcare; towards COVID-19 management." *Materials Today Chemistry* 17 (2020): 100306.
30. Rashid, Md Tahmid and Dong Wang. "CovidSens: A vision on reliable social sensing based risk alerting systems for COVID-19 spread." *arXiv preprint arXiv:2004.04565* (2020).

4 Coronavirus

An Overview of Genetics and Transmission Stages

Neha Gupta and Shayan Ahmed
Jamia Millia Islamia

CONTENTS

4.1 INTRODUCTION

Coronaviruses are a large group of enveloped viruses having positive sense RNA as their genome. These viruses are known to cause infections in mammals and Aves. In humans, coronaviruses are known to cause infections associated with the respiratory tract. These respiratory tract infections can range from mild to lethal [1,2]. Mild infections include common cold (generally caused by rhinoviruses). Lethal infections include outbreaks like Severe Acute Respiratory Syndrome (SARS) as occurred in 2002, Middle East Respiratory Syndrome (MERS) in 2012 and now recently reported Coronavirus Disease-19 (COVID-19) in 2019 [1,3].

The name "coronavirus" is derived from the Latin word "*corona*," meaning "crown" or "wreath." As in the case of coronaviruses, morphologically enveloped glycoproteins exhibit a crown-like appearance containing a genome of about 27–32 kb. The term coronavirus was coined by June Almeida and David Tyrrell who were the first to observe and study the morphology of human coronaviruses [4]. The discovery of coronaviruses was first reported in the 1930s from domesticated chickens having an acute respiratory infection caused by Infectious Bronchitis Virus (IBV) [5]. Six years later, Fred Baudette and Charles Hudson were successful in isolating and cultivating the IBV [6]. Human coronaviruses were discovered in the 1960s. In 1960s, E.C. Kendall, Malcom Byone and David Tyrrell were successful in isolating a novel common cold virus B814. Around the same time, at the University of Chicago, Dorothy Hamre and John Procknow isolated another novel cold virus 229E. The morphological characteristics of these two novel strains B814 and 229E were imaged by a Scottish virologist June Almeida in 1967 with the application of electron microscopy [7]. Since then, other human coronaviruses have been identified such as SARS-CoV in 2002, Human Coronavirus NL63 (HCoV NL63) in 2004, HCoV HKU1 in 2005, MERS-CoV in 2012 and SARS CoV-2 in 2019 [8,9,10].

The emergence of SARS-CoV-2 in December 2019 leading to the COVID-19 pandemic has led to a global health crisis. This outbreak was first identified in December 2019, in Wuhan, China. The World Health Organization (WHO) declared this outbreak as a Public Health Emergency of International Concern on 30 January 2020, and a pandemic on 11 March 2020. As on 23 June 2020, more than 9.1 million cases of COVID-19 have been reported in more than 188 countries and territories, resulting in more than 472,000 deaths; with more than 4.52 million people have recovered. According to WHO, there are neither vaccines nor any specific antiviral treatments for COVID-19 till date. The management involves the treatment of symptoms, supportive care, isolation and experimental measures.

4.2 CLASSIFICATION

Coronaviruses belong to the order Nidovirales, whereby coronaviruses represent the major and largest group in Nidovirales under family Coronaviridae. Order Nidovirales comprises families Coronaviridae, Arteriviridae, Roniviridae and Mesoniviridae. Nidoviruses are enveloped, nonsegmented positive-strand RNA viruses that have certain distinctive characteristics [8,11]. Family Coronaviridae comprises two subfamilies, *Coronavirinae* and *Torovirinae.* Coronaviruses are principally placed and studied under the *Coronavirinae* subfamily; the other subfamily *Torovirinae* comprises genus *Toroviruses* which are known to cause infections in cattle, horses and swine, and *Bafiniviruses* which are known to cause infection in fishes [12]. Members of both *Coronavirinae* and *Torovirinae* differ from each other in the virion structure. Members of subfamily *Coronavirinae* have a flexible nucleocapsid under an envelope whereas members of subfamily *Torovirinae* have a peculiar doughnut-shaped nucleocapsid.

Under the approval of the International Committee on Taxonomy of Viruses, coronaviruses have been sorted into four groups, primarily based on serologic relationships and subsequently on the basis of phylogenetic clustering [13]. These groups are the alpha coronavirus, the beta coronaviruses, and the gamma and delta coronaviruses. ***Figure 4.1 (a, b) represent the taxonomic classification of coronaviruses and their commonly known structure, respectively.***

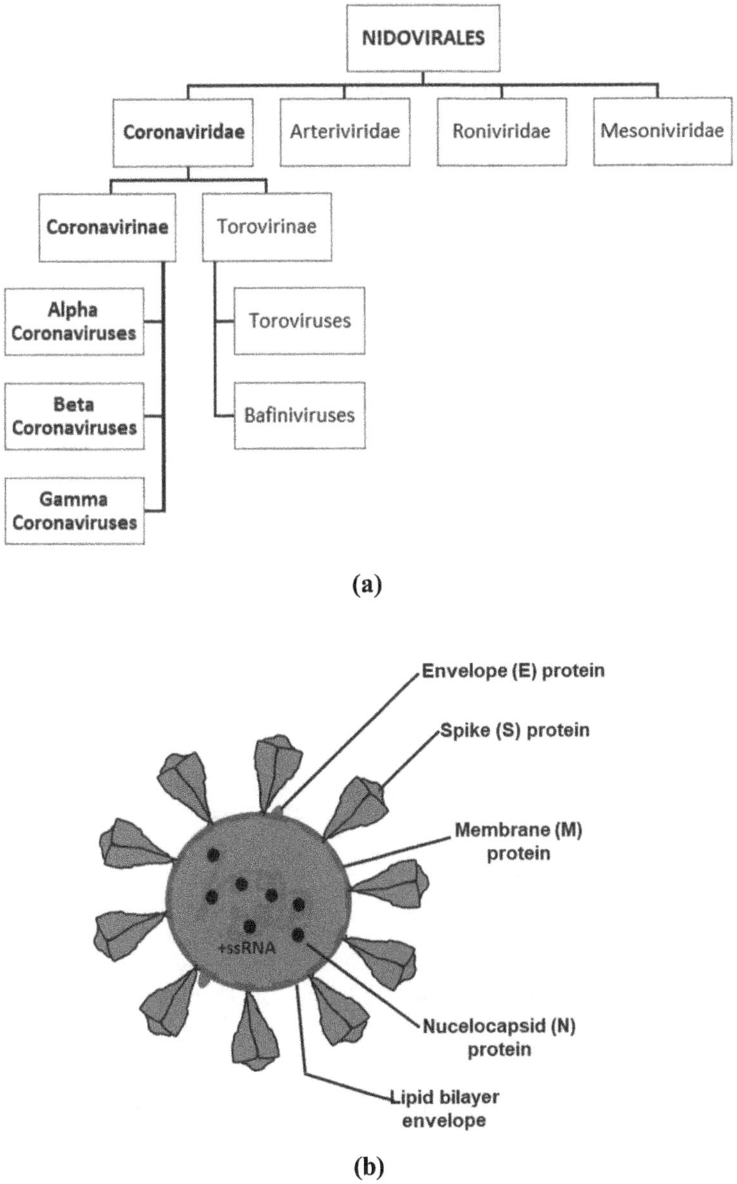

FIGURE 4.1 (a) Taxonomic classification of coronaviruses. (b) A schematic of virion showing structural proteins S, M, E and N.

Almost all alpha and beta coronaviruses are known to have mammalian hosts. Gamma and delta coronaviruses are primarily known to infect avian hosts but not exclusively. ***Examples of coronaviruses species sorted into their respective groups along with their hosts are listed in Table 4.1***. Recent studies have proven that many birds happen to be a rich source of new emerging viruses. Also, it has been proposed

TABLE 4.1
Examples of Different Groups of Coronavirus Species along with Their Host Organisms

Group	Species	Host	References
Alpha Coronaviruses	Transmissible gastroenteritis virus	Pig	[23]
	Porcine respiratory coronavirus	Pig	
	Feline infectious peritonitis virus	Cat	
	Feline enteric coronavirus	Cat	
	Canine coronavirus	Dog	
	Human coronavirus strain 229E	Human	
	Porcine epidemic diarrhea virus	Pig	
	Human coronavirus strain NL63	Human	
	Bat coronavirus strain 61	Bat	
	Bat coronavirus strain HKU2	Bat	
Beta Coronaviruses	Mouse hepatitis virus	Mouse	[23] [62] [63]
	Bovine coronavirus	Cow	
	Rat coronavirus	Rat	
	Sialodacryoadenitis virus	Rat	
	Human coronavirus strain OC43	Human	
	Porcine Hemagglutinating encephalomyelitis virus	Pig	
	Puffinosis coronavirus	Puffin	
	Equine coronavirus	Horse	
	Canine respiratory coronavirus	Dog	
	Human coronavirus strain HKU1	Human	
	Bat SARS coronavirus	Bat	
	SARS coronavirus	Human	
	MERS coronavirus	Human	
	SARS coronavirus 2	Human	
Gamma Coronaviruses	IBV	Chicken	[23]
	Turkey coronavirus	Turkey	
	Pheasant coronavirus	Pheasant	
	Goose coronavirus	Goose	
	Pigeon coronavirus	Pigeon	
	Duck coronavirus	Mallard	
Delta Coronaviruses	Bulbul coronavirus HKU11	ChineseBulbul	[8]
	Munia coronavirus HKU13	Munia Bird	
	White-eye coronavirus HKU16	White-eye Bird	
	Night heron coronavirus HKU19	Night Heron Bird	
	Wigeon coronavirus HKU20	Wigeon	

that both birds and bats act as reservoirs for the incubation and evolution of coronaviruses, owing to their ability to fly and their tendency to roost and lock [14].

4.3 GENOME STRUCTURE AND ORGANIZATION

The coronavirus genome happens to be a nonsegmented, single-stranded and positive-sense RNA genome. The genome size of coronaviruses ranges from 26.4 to 31.7 kilobases [14–16]. Genome size of SARS-CoV is 29,727 nucleotides [17], MERS-CoV is 30,119 nucleotides [18] and SARS CoV-2 is 29,903 nucleotides [19]. In fact, the genome of the coronaviruses is largest among all the RNA viruses, including viruses having segmented RNA genome [20]. The genome of coronaviruses is structurally similar to eukaryotic mRNA in having both 5′ cap [21,22] and 3′ poly-A tail [21]. Although the size of the coronavirus RNA genome is much larger as compared to the length of eukaryotic mRNA, in general, three times more than the eukaryotic mRNA also, in comparison with the eukaryotic mRNA, RNA genome of the coronavirus has a greater number of open reading frames (ORFs) [23].

4.3.1 Common Features of Genome Structure among Coronaviruses

The genome of coronaviruses basically comprises a set of genes in the invariant order ***5′-replicase-S-E-M-N-3′***, with the huge replicase gene that generally occupies the two-third of the coding capacity [23]. Following are the notable common features of genome structure among different coronaviruses:

a. The 5′ end of the genome contains a leader sequence and untranslated region (UTR) ranging from 209 to 528 nucleotides in length. This region also contains similarly positioned short, AUG-initiated ORF, capable of encoding peptides of 3–11 amino acids [20].
b. The 3′ end of the genome also contains the UTR ranging from 288 to 506 nucleotides in length. This region also possesses an octameric sequence of GGAAGAGC at 73–80 upstream from the 3′ poly-A tail [20,24].
c. Every genome has one extremely large gene around 5′ terminal, Gene 1 (replicase gene), covering approximately two-third of the genome separated into two ORFs 1a and 1b. This replica gene is responsible for encoding nonstructural proteins that are known to have a role in proteolytic processing of polyprotein products of replicase gene, replication and transcription [subgenomic mRNA (sg mRNA) synthesis] of the viral genome [20,24,25]. Translation of ORF1a yields polyprotein (pp) 1a (~450 kDa) whereas the translation of ORF1b requires a programmed ribosomal frameshift event that takes place just upstream of the ORF1a stop codon and results in polyprotein (pp) 1ab (~750 kDa) [20,26]. The replicase-transcriptase is the only protein that is translated from the genome; the other products of all downstream ORFs are derived from sg mRNAs [26]. ***Figure 4.2 represents a schematic of the genome organization as found in coronaviruses concerning SARS-CoV, SARS-CoV-2 and MERS-CoV.***

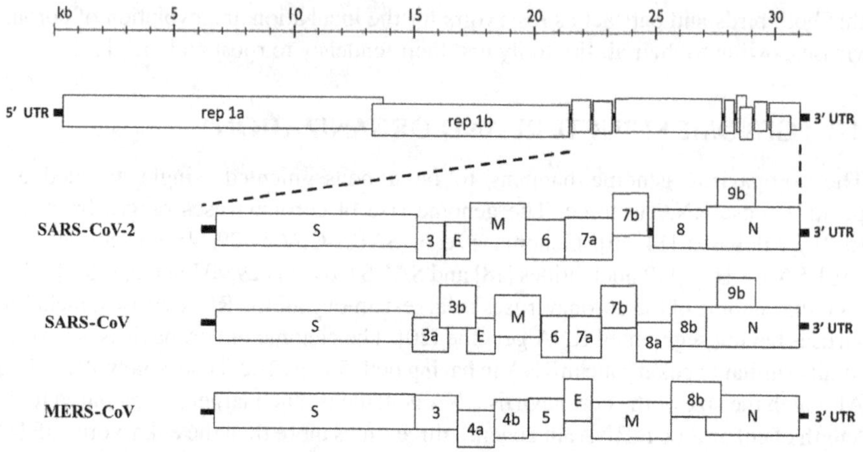

FIGURE 4.2 Coronavirus genome organization.

All coronaviruses encode the structural spike glycoprotein (S), a small envelope protein (E), membrane glycoprotein (M) and nucleocapsid protein (N), covering one-third of the genome around 3′ terminal. These proteins are products of downstream ORFs and are derived translation of the sg mRNAs. A variable number of other virus-specific ORFs that encode for certain proteins called accessory proteins are also found. Generally, these encode nonstructural proteins that are virus-specific in nature [20,24]. ***Table 4.2 enlists the specific ORFs with size of translation products found in certain coronaviruses. The major structural proteins found in coronaviruses are also shown in Figure 4.1 (b).***

4.4 CORONAVIRUS REPLICATION CYCLE

The pathogenicity of a virion depends upon the ability to infect and replicate inside the host cell and then release of the mature virion. This whole sequence of events together constitutes the life cycle of the virus and is discussed below for coronavirus.

4.4.1 Virion Attachment to the Host Cell

Coronaviral infections are initiated when the virion binds to a specific cellular receptor of the host. This is then followed by a series of events culminating in the delivery of the nucleocapsid to the cytoplasm, where the viral genome becomes available for translation. Generally, individual coronaviruses infect only one or a few closely related hosts.

The interaction between the viral S protein and its specific receptor plays the principal role in determining the host species range of coronavirus and tissue tropism. Within the S1 region of a coronavirus S protein, there lie the sites of receptor binding domains (RBD). These RBD sites may further vary depending on the virus, with

TABLE 4.2
Accessory Virus-Specific ORFs along with the Size of Their Respective Translation Products

Virus Species	ORFs	Size of Translation Product	References
SARS Coronavirus	ORF 3a	37 kDa	[64]
	ORF 3b	18 kDa	
	ORF 6	7.5 kDa	
	ORF7a	17.5 kDa	
	ORF 7b	5.5 kDa	
	ORF 8a	14 kDa	
	ORF 8b	5.3 kDa	
	ORF 9b	9.6 kDa	
Transmissible gastroenteritis virus	ORF 3a	7.7 kDa	[65]
	ORF 3b	27.7 kDa	
	ORF 7	0.7 kDa (Hydrophobic Protein)	
Porcine epidemic diarrhea virus	ORF 3	25.3 kDa	[66]
Human coronavirus 229E	ORF 4a	15.3 kDa	[67]
	ORF 4b	10.2 kDa	
Mouse hepatitis virus	ORF 2a	32 kDa	[68]
	ORF 2b	34.6 kDa (Hemagglutinin Esterase Protein)	
	ORF 4	17.8 kDa	
	ORF 5a	13.1 kDa	
	ORF 7	23 kDa (Internal Protein)	
Bovine coronavirus	ORF 2a	32 kDa	[69]
	ORF2b	34.6 kDa (Hemagglutinin Esterase Protein)	
	ORF 4a	4.9 kDa	
	ORF 4b	4.8 kDa	
	ORF 7	23 kDa (Internal Protein)	
IBV	ORF 3a	6.7 kDa	[70]
	ORF 3b	7.4 kDa	
	ORF 5a	7.5 kDa	
	ORF 5b	9.5 kDa	

some having the RBD at the N-terminus of S1 like Mouse Hepatitis Virus (MHV) [27] while others like SARS-CoV have the RBD at the C-terminus of S1 [28].

Angiotensin-converting enzyme 2 (ACE2) happens to be the cellular receptor for both SARS-CoV and SARS-CoV-2. ACE2 is present abundantly in the lung tissue particularly in the alveolar epithelial cells and guides the entry of SARS-CoV and SARS-CoV-2 inside the lung epithelial tissue leading to infection [29]. Similarly, dipeptidyl peptidase 4 (DPP4) which is found upregulated in people with obstructive pulmonary disorder acts as a receptor for MERS-CoV and is responsible for guiding the entry of MERS-CoV in the lung tissue leading to infection [30].

A majority of the coronaviruses use peptidases as their cellular receptor. ***The known cellular receptors for alpha- and beta coronaviruses (members of these two groups cause infections in the mammals) are presented in Table 4.3.***

TABLE 4.3
Cellular Receptors of Alpha Coronaviruses and Beta Coronaviruses

Group	Viruses	Cellular Receptors	References
Alpha coronaviruses	Transmissible gastroenteritis virus	Porcine aminopeptidase N	[71]
	Porcine epidemic diarrhea virus	Porcine aminopeptidase N	[72]
	Feline infectious peritonitis virus	Feline aminopeptidase N	[73]
	Feline enteric coronavirus	Feline aminopeptidase N	[73]
	Canine coronavirus	Canine aminopeptidase N	[73]
	Human coronavirus strain 229E	Human aminopeptidase N	[74]
	Human coronavirus strain NL63	ACE2	[75]
Beta coronaviruses	Mouse hepatitis virus	Murine carcinoembryonic antigen–related adhesion molecule 1	[76]
	Bovine coronavirus	N-acetyl-9-O-acetylneuraminic acid	[77] [78]
	Human coronavirus strain OC43	N-acetyl-9-O-acetylneuraminic acid	[79] [80]
	SARS coronavirus	ACE2	[29]
	MERS coronavirus	DPP4	
	SARS coronavirus 2	ACE2	

4.4.2 Viral Entry and Uncoating

Large-scale rearrangements of the S protein result in the entry of the virion into the host cell, which further leads to the fusion of viral and cellular membranes. These rearrangements are the result of some combination of receptor binding, proteolytic cleavage of S protein and exposure to acidic pH [31]. In mature virions, these S proteins of many coronaviruses are uncleaved and require an encounter with a protease (general *cathepsin* at acidic pH), at the entry step of infection to separate the receptor-binding and fusion components of the spike [32]. Cleavage of S protein occurs at two sites within the S2 portion of the protein, the first cleavage separates the RBD and fusion domains of the S protein and second cleavage at S2′ exposes the fusion peptide [33]. This exposed fusion peptide inserts into the cellular membrane. Fusion is generally known to occur within the acidified endosomes, but in some coronaviruses, fusion also occurs at the plasma membrane, as in the case of MHV [32].

4.4.3 Expression of Replicase-Transcriptase Complex

Following the delivery of the viral nucleocapsid to the cytoplasm, the next event that takes place is the translation of the replicase gene from the genomic RNA of the virus. This gene consists of two large ORFs, rep 1a and rep 1b; both of these share a small region of overlap between them. These rep 1a and rep 1b are known to express two co-terminal polyproteins, pp1a and pp1ab [20]. To express both polyproteins, the virus utilizes an RNA pseudoknot which is involved in ribosomal frameshifting from the rep1a reading frame into the rep1b ORF and a slippery sequence

(5′-UUUAAAC-3′) [26]. In most cases, the ribosome shows the capability of unwinding the pseudoknot structure and continues the process of translation until it encounters the rep1a stop codon [26]. Indeed, the pseudoknot also blocks the ribosome from continuing elongation process, this leading it to pause on the slippery sequence, resulting in changes in the reading frame by moving back one nucleotide, a, -1(minus one) frameshift, before the ribosome leads to melting of pseudoknot structure and extend translation into rep1b, resulting in the translation of pp1ab [26]. Reporter gene expression studies suggest that the coronaviral ribosomal frameshifting event is as high as 25%–30%; however, the *in vivo* frequency in infected cells remains to be quantitated [34]. It is concluded that the role of programmed frameshifting is to provide a fixed ratio of translation products for assembly into a macromolecular complex [35]. The autoproteolytically processing of polyproteins pp1a (440–500 kDa) and pp1ab (740– 810 kDa) results in mature products that are designated nsp1 to nsp16 (except for the gamma coronaviruses, which do not have a counterpart of nsp1). Many partial proteolytic products are also generated as a result of processing, which may have functional importance [36]. There are two types of polyprotein cleavage activity. Papain-like proteases (PLpro), which are generally one or two situated in nsp3, carry out the relatively specialized separation of nsp1, nsp2 and nsp3. The main protease (Mpro) present in nsp5 is known to perform the remaining 11 cleavage events [37]. Several crystal structures have been determined for PLpro and Mpro of SARS-CoV and other coronaviruses, and these enzymes present attractive targets for antiviral drug design [36].

The processed nsps assemble to form the coronavirus replicase, which is also referred to as the replicase-transcriptase complex (RTC). Along with the PLpro and Mpro, the products of rep1a perform several other activities that are important in establishing cellular conditions favorable for infection. While some of these activities are directly linked to RNA synthesis others are nonessential for viral replication, but may have major effects on virus-host interactions [36]. Assembly of the nsps into RTC provides a suitable environment for RNA synthesis and ultimately leading to RNA replication and transcription of the subgenomic RNAs. ***A list of nonstructural proteins and their functions is presented in Table 4.4.***

4.4.4 Replication and Transcription

Viral RNA synthesis follows the translation and assembly of the viral RTCs. Viral RNA synthesis is responsible for producing both genomic and subgenomic RNAs. Subgenomic RNAs are known to serve as mRNAs for the genes downstream of the replicase polyproteins called structural and accessory genes. All the positive-sense subgenomic RNAs form a set of nested RNAs because these are 3′ co-terminal with the full-length viral genome, and thus, this happens to be a distinctive property of the order Nidovirales.

Each sgRNA consists of a leader RNA of 70–100 nucleotides, which is identical to the 5′ end of the genome. This leader RNA is found to be fused with a body RNA, which is identical to a segment of the 3′ end of the genome. This fusion of the leader RNA to body RNAs is known to occur at short motifs known as Transcriptional Regulatory Sequences (TRSs) [38].

TABLE 4.4
Nonstructural Proteins and Their Respective Functions Present in Coronaviruses

Nonstructural Proteins	Functions	References
nsp1	Promotes cellular mRNA degradation and blocks host cell translation, results in blocking innate immune response.	[81]
nsp2	No known function, binds to prohibitin proteins	[82]
nsp3	Large multidomain transmembrane protein, interacts with N protein, cleaves viral polyprotein, responsible for promoting cytokine expression and blocks host cell immune response.	[83] [84]
nsp4	Transmembrane scaffold protein, important for proper structure of double membrane vesicles (DMVs)	[85]
nsp5	Mpro, cleaves viral polyprotein	[86]
nsp6	Potential transmembrane scaffold protein	[87]
nsp7	Forms a complex (hexadecameric) with nsp8, may serve as processivity clamp for RNA polymerase	[88]
nsp8	Forms hexadecameric complex with nsp7, may act as processivity clamp for RNA polymerase, may act as primase	[89]
nsp9	RNA binding protein	[90]
nsp10	Cofactor for nsp16 and nsp14, forms heterodimer with both and stimulates exoribonuclease (ExoN) and 2′-O-methyltransferase (2-O-MT) activity.	[91]
nsp12	RdRp	[92]
nsp13	RNA helicase, 5′ triphosphatase	[93]
nsp14	N7-methyltransferase (N7 MTase) and 3′-5′ exoribonuclease, ExoN; N7 MTase adds 5′ cap to viral RNAs, ExoN activity is important for proofreading of viral genome	[94] [95]
nsp15	Viral endoribonuclease, NendoU	[96]
nsp16	2′-O-MT; shields viral RNA from melanoma differentiation associated protein 5 (MDA5 recognition)	[97] [98]

Like the genome, the sgRNAs have 5′ caps and 3′ polyadenylate tails. The negative-strand intermediates are responsible for producing both genomic and subgenomic RNAs [38]. These negative-strand intermediates contain both poly-uridylate and antileader sequences and are only about 1% as abundant as their positive-sense counterparts [39].

Coronavirus genomes contain *cis*-acting RNA elements at both the 5′ and 3′ ends that allow selective recognition of genomes as templates for the RTC and play essential roles in RNA synthesis. There are seven stem-loop structures within the 5′ UTR of the genome that may extend into the replicase 1a gene [40]. One of the stem loops present at the 5′ end of these displays the leader copy of the TRS (TRS-L) in its loop, and another contains the start codon of the rep 1a gene within its stem. Many of these defined stem looped structures can be exchanged among the genomes of different beta coronaviruses [41]. The 3′ UTR contains a pseudoknot (a bulged stem-loop) and a hypervariable region [38,42].

During the production of subgenomic RNAs, the fusion of the leader and body TRS segments happens to be one of the most novel aspects of the coronavirus replication process. It is largely believed to occur during the discontinuous extension of negative-strand RNA, but originally it was thought to occur during the synthesis of the positive strand [38]. In this model, both genomic and subgenomic negative-strand RNAs are initiated by the RTC at the 3′ end of the positive-strand genome template. This model proposes that the RdRp pauses at any one of the body-TRS sequences (TRS-B). At this point, either the growing negative strand is elongated by the RdRp, or alternatively, it may switch to the leader at the 5′ end of the genome template. This switch occurs as a result of complementarity between the 3′ end of the nascent negative strand and the TRS-L of the genome. The resulting negative-strand sgRNA, in a partial duplex with positive-strand RNA, then serves as a template for the synthesis of multiple copies of the corresponding positive-strand sgRNA [39].

The fusion of leader-to-body during negative-strand synthesis is highly supported by accumulated experimental results with coronaviruses. First, negative-strand sgRNAs contain antileaders at their 3′ ends. Secondly, in infected cells, transcription intermediates containing negative-strand sgRNAs are found to exist in association with the genome. These complexes are known for their active participation in transcription and these can be biochemically separated from replication intermediates containing genome-length negative-strand RNAs [43]. For discontinuous negative-strand synthesis, engineered or naturally occurring variant nucleotides incorporated into the TRS-B, rather than the TRS-L, end up in the leader–body junction of the resulting sgRNA [44]. There remains, however, considerable further work to be done to elucidate the details of the model. It is not known as how the transcribing RdRp (RNA dependent RNA polymerase) might continuously monitor the ability of its nascent product to base pair to the TRS-L. Additionally, the synthesis of genome-length negative strands would require the RdRp to bypass all of the TRS-B sites in the genome template.

Several cellular proteins, including hnRNP A1, polypyrimidine tract binding protein, mitochondrial aconitase and polyadenylate-binding protein, are known to participate in coronavirus RNA synthesis, mainly based on their ability to bind in vitro to genomic RNA segments.

4.4.5 RNA Recombination

Coronaviruses are capable of recombining their genome elements using both homologous and nonhomologous recombination. The ability of these viruses to show recombination is principally linked with the strand switching ability of the RdRp. Recombination plays a prominent role in viral evolution and the accumulation of mutations. Recombination serves as the basis for targeted RNA recombination, which happens to be a reverse genetics tool. This finds applications in engineering viral recombinants at the 3′ end of the genome [45].

4.4.6 Assembly and Release of Virions

Replication and subgenomic RNA synthesis are followed by the translation of the structural proteins of the virus; S, E, and M, and these are inserted into the endoplasmic reticulum. These proteins move along the secretory pathway into the endoplasmic reticulum–Golgi intermediate compartment (ERGIC). Viral genomes then undergo encapsidation by N protein leading to the formation of mature virions by budding into membranes of the ERGIC containing viral structural proteins [46].

Most of the protein-protein interactions required for assembly of coronaviruses are directed by M protein. However, M protein expression alone is not sufficient for virion formation, as M protein expression alone cannot result in the formation of virus-like particles (VLPs). When M protein is expressed along with E protein it results in the formation of VLPs; therefore these two proteins are known to function together to produce coronavirus envelopes [47]. N protein has roles in enhancing the VLP formation, thus the fusion of the encapsidated genomes into the ERGIC enhances the formation of the viral envelope. The S protein is incorporated into virions at this step, but is not required for assembly. The ability of the S protein to interact with the M protein in the ERGIC is critical for its incorporation into virions [48]. M protein interactions are responsible for promoting the envelope maturation. E protein assists M protein in assembly of the virion; E protein is known to induce membrane curvature. Additionally, the E protein is known to play a role in altering the secretory pathway of the host organism and promotes the release of virions [49]. Binding of the M protein to the nucleocapsid plays a role in promoting the completion of virion assembly. The interaction between M protein and nucleocapsid has been mapped to the C-terminus of the endodomain of M with C-terminal domain of the N protein [50]. Among different RNA species produced during infection, the N protein selectively packages only positive-sense full-length genomes. In the coding sequence of nsp15 a packaging signal has been identified for MHV. It was concluded that any mutation in this signal does not really affect the multiplication of the virus. Also, the underlying mechanism as to how this packaging signal works has not been determined [51]. Following assembly of components and budding, virions are transported to the cell surface in vesicles and released by exocytosis. In many coronaviruses, the S protein does not get assembled into virions, it moves to the cell surface where it is known to play a role in mediating cell-cell fusion between infected cells and uninfected cells. This results in the formation of large multinucleated cells, which provides an advantage to the virus and it spreads within the infected organism being undetected or without getting neutralized by the action of virus-specific antibodies. ***Figure 4.3 summarizes the sequence of events involved from infection to release of virions.***

4.5 CORONAVIRUS REVERSE GENETICS

Reverse genetics is a molecular method that is known to determine the functions of a gene. This is done by analyzing the phenotypic effects that are results of genetically engineered specific nucleic acid sequences within the gene.

Classical coronavirus genetics mainly deals with two types of mutants. The first is particularly deletion mutants that are known to occur naturally; these provide indications for genetic changes responsible for different pathogenic traits. The second is

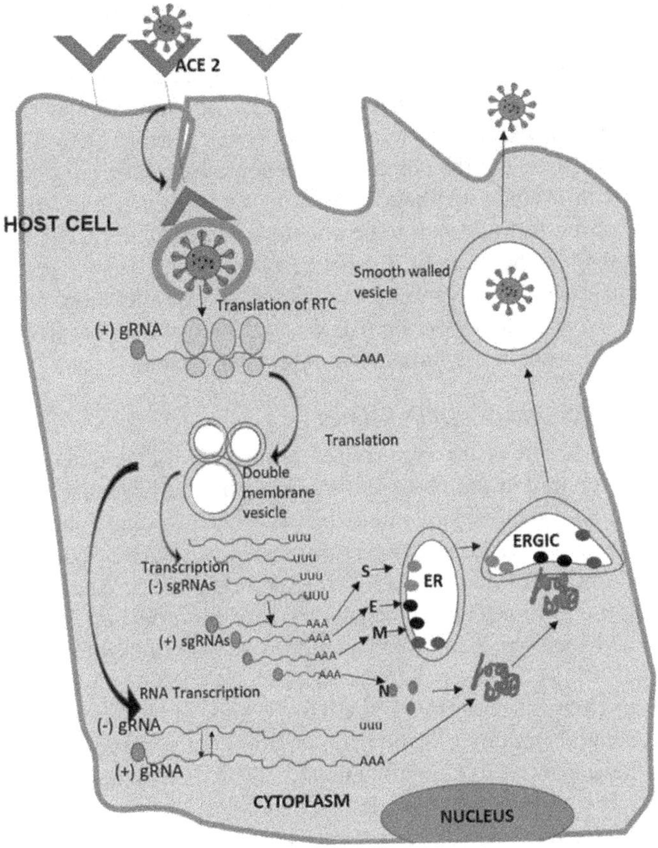

FIGURE 4.3 Replication cycle of coronavirus.

Figure 4.3 shows the complete replication cycle of a coronavirus, from entry of virion through receptor binding followed by uncoating, translation of replicase-transcriptase complex (RTC) and transcription, leading to formation of genomic and subgenomic RNAs, and ultimately to assembly and release of the mature virions.

temperature-sensitive (ts) mutants that were isolated from MHV following chemical mutagenesis. Some of these mutations are valuable in the analysis of the functions of structural proteins [52]. The study of coronavirus genetics was broadly restricted to the analysis of ts mutants [20], defective RNA templates that depend on replicase proteins provided in trans by a helper virus [53] and recombinant viruses generated by targeted recombination.

4.5.1 Tools of Reverse Genetics

4.5.1.1 Targeted RNA Recombination

Targeted RNA recombination was developed as a tool for a reverse genetic system for coronaviruses when the construction of full-length infectious cDNA clones was not thought to be feasible. RNA recombination in coronaviruses came to knowledge due to the accumulation of experimental evidences, ever since its first description in the mid-1980s [22]. Recombination of coronaviruses happens to be a process of vital importance in contributing to the natural evolution of many coronaviruses. Targeted RNA recombination was devised as a tool for reverse genetics by means of introducing specified changes into the coronavirus genome by taking advantage of recombination between a donor synthetic RNA and a recipient parent virus possessing some characteristic that allows it to be counter-selected. After RNA recombination in infected cells, on the basis of distinctive phenotypic property not found in the original recipient virus, viral progeny bearing the desired alterations are selected. An abundance of experimental work on MHV had demonstrated that RNA recombination is a frequent event in the coronavirus infectious cycle.

4.5.1.2 Construction of cDNA Clones

To access a major part of the coronavirus genome, full-length cDNA clones were constructed, overcoming the obstacles presented by the huge size of the replicase gene and the high instability of various regions when propagated in bacterial clones [54,55]. Two methods are known to perform reverse genetics by constructing cDNA clones. In the first, a full-length cDNA copy of a coronavirus genome is assembled downstream of a cytomegalovirus promoter in a bacterial artificial chromosome (BAC) vector, which has a low copy number that contributes to its stability [55]. Transcription of infectious coronavirus RNA by host RNA polymerase II leads to infection through transfected BAC DNA. This method of introducing infection eliminates the potential limitation of *in vitro* capping and synthesis of genomic RNA [34,54]. In the second, a full-length genomic cDNA is assembled through *in vitro* ligation of smaller cloned cDNA fragments; ligation is known to occur in a directed order that is dictated by the use of asymmetric restriction sites. Transcription of the infectious genomic RNA obtained *in vitro* and is used to transfect susceptible host cells [55]. Genome of coronavirus can be constructed entirely from cDNAs synthetically by extending this method [55].

4.6 TRANSMISSION OF CORONAVIRUSES

Most of the coronaviruses find their ways to spread to susceptible hosts by respiratory or fecal-oral routes of infection, where the replication process first occurs in epithelial cells. Some of the coronaviruses replicate principally in respiratory epithelial cells like HCoV OC43, HCoV 229E and Porcine Respiratory Coronavirus (PRCoV); these produce more virions and cause local respiratory symptoms. Other coronaviruses, including Transmissible Gastroenteritis Virus (TGEV), Bovine Coronavirus (BCoV), Porcine Hemagglutinating Encephalomyelitis Virus (PHEV), Canine Coronavirus (CCoV) and Feline Enteric Coronavirus (FECoV), and some enteric strains of MHV are known to infect epithelial cells of the enteric tract. Some of these viruses, such as TGEV, cause diarrhea as a result of an infection that is severe and sometimes proves to be fatal for young animals. These enteric infections in adult animals maintain the virus in the population [56]. Along with the localized infections of the respiratory or enteric tracts, several coronaviruses are also known to cause severe diseases in their respective hosts, for example, SARS-CoV spreads from the upper airway to cause a severe lower respiratory tract infection; PHEV of swine is known to cause enteric infection but is also found to be neurotropic [57]. Birds have proven to be a rich source of new viruses. It has also been proposed that bats and birds are ideally suited as reservoirs for the incubation and evolution of coronaviruses, owing to their common ability to fly and their propensity to roost and lock [14].

4.6.1 Transmission of Human Coronaviruses

Before the SARS-CoV, MERS-CoV and SARS-CoV-2 outbreak, coronaviruses HCoV 229E, HCoV NL63, HCoV OC43 and HCoV HKU1 were only known to cause mild and self-limiting respiratory infections in humans.

SARS-CoV spreads only through direct contact with infected individuals after the onset of illness [58]. Bats are considered as the ultimate source of SARS-CoV [16]. Although bat was the original source of SARS-CoV, its global spread occurred by human-to-human transmission. Transmission occurred through close contact through infectious droplets and probably through aerosols in some instances [58].

MERS-CoV transmits mainly through human-camel (zoonotic transmission) contact or occasionally through human-human contact. MERS outbreak is also known to be a result of healthcare-associated transmission in Saudi Arabia and South Korea [59].

With currently available naïve information, it has been discovered that SARS-CoV-2 spreads through close contact with the infected individual, generally through droplet transmission. Based on the results of the viral genome sequencing as well as the evolutionary analysis, the suspected natural host of virus origin is the bat, and SARS-CoV-2 might be transmitted from bats via unknown intermediate hosts to infect humans [29]. When an infected person coughs, sneezes or talks, the virus is released in the respiratory secretions and these droplets are capable of infecting other individuals if they make direct contact with the mucous membranes on the uninfected individuals. Infections can also occur by touching an infected surface and followed by eyes, nose or mouth. Extensive environmental contamination from infected patients, like contamination of air vents, door handles, toilet bowl, sink, and on personnel protective equipment also contributes to the transmission of coronavirus [60]. Nosocomial mode of transmission is also one important source for infection in healthcare professionals during the examination, transport, noninvasive ventilation, intubation and bronchoscopy of patients. ***Figure 4.4 represents the key reservoir and modes of transmission of coronaviruses.***

4.6.2 Stages of Transmission

According to WHO, following four stages of transmission of coronavirus particularly in case of SARS-CoV-2 are known [61]:

- **Stage 1:** This stage of transmission prevails when the infected people from the country of disease origin or from any affected country travel to another unaffected country. People who test positive for respiratory illness

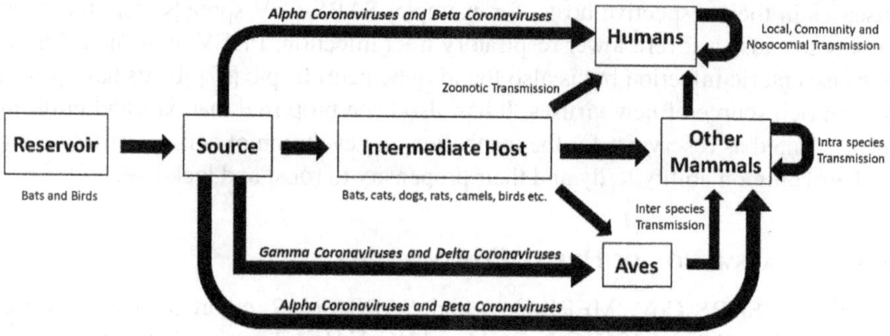

FIGURE 4.4 Key reservoir and modes of transmission of coronaviruses.

Figure 4.4 shows a diagrammatic representation for the reservoirs, sources, intermediate hosts and target hosts of coronaviruses. Possible modes of transmission are mentioned, viz., intra-species transmission, inter-species transmission, zoonotic transmission, local transmission, community transmission and nosocomial transmission.

at this stage are the ones who have a travel history. At this stage, the local transmission does not occur. Therefore, quite a few number of people are infected at this stage.

- **Stage 2:** This is the stage where the local transmission from source individual, i.e., the infected individual having travel history to other countries that are already affected, is known to occur. At this stage of transmission, people generally family and friends who have come in close contact with the infected source individual having travel history test positive. At this stage, the number of positive cases of infected people increases, but it is still manageable to treat the infected individuals by locating them and providing them with appropriate medical care. Social distancing and home quarantine measures help prevent the progression of the disease leading to stage 3 of community transmission.
- **Stage 3:** At this stage community transmission is known to occur. This happens when the actual source of the infection in an individual cannot be traced and isolated. At this stage, individuals test positive for the infection even if they have not come in contact with an infected person or despite not having any travel history to the already affected countries. Once the community transmission begins the outbreak may spread at a fast rate in clusters and transmission becomes uncontrollable.
- **Stage 4:** At this stage of transmission, the infectious disease spreads all over within the country. It soon becomes an epidemic due to several clusters of the infection, affecting many different parts of the country. This stage is characterized by massive numbers of individuals getting infected, across all the different age groups, a large number of deaths are reported among the immune-compromised and elderly patients, it becomes very difficult to control the rate of spread of the disease at this stage. In such a situation, the government officials and health ministry of the affected country are forced to

announce and implement certain strict legislation, for all the citizens to stay indoors by enforcing a lockdown, banning any gathering of large groups of people and shutting down all the public places. To avoid further fatalities that may arise due to the disease and to save the healthy population form getting infected, these extreme steps are undertaken.

4.7 FUTURE DIRECTIONS

Considering the major global health emergencies that have known to occur in recent years like the SARS-CoV outbreak, MERS-CoV outbreak and COVID-19, it becomes necessary to study and understand the genetics and pathogenesis of coronaviruses. Emergence of novel coronaviruses relies on the mutations that might occur, leading to alterations in the genome of the virus. Also, genomic characterization is useful in accurately identifying the origin and evolution of the virus. Understanding the molecular biology of coronaviruses helps in studying the pathogenicity of the virus and identifying certain targets to develop specific and effective antiviral drugs. Further, tools of reverse genetics like targeted RNA recombination facilitate the examination of the genome and its elements. Defining the genetic organization, molecular biology, mechanism of pathogenesis and tools of reverse genetics can significantly contribute to vaccine development and controlling coronavirus outbreaks that lead to global socio-economic health emergencies.

ACKNOWLEDGEMENTS

Dr. Neha Gupta acknowledges the Department of Science and Technology (DST), Government of India, for providing support through the INSPIRE FACULTY Fellowship scheme.

REFERENCES

1. Cascella M, Rajnik M, Cuomo A, Dulebohn SC, Di Napoli R (2020) Features, evaluation and treatment coronavirus (COVID-19). In: Statpearls [internet]. StatPearls Publishing.
2. Payne S (2017) Chapter 17: Family coronaviridae. In: Payne S (ed.) *Viruses*. Academic Press, pp. 149–158. doi: 10.1016/B978–0–12–803109–4.00017–9
3. Sahu K, Mishra A, Lal A (2020) Coronavirus disease-2019: An update on third coronavirus outbreak of 21st century. *QJM: An International Journal of Medicine* 113 (5): 384–386.
4. Tyrrell DAJ, Fielder M (2002) *Cold Wars: The Fight against the Common Cold*. Oxford: Oxford University Press.
5. Estola T (1970) Coronaviruses, a new group of animal RNA viruses. *Avian Diseases* 14 (2): 330–336.
6. Decaro N (2011) Gammacoronavirus‡: Coronaviridae. *The Springer Index of Viruses*. pp. 403–413. doi: 10.1007/978-0-387–95919-1_58
7. Almeida J (2008) June Almeida (née Hart). *British Medical Journal* 336 (7659): 1511–1511. doi: 10.1136/bmj.a434.
8. de Groot RJ, Baker S, Baric R, Enjuanes L, Gorbalenya A, Holmes K, Perlman S, Poon L, Rottier P, Talbot P (2012) Family coronaviridae. Virus taxonomy. pp. 806–828.

9. Woo PC, Lau SK, Lam CS, Lai KK, Huang Y, Lee P, Luk GS, Dyrting KC, Chan K-H, Yuen K-Y (2009) Comparative analysis of complete genome sequences of three avian coronaviruses reveals a novel group 3c coronavirus. *Journal of Virology* 83 (2): 908–917.
10. Su S, Wong G, Shi W, Liu J, Lai AC, Zhou J, Liu W, Bi Y, Gao GF (2016) Epidemiology, genetic recombination, and pathogenesis of coronaviruses. *Trends in Microbiology* 24 (6): 490–502.
11. Enjuanes L, Brian D, Cavanagh D, Holmes K, Lai M, Laude H, Masters P, Rottier P, Siddell S, Spaan W (2000) Coronaviridae. In: Murphy MA Fauquet CM, Bishop DHL, Ghabrial SA, Jarvis AW, Martelli GP, Mayo MA and Summers MD (eds.) *Virus Taxonomy, Classification and Nomenclature of Viruses.* New York: Academic Press. pp. 835–849.
12. Snijder EJ, Horzinek MC (1993) Toroviruses: Replication, evolution and comparison with other members of the coronavirus-like superfamily. *Journal of General Virology* 74 (Pt 11): 2305–2316. doi: 10.1099/0022–1317–74–11–2305
13. Gorbalenya AE, Enjuanes L, Ziebuhr J, Snijder EJ (2006) Nidovirales: Evolving the largest RNA virus genome. *Virus Research* 117 (1): 17–37. doi: 10.1016/j.virusres. 2006.01.017
14. Woo PC, Lau SK, Huang Y, Yuen K-Y (2009) Coronavirus diversity, phylogeny and interspecies jumping. *Experimental Biology and Medicine* 234 (10): 1117–1127.
15. Woo PC, Huang Y, Lau SK, Yuen K-Y (2010) Coronavirus genomics and bioinformatics analysis. *Viruses* 2 (8): 1804–1820.
16. Lau SK, Woo PC, Li KS, Huang Y, Tsoi H-W, Wong BH, Wong SS, Leung S-Y, Chan K-H, Yuen K-Y (2005) Severe acute respiratory syndrome coronavirus-like virus in Chinese horseshoe bats. *Proceedings of the National Academy of Sciences* 102 (39): 14040–14045.
17. Rota PA, Oberste MS, Monroe SS, Nix WA, Campagnoli R, Icenogle JP, Penaranda S, Bankamp B, Maher K, Chen M-h (2003) Characterization of a novel coronavirus associated with severe acute respiratory syndrome. *Science* 300 (5624): 1394–1399.
18. Yin Y, Wunderink RG (2018) MERS, SARS and other coronaviruses as causes of pneumonia. *Respirology* 23 (2): 130–137.
19. Khailany RA, Safdar M, Ozaslan M (2020) Genomic characterization of a novel SARS-CoV-2. Gene reports: 100682.
20. Lai MM, Cavanagh D (1997) The molecular biology of coronaviruses. *Advances in Virus Research* 48: 1–100 doi: 10.1016/s0065–3527(08)60286–9
21. Lai M, Stohlman SA (1978) RNA of mouse hepatitis virus. *Journal of Virology* 26 (2): 236–242.
22. Lai MM, Stohlman SA (1981) Comparative analysis of RNA genomes of mouse hepatitis viruses. *Journal of Virology* 38 (2): 661–670.
23. Masters PS (2006) The molecular biology of coronaviruses. *Advances in Virus Research* 66: 193–292 doi: 10.1016/s0065–3527(06)66005–3
24. Chen Y, Liu Q, Guo D (2020) Emerging coronaviruses: Genome structure, replication, and pathogenesis. *Journal of Medical Virology* 92 (4): 418–423. doi: 10.1002/ jmv..25681
25. Fehr AR, Perlman S (2015) Coronaviruses: An overview of their replication and pathogenesis. *Methods in Molecular Biology* 1282: 1–23. doi: 10.1007/978-1-4939-2438-7_1
26. Brierley I, Boursnell M, Binns M, Bilimoria B, Blok V, Brown T, Inglis S (1987) An efficient ribosomal frame-shifting signal in the polymerase-encoding region of the coronavirus IBV. *The EMBO Journal* 6 (12): 3779–3785.
27. Kubo H, Yamada YK, Taguchi F (1994) Localization of neutralizing epitopes and the receptor-binding site within the amino-terminal 330 amino acids of the murine coronavirus spike protein. *Journal of Virology* 68 (9): 5403–5410.

28. Cheng PKC, Wong DA, Tong LKL, Ip S-M, Lo ACT, Lau C-S, Yeung EYH, Lim WWL (2004) Viral shedding patterns of coronavirus in patients with probable severe acute respiratory syndrome. *Lancet* 363 (9422): 1699–1700. doi: 10.1016/s0140–6736(04)16255–7
29. Zhou P, Yang X-L, Wang X-G, Hu B, Zhang L, Zhang W, Si H-R, Zhu Y, Li B, Huang C-L, Chen H-D, Chen J, Luo Y, Guo H, Jiang R-D, Liu M-Q, Chen Y, Shen X-R, Wang X, Zheng X-S, Zhao K, Chen Q-J, Deng F, Liu L-L, Yan B, Zhan F-X, Wang Y-Y, Xiao G-F, Shi Z-L (2020) A pneumonia outbreak associated with a new coronavirus of probable bat origin. *Nature* 579 (7798): 270–273. doi: 10.1038/s41586-020-2012-7
30. Mou H, Raj VS, Van Kuppeveld FJ, Rottier PJ, Haagmans BL, Bosch BJ (2013) The receptor binding domain of the new Middle East respiratory syndrome coronavirus maps to a 231-residue region in the spike protein that efficiently elicits neutralizing antibodies. *Journal of Virology* 87 (16): 9379–9383.
31. Bosch BJ, Rottier PJ (2007) Nidovirus entry into cells. *Nidoviruses.* pp. 157–178.
32. Bosch BJ, Van der Zee R, De Haan CA, Rottier PJ (2003) The coronavirus spike protein is a class I virus fusion protein: Structural and functional characterization of the fusion core complex. *Journal of Virology* 77 (16): 8801–8811.
33. Belouzard S, Chu VC, Whittaker GR (2009) Activation of the SARS coronavirus spike protein via sequential proteolytic cleavage at two distinct sites. *Proceedings of the National Academy of Sciences of the United States of America* 106 (14): 5871–5876. doi: 10.1073/pnas..0809524106
34. Genetics and Reverse Genetics of Nidoviruses. In: *Nidoviruses.* pp. 47–64. doi: 10.1128/9781555815790.ch4
35. Araki K, Gangappa S, Dillehay DL, Rouse BT, Larsen CP, Ahmed R (2010) Pathogenic virus-specific T cells cause disease during treatment with the calcineurin inhibitor FK506: Implications for transplantation. *Journal of Experimental Medicine* 207 (11): 2355–2367. doi: 10.1084/jem.20100124
36. Ziebuhr J, Snijder EJ, Gorbalenya AE (2000) Virus-encoded proteinases and proteolytic processing in the Nidovirales. *Journal of General and Molecular Virology* 81 (Pt 4): 853–879. doi: 10.1099/0022–1317–81–4–853
37. Mielech AM, Chen Y, Mesecar AD, Baker SC (2014) Nidovirus papain-like proteases: Multifunctional enzymes with protease, deubiquitinating and deISGylating activities. *Virus Research* 194: 184–190.
38. Sawicki SG, Sawicki DL, Siddell SG (2007) A contemporary view of coronavirus transcription. *Journal of Virology* 81 (1): 20–29. doi: 10.1128/jvi.01358–06
39. Sethna PB, Hofmann MA, Brian DA (1991) Minus-strand copies of replicating coronavirus mRNAs contain antileaders. *Journal of Virology* 65 (1): 320–325.
40. Raman S, Bouma P, Williams GD, Brian DA (2003) Stem-loop III in the 5' untranslated region is a cis-acting element in bovine coronavirus defective interfering RNA replication. *Journal of Virology* 77 (12): 6720–6730. doi: 10.1128/jvi.77.12.6720–6730.2003
41. Guan BJ, Wu HY, Brian DA (2011) An optimal cis-replication stem-loop IV in the 5' untranslated region of the mouse coronavirus genome extends 16 nucleotides into open reading frame 1. *Journal of Virology* 85 (11): 5593–5605. doi: 10.1128/jvi.00263–11
42. Liu Q, Johnson RF, Leibowitz JL (2001) Secondary structural elements within the 3' untranslated region of mouse hepatitis virus strain JHM genomic RNA. *Journal of Virology* 75 (24): 12105–12113. doi: 10.1128/JVI..75.24.12105–12113.2001
43. Sawicki SG, Sawicki DL (1990) Coronavirus transcription: Subgenomic mouse hepatitis virus replicative intermediates function in RNA synthesis. *Journal of Virology* 64(3): 1050–1056.
44. Pasternak AO, van den Born E, Spaan WJ, Snijder EJ (2001) Sequence requirements for RNA strand transfer during nidovirus discontinuous subgenomic RNA synthesis. *Embo Journal* 20 (24): 7220–7228. doi: 10.1093/emboj/20.24.7220

45. Keck JG, Makino S, Soe LH, Fleming JO, Stohlman SA, Lai MM (1987) RNA recombination of coronavirus. *Advances in Experimental Medicine and Biology* 218: 99–107. doi: 10.1007/978-1-4684-1280-2_11
46. Krijnse-Locker J, Ericsson M, Rottier PJ, Griffiths G (1994) Characterization of the budding compartment of mouse hepatitis virus: Evidence that transport from the RER to the Golgi complex requires only one vesicular transport step. *Journal of Cell Biology* 124 (1–2): 55–70. doi: 10.1083/jcb.124.1.55
47. Bos EC, Luytjes W, van der Meulen HV, Koerten HK, Spaan WJ (1996) The production of recombinant infectious DI-particles of a murine coronavirus in the absence of helper virus. *Virology* 218 (1): 52–60. doi: 10.1006/viro.1996.0165
48. Siu YL, Teoh KT, Lo J, Chan CM, Kien F, Escriou N, Tsao SW, Nicholls JM, Altmeyer R, Peiris JSM, Bruzzone R, Nal B (2008) The M, E, and N structural proteins of the severe acute respiratory syndrome coronavirus are required for efficient assembly, trafficking, and release of virus-like particles. *Journal of Virology* 82 (22): 11318–11330. doi: 10.1128/jvi..01052–08
49. Ye Y, Hogue BG (2007) Role of the coronavirus E viroporin protein transmembrane domain in virus assembly. *Journal of Virology* 81 (7): 3597–3607. doi: 10.1128/jvi..01472–06
50. Hurst KR, Kuo L, Koetzner CA, Ye R, Hsue B, Masters PS (2005) A major determinant for membrane protein interaction localizes to the carboxy-terminal domain of the mouse coronavirus nucleocapsid protein. *Journal of Virology* 79 (21): 13285–13297. doi: 10.1128/JVI..79.21.13285–13297.2005
51. Kuo L, Masters PS (2013) Functional analysis of the murine coronavirus genomic RNA packaging signal. *Journal of Virology* 87(9): 5182–5192. doi: 10.1128/jvi.00100–13
52. Koolen MJ, Osterhaus AD, Van Steenis G, Horzinek MC, Van der Zeijst BA (1983) Temperature-sensitive mutants of mouse hepatitis virus strain A59: Isolation, characterization and neuropathogenic properties. *Virology* 125(2): 393–402. doi: 10.1016/0042–6822(83)90211–8
53. Izeta A, Smerdou C, Alonso S, Penzes Z, Mendez A, Plana-Durán J, Enjuanes L (1999) Replication and packaging of transmissible gastroenteritis coronavirus-derived synthetic minigenomes. *Journal of Virology* 73(2): 1535–1545. doi: 10.1128/jvi..73.2.1535–1545.1999
54. Deming DJ, Baric RS (2008) Genetics and reverse genetics of Nidoviruses. In: Nidoviruses. American Society of Microbiology. doi: 10.1128/9781555815790..ch4
55. Becker MM, Graham RL, Donaldson EF, Rockx B, Sims AC, Sheahan T, Pickles RJ, Corti D, Johnston RE, Baric RS, Denison MR (2008) Synthetic recombinant bat SARS-like coronavirus is infectious in cultured cells and in mice. *Proceedings of the National Academy of Sciences of the United States of America* 105 (50): 19944–19949. doi: 10.1073/pnas..0808116105
56. Saif LJ (2004) Animal coronavirus vaccines: Lessons for SARS. *Developmental Biology (Basel)* 119: 129–140.
57. McIntosh K (1974) Coronaviruses: A comparative review. In: Arber W, Haas R, Henle W et al. (eds.) *Current Topics in Microbiology and Immunology / Ergebnisse der Mikrobiologie und Immunitätsforschung*. Springer Berlin Heidelberg. pp. 85–129.
58. Peiris JS, Yuen KY, Osterhaus AD, Stöhr K (2003) The severe acute respiratory syndrome. *The New England Journal of Medicine* 349 (25): 2431–2441. doi: 10.1056/NEJMra032498
59. Balkhy HH, Alenazi TH, Alshamrani MM, Baffoe-Bonnie H, Arabi Y, Hijazi R, Al-Abdely HM, El-Saed A, Al Johani S, Assiri AM, Bin Saeed A (2016) Description of a hospital outbreak of Middle East respiratory syndrome in a large tertiary care hospital in Saudi Arabia. *Infection Control and Hospital Epidemiology* 37 (10): 1147–1155. doi: 10.1017/ice..2016.132

60. Casanova LM, Jeon S, Rutala WA, Weber DJ, Sobsey MD (2010) Effects of air temperature and relative humidity on coronavirus survival on surfaces. *Applied and Environmental Microbiology* 76 (9): 2712–2717. doi: 10.1128/aem..02291–09
61. World Health Organization (2020) Coronavirus disease 2019 (COVID-19): Situation report, 72. Geneva: World Health Organization.
62. Zaki AM, van Boheemen S, Bestebroer TM, Osterhaus AD, Fouchier RA (2012) Isolation of a novel coronavirus from a man with pneumonia in Saudi Arabia. *The New England Journal of Medicine* 367 (19): 1814–1820. doi: 10.1056/NEJMoa1211721
63. Xu X, Chen P, Wang J, Feng J, Zhou H, Li X, Zhong W, Hao P (2020) Evolution of the novel coronavirus from the ongoing Wuhan outbreak and modeling of its spike protein for risk of human transmission. *Science China Life Sciences* 63 (3): 457–460. doi: 10..1007/s11427-020-1637–5
64. Schaecher SR, Pekosz A (2009) SARS coronavirus accessory gene expression and function. *Molecular Biology of the SARS-Coronavirus*: 153–166. doi: 10..1007/978-3-642–03683-5_10
65. Almazán F, González JM, Pénzes Z, Izeta A, Calvo E, Plana-Durán J, Enjuanes L (2000) Engineering the largest RNA virus genome as an infectious bacterial artificial chromosome. *Proceedings of the National Academy of Sciences* 97 (10): 5516–5521. doi: 10.1073/pnas..97.10.5516
66. Kocherhans R, Bridgen A, Ackermann M, Tobler K (2001) Completion of the porcine epidemic diarrhoea coronavirus (PEDV) genome sequence. *Virus Genes* 23 (2): 137–144. doi: 10.1023/a:1011831902219
67. Herold J, Raabe T, Siddell S.G. (1993) Molecular analysis of the human coronavirus (strain 229E) genome. In: *Unconventional Agents and Unclassified Viruses.* Vienna: Springer, pp. 63–74.
68. Sarma JD, Fu L, Hingley ST, Lai MM, Lavi E (2001) Sequence analysis of the S gene of recombinant MHV-2/A59 coronaviruses reveals three candidate mutations associated with demyelination and hepatitis. *Journal of Neurovirology* 7 (5): 432–436.
69. Chouljenko VN, Lin XQ, Storz J, Kousoulas KG, Gorbalenya AE (2001) Comparison of genomic and predicted amino acid sequences of respiratory and enteric bovine coronaviruses isolated from the same animal with fatal shipping pneumonia. *Journal of General and Molecular Virology* 82 (Pt 12): 2927–2933. doi: 10.1099/0022–1317–82–12–2927
70. Boursnell ME, Brown TD, Foulds IJ, Green PF, Tomley FM, Binns MM (1987) Completion of the sequence of the genome of the coronavirus avian infectious bronchitis virus. *Journal of General and Molecular Virology* 68 (Pt 1): 57–77. doi: 10.1099/0022–1317–68–1–57
71. Delmas B, Gelfi J, L'Haridon R, Vogel, Sjöström H, Norén, LH (1992) Aminopeptidase N is a major receptor for the enteropathogenic coronavirus TGEV. *Nature* 357 (6377): 417–420. doi: 10.1038/357417a0
72. Li BX, Ge JW, Li YJ (2007) Porcine aminopeptidase N is a functional receptor for the PEDV coronavirus. *Virology* 365 (1): 166–172. doi: 10.1016/j.virol.2007.03.031
73. Tresnan DB, Levis R, Holmes KV (1996) Feline aminopeptidase N serves as a receptor for feline, canine, porcine, and human coronaviruses in serogroup I. *Journal of Virology* 70 (12): 8669–8674.
74. Yeager CL, Ashmun RA, Williams RK, Cardellichio CB, Shapiro LH, Look AT, Holmes KV (1992) Human aminopeptidase N is a receptor for human coronavirus 229E. *Nature* 357 (6377): 420–422. doi: 10.1038/357420a0
75. Hofmann H, Pyrc K, van der Hoek L, Geier M, Berkhout B, Pöhlmann S (2005) Human coronavirus NL63 employs the severe acute respiratory syndrome coronavirus receptor for cellular entry. *Proceedings of the National Academy of Sciences of the United States of America* 102 (22): 7988–7993. doi: 10.1073/pnas..0409465102

76. Dveksler GS, Pensiero MN, Cardellichio CB, Williams RK, Jiang GS, Holmes KV, Dieffenbach CW (1991) Cloning of the mouse hepatitis virus (MHV) receptor: Expression in human and hamster cell lines confers susceptibility to MHV. *Journal of Virology* 65 (12): 6881–6891.
77. Schultze B, Herrler G (1993) Recognition of N-acetyl-9-O-acetylneuraminic acid by bovine coronavirus and hemagglutinating encephalomyelitis virus. *Advances in Experimental Medicine and Biology* 342: 299–304 doi: 10.1007/978-1-4615-2996-5_46
78. Vlasak R, Luytjes W, Spaan W, Palese P (1988) Human and bovine coronaviruses recognize sialic acid-containing receptors similar to those of influenza C viruses. *Proceedings of the National Academy of Sciences of the United States of America* 85 (12): 4526–4529. doi: 10.1073/pnas..85.12.4526
79. Li W, Moore MJ, Vasilieva N, Sui J, Wong SK, Berne MA, Somasundaran M, Sullivan JL, Luzuriaga K, Greenough TC, Choe H, Farzan M (2003) Angiotensin-converting enzyme 2 is a functional receptor for the SARS coronavirus. *Nature* 426 (6965): 450–454. doi: 10.1038/nature02145
80. Raj VS, Mou H, Smits SL, Dekkers DH, Müller MA, Dijkman R, Muth D, Demmers JA, Zaki A, Fouchier RA, Thiel V, Drosten C, Rottier PJ, Osterhaus AD, Bosch BJ, Haagmans BL (2013) Dipeptidyl peptidase 4 is a functional receptor for the emerging human coronavirus-EMC. *Nature* 495 (7440): 251–254. doi: 10.1038/nature12005
81. Kamitani W, Huang C, Narayanan K, Lokugamage KG, Makino S (2009) A two-pronged strategy to suppress host protein synthesis by SARS coronavirus Nsp1 protein. *Nature Structural & Molecular Biology* 16 (11): 1134.
82. Cornillez-Ty CT, Liao L, Yates JR, Kuhn P, Buchmeier MJ (2009) Severe acute respiratory syndrome coronavirus nonstructural protein 2 interacts with a host protein complex involved in mitochondrial biogenesis and intracellular signaling. *Journal of Virology* 83 (19): 10314–10318. doi: 10.1128/jvi..00842–09
83. Neuman BW, Joseph JS, Saikatendu KS, Serrano P, Chatterjee A, Johnson MA, Liao L, Klaus JP, Yates JR, Wüthrich K, Stevens RC, Buchmeier MJ, Kuhn P (2008) Proteomics analysis unravels the functional repertoire of coronavirus nonstructural protein 3. *Journal of Virology* 82 (11): 5279–5294. doi: 10.1128/jvi..02631–07
84. Frieman M, Ratia K, Johnston RE, Mesecar AD, Baric RS (2009) Severe acute respiratory syndrome coronavirus papain-like protease ubiquitin-like domain and catalytic domain regulate antagonism of IRF3 and NF-κB signaling. *Journal of Virology* 83 (13): 6689–6705. doi: 10.1128/jvi..02220–08
85. Gadlage MJ, Sparks JS, Beachboard DC, Cox RG, Doyle JD, Stobart CC, Denison MR (2010) Murine hepatitis virus nonstructural protein 4 regulates virus-induced membrane modifications and replication complex function. *Journal of Virology* 84 (1): 280–290. doi: 10.1128/jvi.01772–09
86. Lu Y, Lu X, Denison MR (1995) Identification and characterization of a serine-like proteinase of the murine coronavirus MHV-A59. *Journal of Virology* 69 (6): 3554–3559.
87. Oostra M, Hagemeijer MC, van Gent M, Bekker CPJ, te Lintelo EG, Rottier PJM, de Haan CAM (2008) Topology and membrane anchoring of the coronavirus replication complex: Not all hydrophobic domains of nsp3 and nsp6 are membrane spanning. *Journal of Virology* 82 (24): 12392–12405. doi: 10.1128/jvi..01219–08
88. Zhai Y, Sun F, Li X, Pang H, Xu X, Bartlam M, Rao Z (2005) Insights into SARS-CoV transcription and replication from the structure of the nsp7-nsp8 hexadecamer. *Nature Structural & Molecular Biology* 12 (11): 980–986. doi: 10..1038/nsmb999
89. Imbert I, Guillemot J-C, Bourhis J-M, Bussetta C, Coutard B, Egloff M-P, Ferron F, Gorbalenya AE, Canard B (2006) A second, non-canonical RNA-dependent RNA polymerase in SARS coronavirus. *The EMBO Journal* 25 (20): 4933–4942. doi: 10.1038/sj.emboj..7601368

90. Egloff M-P, Ferron F, Campanacci V, Longhi S, Rancurel C, Dutartre H, Snijder EJ, Gorbalenya AE, Cambillau C, Canard B (2004) The severe acute respiratory syndrome-coronavirus replicative protein nsp9 is a single-stranded RNA-binding subunit unique in the RNA virus world. *Proceedings of the National Academy of Sciences of the United States of America* 101 (11): 3792–3796. doi: 10.1073/pnas..0307877101
91. Decroly E, Debarnot C, Ferron F, Bouvet M, Coutard B, Imbert I, Gluais L, Papageorgiou N, Sharff A, Bricogne G (2011) Crystal structure and functional analysis of the SARS-coronavirus RNA cap 2′-O-methyltransferase nsp10/nsp16 complex. *PLoS Pathogens* 7 (5): e1002059.
92. Xu X, Liu Y, Weiss S, Arnold E, Sarafianos SG, Ding J (2003) Molecular model of SARS coronavirus polymerase: Implications for biochemical functions and drug design. *Nucleic Acids Research* 31 (24): 7117–7130. doi: 10..1093/nar/gkg916
93. Ivanov KA, Ziebuhr J (2004) Human coronavirus 229E nonstructural protein 13: Characterization of duplex-unwinding, nucleoside triphosphatase, and RNA 5'-triphosphatase activities. *Journal of Virology* 78 (14): 7833–7838. doi: 10.1128/jvi.78.14.7833–7838.2004
94. Minskaia E, Hertzig T, Gorbalenya AE, Campanacci V, Cambillau C, Canard B, Ziebuhr J (2006) Discovery of an RNA virus 3′→5′ exoribonuclease that is critically involved in coronavirus RNA synthesis. *Proceedings of the National Academy of Sciences of the United States of America* 103 (13): 5108–5113. doi: 10.1073/pnas..0508200103
95. Eckerle LD, Lu X, Sperry SM, Choi L, Denison MR (2007) High fidelity of murine hepatitis virus replication is decreased in nsp14 exoribonuclease mutants. *Journal of Virology* 81 (22): 12135–12144. doi: 10.1128/jvi.01296–07
96. Bhardwaj K, Sun J, Holzenburg A, Guarino LA, Kao CC (2006) RNA recognition and cleavage by the SARS coronavirus endoribonuclease. *Journal of Molecular Biology* 361 (2): 243–256. doi: 10.1016/j.jmb.2006.06.021
97. Decroly E, Imbert I, Coutard B, Bouvet M, Selisko B, Alvarez K, Gorbalenya AE, Snijder EJ, Canard B (2008) Coronavirus nonstructural protein 16 is a cap-0 binding enzyme possessing (nucleoside-2′O)-methyltransferase activity. *Journal of Virology* 82 (16): 8071–8084. doi: 10.1128/jvi..00407–08
98. Züst R, Cervantes-Barragan L, Habjan M, Maier R, Neuman BW, Ziebuhr J, Szretter KJ, Baker SC, Barchet W, Diamond MS, Siddell SG, Ludewig B, Thiel V (2011) Ribose 2′-O-methylation provides a molecular signature for the distinction of self and non-self mRNA dependent on the RNA sensor Mda5. *Nature Immunology* 12 (2): 137–143. doi: 10.1038/ni.1979

90. [illegible] (2010) [illegible] coronavirus replicative system [illegible] RNA virus world. [illegible]
91. Decroly E, [illegible] (2011) Crystal structure and functional analysis of the SARS-coronavirus [illegible] complex. [illegible]
92. [illegible] SARS-coronavirus [illegible]
93. [illegible]
94. [illegible] (2006) [illegible] SARS-coronavirus [illegible] RNA [illegible]
95. [illegible]
96. [illegible] SARS [illegible]
97. [illegible] (2020) [illegible]
98. [illegible]

5 Presumable Strategies to Combat the COVID-19 Deleterious Effects

Harshit Agarwal, Shakti Bhushan, Pooja Pant, and Vandana Meena
Banaras Hindu University

Nupur Gupta
Guru Nanak Dev University

CONTENTS

5.1 INTRODUCTION

In the present scenario of viral diseases, there are more than 200 different viral types that are associated with cold-like symptoms including fever. Coronaviruses are the most common virus family behind colds, with seven known species that affect humans namely Severe Acute Respiratory Syndrome Coronavirus 2 (SARS-CoV-2), Severe Acute Respiratory Syndrome Coronavirus (SARS-CoV), the Middle East Respiratory Syndrome Coronavirus (MERS-CoV), Human Coronavirus (HKU1), Human Coronavirus NL63 (HCoV-NL63), Human Coronavirus OC43 (HCoV-OC43), and Human Coronavirus 229E (HCoV-229E) (Andersen et al. 2020;

Chen 2020). There had been some high-profile cases, including the SARS outbreak several years ago, and the current coronavirus outbreak in China. The studies show that most of the coronavirus infections find their origin in animals, specifically bats, or rodents (Zhou et al. 2020). A pandemic will lead to massive deaths. In the last 100 years, the worst pandemic was in 1918; an Influenza Global Pandemic with an estimated 50–100 million deaths while no global coronavirus pandemic was born. Two global epidemics were triggered however by the family of viruses, including extreme acute respiratory syndrome or SARS (2001–2003) as well as the Middle East or MERS (2012–2015) causing 15,000 deaths. Their name comes from the Latin for a crown, "corona," for their crown-like surface projections. SARS-CoV-2 is reportedly responsible for a respiratory disease epidemic known as the COVID-19, which has spread to several countries around the world and is declared as the biggest pandemic of the 21st century (Yu et al. 2020). SARS-CoV-2 is an RNA virus of a single strand, whose infection depth depends on the polarity of the RNA. Since the RNA can be negative or positive, the studies show that the positive-sense RNA is contagious in humans, and the negative sense RNA being the complex ones does not replicate by itself and therefore are not infectious, unless transformed into positive-sense RNA. The newly discovered 2019 novel coronavirus causes respiratory disease wherein the patients initially exhibit fever and dry cough. This is a viral infection that spreads through consumption of contaminated person's virus droplets. The microdroplets from the infected person make their way into the epithelial cells of the person in contact. However, the virus is tested to be active in the air for long hours too.

COVID-19 was first recorded in December 2019 at Wuhan, in China, however, by 30th January a total of 7818 cases were confirmed globally, and most cases were in China with 82 registered outside China in 18 countries. As per the reports of WHO on 30 June 2020, concerning Corona cases, there were a total of 10,421,490 confirmed cases of COVID-19 among which 5,679,143 cases were recovered and 508,419 people have been reported dead. These numbers are increasing hourly which is a point of concern (Webmeter 2020).

5.2 COVID-19 TEST MECHANISM

Since it is an RNA virus, it possesses the capability of mutating and generating subspecies. The first and foremost job is to identify whether a person is infected or not, meaning, a person is COVID-19 positive or negative. According to World Health Organization, "reverse transcriptase-polymerase chain reaction" tests (rTR-PCR test) can be helpful in the detection of a single RNA sequence of coronavirus by rTR-PCR and confirmation of the disease. The test includes the study of the patient's nasal and throat swabs. The DNA and certain forms of the virus from our genetic material; COVID-19 is a single-stranded RNA, but the PCR check can only copy DNA, as discussed above. From now on, one-stranded RNA must be translated into two-stranded DNA, to use in the PCR test. The DNA sample is applied to a test tube with the aid of primers which is helpful in the creation of copies of DNA. After the primers have been connected to DNA, they provide a foundation for copying the DNA enzyme. It is achieved by the reheating and the refrigeration of millions of DNA copies. The fluorescent test applied to the test tube by adding the fluorescent dyes while the

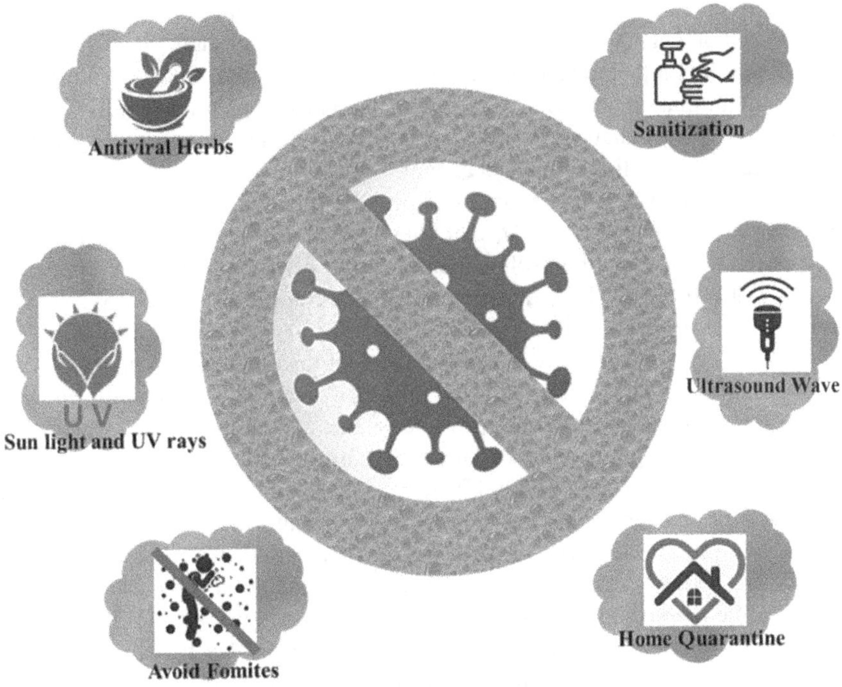

FIGURE 5.1 Possible ways to decrease the effect of virus.

copying of the DNA enhances the fluorescence. As more copies of the DNA virus are made, fluorescence increases. When a certain degree of fluorescence is reached, the result is positive. If the virus did not appear in the sample, the PCR test would not have copied the fluorescence threshold, and the result is null (Tang et al. 2020; Lan et al. 2020). Recently, several drugs as hydroxychloroquine and azithromycin medicine have been tested for the effective treatment of COVID-19 (Gautret et al. 2020); however, this medication would be harmful to patients with cardiovascular problems as the drugs cause arrhythmia. "Arrhythmia" in simple terms implies abnormality in heart beating rhythm. Where scientists and researchers worldwide are in the continuous development of a proper vaccine for this deadly viral infection, we on our hand can make our contribution in combating this pandemic. The present article is important to understand the possible modalities to decrease the effect of coronavirus which is also shown in Figure 5.1.

5.3 POSSIBLE WAYS FOR DECREASING THE EFFECT OF COVID-19

5.3.1 Sunlight/Ultra Violet Light as a Natural Sanitization

Viral reactions are affected by the time and amount of exposure to sunlight. Light therapy has been found to hold great potential in the healing of wounds and hair regrowth in a few animals. The highest rate of fast viral response observed during the season with full daily sunlight and Vitamin D affects the viral reaction to the therapy based on peginterferon/ribavirin (Hernández-Alvarez et al. 2017). For years, photobiomodulation therapy has found great potential in repairing worn-out tissues, reducing inflammation and relieving of acute and chronic pain using infrared (IR)/near-infrared (NIR) radiations; however, it only helps if the correct intensity and the exposure time of laser light are taken care of.

"Photobiomics" may affect a large number of living organisms, referring to the cumulative effects on metabolism, cytokines, transcription factors, microbiomes, and co-ordination factors of photobiomodulation (Liebert et al. 2019). However, studies show that ultraviolet (UV) radiation sensitivity can inhibit immunity from cell regulation and can thus adversely affect virus infections (Norval 2006). Far-UVC (Ultraviolet C) has successfully inactivated aerosol viruses with an amazingly low dose of over 95% of the H1N1 aerosolized influenza virus, which allows the dissemination of transmitted infectious diseases safe, stable, and cost-effective (Welch et al. 2018). Vacuum-UV light is an essential method for disinfecting microorganisms in the environment (Szeto et al. 2020). UV light is beneficial for lessening pathogenic viruses on fruit and vegetables without making any visible changes (Fino and Kniel 2008). In the present scenario when COVID-19 pandemic is at an alarming rate, scientists are trying to study ultraviolet germicidal radiation for the detection of virus in public places such as schools and restaurants and use the UV light for the disinfection of these contaminated public places to stop further transmission of the virus. These may act as a useful tool at very low-cost and can be used by supermarket facilities or restaurants where the health-related viral disease is associated with food handling. Hence we assume that frequent exposure to UV and IR/NIR but with specific calculated intensity is required in one or the other way to minimize the severity of COVID-19.

5.3.2 Antiviral Herbs of the Indian Subcontinent

Ayurveda (an Indian traditional system of medicine practiced in the Indian subcontinent) mentions numerous plants of herbal origin which may act effectively (prophylactically or treatment) against the symptoms of respiratory tract infections and enhances respiratory health. These symptoms are also expressed in the COVID-19 so these plants may support the restoration of respiratory health. Pippali (*Piper longum*), Kantkari (*Solanum xanthocarpum*), Karkatshringi (*Pistacia integerrima*), Kasmard (*Cassia occidentalis*), Agastya (*Sesbania grandiflora*), Shati (*Hedychium spicatum*), Karchura (*Curcuma zedoaria*), Pushkarmula (*Inula racemosa*), Bharangi (*Clerodendrum serratum*), Dugdhika (*Euphorbia thymifolia*) are some herbs mentioned in Ayurveda specifically for respiratory ailments. Haritaki (*Terminalia chebula*), Amlaki (*Emblica officinalis*), Guduchi (*Tinospora cordifolia*), Ashwagandha (*Withania somnifera*), Vridhadaru (*Argyreia nervosa*), Nagbala (*Grewia hirsuta*) are established herbs for enhancing immunity and can be used on preventive aspect against COVID-19. These herbs need more scientific validation and trials to be used against treatment or in the prevention of COVID-19. Some plants like Neem (*Azadirachta indica*), Tulsi (*Ocimum sanctum*), Karpoora (*Cinnamomum camphora*), Chandana (*Santalum album*) boiled in water and whole content may be used for bathing purpose and also in home fumigation by controlled burning to generate fumes (Savan and Mohanlal 2019).

Some antiviral and immunomodulatory processes are recognized for their use of food and herbs. The disease is known to have immunomodulatory properties, including Aloe, Angelica gigas, Astragalus membranaceus, Ganoderma lucidum, Scutellaria baicalensis, and Panax ginseng. It has been shown that antiviral activity against influenza was present in extracts and bioactive compounds from garlic, ginger, Korea's red ginseng, eucalyptus, tea tree, Tianmingjing, Machixian, fish mint, Chinese cabbage, capes, zhebeimu (Panyod, Ho, and Sheen 2020). A large number of herbal phytoconstituents have been researched widely for its antiviral function. Androgragrapholide (*Diterpene lactone*) has immuno and anti-inflammatory features (Tan et al. 2017) and dehydroandrographolide succinic acid monodester, which is derived from Andrographispaniculata, and it prevents HIV in vitro (Puri et al. 1993). Allium sativum (garlic)'s biological actions and its organosulfur compounds and allicin are expected to be acted in humans (Ankri and Mirelman 1999; Chang et al. 1991). Punica granatum, Ocimum sanctum, Azadirachta indica, and Carica papaya are to be further studied for potential Herpes simplex virus (HSV)-resistant practices. The plants are chosen based on their conventional prescribed use in the texts of Ayurveda (Weber et al. 1992; Thomson and Ali 2005). Herbs described in the context are abundantly available in the Indian subcontinent and are primarily used against contagious ailments since ancient times in Indian History. These herbs or their photo-constituents may act as a potential source of antivirals.

5.3.3 Sanitization

Medical science has believed for years that the process of washing hands is important to avoid illness. Remember that if the hand sanitizer is 60% alcohol, then wash hands with soap or else. It is advantageous to clean the hands because even if there is an infection in the hands, it is removed (Jing et al. 2020). Also, let us know that washing hands with soap and lukewarm water is considered better than a sanitizer. According to research, 40% of the people do not wash their hands before eating food, whereas more than these people do not wash their hands with bowel movements. According to this study, if the process of washing hands is carried out honestly, then every year millions of people can be saved. It is one of the most powerful weapons against the coronavirus. If the hands are infected and not washed/cleaned, then the germs can easily transfer to our co-workers, relatives, and friends. Door handles, toilets or bathrooms, tap spout, railing, lift button, and many other places that are in public contact can get you infected by merely a small physical contact. Remember that millions of germs are present in 1 g of human feces. These germs reach the body easily through contact. If you have a habit of repeatedly reaching hands to the mouth, nose, and eyes, then inadvertently your hand infection will affect the mouth and eyes. These infections can make you seriously ill. Remember that washing your hands with soap is more effective than a sanitizer. For proper cleaning, always apply soap on your hands for a long time; wait for the foam formation, clean the palms and the fingers along the back. Remove all foam from excess water. Now wipe with a clean and dry cloth. By doing this, you will not only be able to avoid infections of respiratory diseases but will also be able to avoid stomach diseases. Among the germs that enter the body through hands, stomach infection is the most painful after coronavirus. These germs reach the body with contaminated water and infected hands, due to which the deadly infections like typhoid and cholera are born. On average, 4 out of 10 cases of diarrhea can be avoided by washing hands properly. Another important advantage of hand washing is the protection from eye infections. Infectious diseases of the eye like conjunctivitis and trachoma are spread only through infected hands. In such a situation, the correct procedure of washing hands can prove to be very effective. Staphylococcus is a bacteria found in the skin and nostrils of the body. If these germs reach open wounds, skin infection occurs. By descending to the soft tissue inside the skin, these bacteria reach vital components including the joints and bones of the body. This infection can be avoided to a great extent by hand washing.

Sanitizer is usually provided in two forms: alcohol and nonalcohol (Jing et al. 2020; Fleur and Jones 2017). The sanitizer typically contains alcohol; the same has been found in beer and wine. Isopropanol (usually known as alcohol rubbing) and, less frequently, propanol are other alcohols that are used. Hand sanitizers are usually made of alcohol and contain between 60% and 95% alcohol, and for specific purposes, suppliers add extra ingredients. Specific agents, including chlorhexidine or benzalkonium chloride, are also active against virus or bacteria and also in nonalcoholic sanitizers, these ingredients are also essential. Ingredients like glycerol do not dry out hands. Used in minimal amounts, hydrogen peroxide avoids bacterial degradation of the sanitizer. Alcohols destroy most bacteria and viruses successfully. They affect the protein structure, which allows it to "denature" the outer shells of viruses and bacteria, to kill them, and to avoid infections. While in many cases, alcohols are found to be the most effective method of virus suppression, few viruses do not have an outer layer (known as an envelope) and are, therefore, unaffected by the use of alcohol. The nonenveloped viruses, for example, noroviruses, are not alcohol-killed. Often chlorhexidine is added to a sanitizer containing alcohol to kill certain bacteria and viruses. There are some reports that it improves the performance of alcohol-based sanitizers. However, coronavirus is a virus enveloped and alcohols are also successful in fighting it. Benzalkonium chloride is mostly used in hand sanitizers with no alcoholic content, so bacteria are easily regulated and virus activities are reduced. Sanitizers based on nonalcohol are usually less effective than those based on alcohol. The proportion of alcohol per volume will, however, be at least 60% and above that, bacteria and viruses are less likely to kill you. With the increased level of alcohol, the capacity of drug-based hand sanitizers rises. But very high (above 95%) concentrations are less effective. Proteins are not so easily denatured when no water is present in the environment. The dirt on the palms is also a contributing factor. Hand sanitizers do not absorb this unless they are coated with dirt or grate. Bacteria or viruses will stay in the soil on your hands (Egwari et al. 2016).

5.3.4 Fomites

Fomites are usually objects such as clothes, utensils, furniture, improperly sanitized medical equipment through which viral and bacterial infections are often transferred from one carrier to another (Kraay et al. 2018). Viruses account for 60% of human infection, and most infectious infections are caused by respiratory and enteric viruses. Fomites are infected with the virus by the direct body or body secretions, by soiled hands, the aerosol virus (droplet propagations) that is produced through voice, sneezing, coughing or vomiting, or by contact with an airborne virus that settles after a tainted fomite has been disrupted, i.e., through the shaking of a tainted blanket. In the present context of coronavirus issue, it is important to avoid the direct use of any fomites. Before using any object, it is necessary to clean the object with proper sanitization. Stephanie A. Boone published an article on the role of Fomites in the dissemination of Respiratory and Enteric Viral Disease, where he addressed the importance of Fomites in spreading most of the virus including coronavirus. Also, they have explained clearly how the virus stays on the objects (Pan et al. 2020; Boone and Gerba 2007).

5.3.5 Sound Vibration

Depending upon the range of frequencies, sound waves are classified into three regimes based on frequency: infrasound having the frequency range 104<–20 Hz, audible sound having the frequency range (20–10^4 Hz), and ultrasonic sound (10^4–10^{12} Hz) which has been shown in Figure 5.2. Sound waves have a long history in biomedical applications (Sarvazyan et al. 2010). Indian Vedic history reveals the use of shanks and temples bells (metallic) to clean the temple premises since it was believed that the sound waves generated by these bells and shankhs could remove the negative energies and maintain a positive aura. According to the physics of sound vibration, the metallic sound produces a spatial harmonic wave and its wavelength matches the dimension of many bacteria and viruses. Ultrasound is used on a big scale in the field of medical science. Since the metallic sound frequency matches

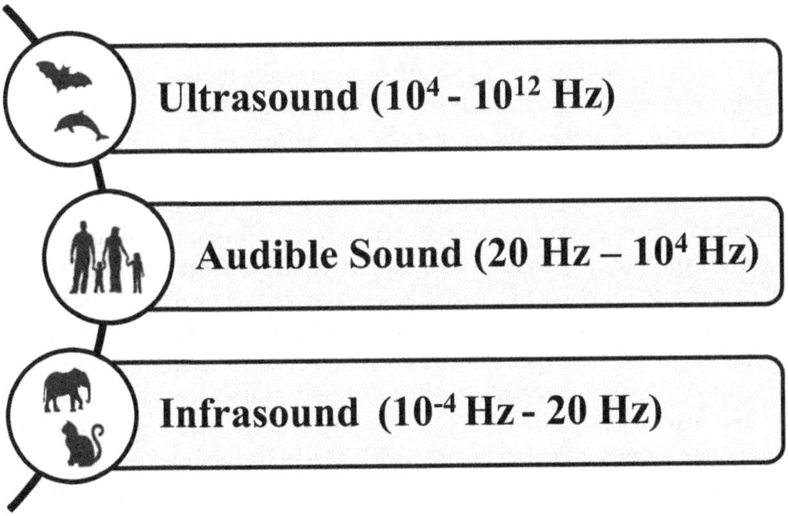

FIGURE 5.2 Classification of sound waves based on frequencies.

the resonance condition; the natural frequency of the ultrasonic sound wave matches the natural frequency vibration of the corresponding bacteria and virus membrane. For resonance with microbes less than 2 micrometers, the ultrasound frequency will be required. Ultrasonic sound waves are useful in various fields as in disrupting and lysate cells; inactivation of viral and bacterial cells; release of intracellular material, e.g., proteins, antigens; stimulation of bacterial activity; preparation of nano-drug carriers, etc. (Xing et al. 2017).

Ultrasonic sound waves have been widely used in the food processing industries to kill the bacteria present in the food units. Scientifically, the ultrasound waves create alternate expansion and compression sites that lead to the formation of vapor bubbles. This formation of pressure regions and varying pressure/temperature causes cavitation which is responsible for the killing of bacteria. This implies that not ultrasound alone, but it is the combined effect of longitudinal waves and the pressure/heat that emerges as a promising candidate for disrupting the bacterial cells (Piyasena, Mohareb, and McKellar 2003).

We have already discussed that DNA or RNA of the virus is enclosed within a protective protein shell called "capsid." Once this shell is rendered inactive, the virus itself becomes inactive. Hence the major role of researchers is to find an active way of rupturing this protective sheath around the virus. There is a set of natural frequencies at which our body cells along with the localized viruses vibrate freely. It can be easily correlated with every physical system around us. This implies that our physical systems also possess a set of the natural frequency of their own at which they can vibrate, referring these as free vibrations. However, just as the lasers have an external pumping source, these systems can also be excited using the mechanical vibrations/forced vibrations. When the excited frequency matches any of the naturally occurring frequencies of the system, it can successfully damage the cells around it. An opera singer shattering the wine glass is an appropriate example of wonders that sound energy can do.

The matching of the external frequency of the ultrasound waves with the natural frequency results in the resonance condition, which leads to the formation of alternate rarefactions and compressions. These alternate pressure regions impose mechanical stress which is intense enough to shatter the viral encapsulation, and therefore, once the outer covering is broken, the virus becomes ineffective.

Since researchers are in continuous process of making vaccinations and equipment to combat the deadly coronavirus, it is highly required to consider the science of ultrasound waves into saving mankind from further turmoil of the disease. Long-term sound exposure can change the physical structure of these microbes as it will be useful to decrease the effect of bacteria and viruses in the local region because ultrasonic sound vibration played an important role to distort the structure of the virus.

Moreover, in all these possible ways, if someone is feeling the symptoms of COVID-19, it is most important to follow the home quarantine for at least 15 days where one must avoid getting involved in any kind of meetings and gatherings. Home quarantine is not the treatment of COVID-19 but definitely a way to prevent the transmission of an infectious disease. In the current context of the pandemic, it is important to distract the person from the negative emotions and psychologically mental disorder during isolation.

5.4 SCIENTIFIC ADVANCEMENT IN CASE OF COMBATING THE COVID-19

5.4.1 AI Sensors

Sensors refer to the devices that are involved in the detection and measuring of the response of any optical, electrical, or chemical signal by converting any physical parameter such as temperature, pressure, or humidity into a signal that can be measured electrically. Machine Intelligence (MI) has proved to be a remarkable feature for medical/precise diagnosis and treatment, in which new resources for researchers, physicians, and patients are open have been developed that allow them to achieve better outcomes (Jordan and Mitchell 2015). Upon the incorporation of MI into healthcare fields, health research and the quality of patient care can be significantly improved. Just like sense organs are to the human body, electronic sensors are to artificial intelligence (AI). AI includes programming computers to make them behave as if humans do such as in the field of robotics, where no sensors employed imply a deaf and dumb robot incapable of functioning. MI/AI involves combining advanced technologies to make a machine capable enough to adapt to the human actions of reading, learning, decision-making or occasionally changing behaviors (Ho 2020). This is attained by bringing in the use of certain technological systems such as neural networks, expert systems, logics, and genetic algorithms. MI technology has already been applied to many areas that required the interpretation of data acquired from the sensors processed. AI algorithms are also actively used in Google Assistant, Amazon Alexa, and Apple Siri. In various parts of China, robots are being used by healthcare systems to disinfect the contaminated rooms using UV rays. Recently, the AI and data analysis techniques acknowledged the prospective for AI technology and data science to provision the work of epidemiologists and government crisis response teams. But considering the diversity of fields, the academics, researchers, and health professionals are in the process of establishment of sectors where AI could be brought into use in other forms.

In the present scenario wherein the world is dealing with a dangerous pandemic "COVID-19," the AI sensors are being actively used in "contact mapping" for detecting the infected/noninfected people. These are also being used to observe the spread of disease and that way makes it possible for the government to manage the priority population. These are also being used to detect the fever by mere scanning of the face even with their faces covered (Kricka et al. 2020). During an emergency, the administration of medications approved for human use based on medication systems that can also successfully treat COVID-19 can be used easily. The integrating AI algorithm for early and accurate detection of COVID-19 is shown in Figure 5.3.

To optimize combination therapy design, the use of AI is needed more than traditional medicine screening and recasting. Instead of waiting for the outcome of the finished testing to inform the subsequent design of the procedure, AI-driven drug combination design is possible primary security in deciding the appropriate combination therapy to initiate. Combination therapy with AI can even deter the use and can take on a task in avoiding downstream prescriptional medicine defective in the application of suboptimal clinical care, patient death, and morbidity.

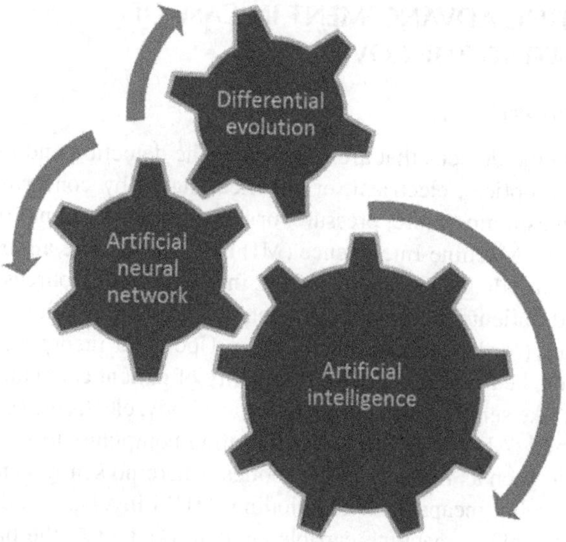

FIGURE 5.3 AI algorithm for COVID-19 detection.

This can lead to the rapid identification of regimes that mediate unexpected and significantly improved treatment outcomes (Shi et al. 2020). Data sets focused on COVID-19 (e.g., CORD-19), effective search tools (e.g., WellAI, SciSight), and mobile communications-based touch monitoring might be considered as efficient tools for disease research in COVID-19. Specifically in the context of urgency of diagnostic and treatment challenges are arising from the pandemic, where easily accessible AI-enabled tools and database material are valuable. The AI system may contribute to the rapid diagnosis of COVID-19 patients if computed tomography scanning is available and along with associated clinical history (Mei et al. 2020). A global effect of AI technologies involves flexible ways of exchanging knowledge, models, and code which may require adaptation to local environments and cross-border collaboration (Luengo-Oroz et al. 2020).

The new AI-enabled search tools are hoped to promote research and development as the world tries to find a reliable, timely, and effective treatment for this devastating pandemic. Solidarity will serve as the guiding concept to help the creation and implementation of the global sustainable development agenda of the creative and ethical frameworks for the AI to tackle COVID-19 pandemics. Governments worldwide are making use of thermal scanners that use AI and big data to combat the novel virus. There is contactless temperature-detection software that has been employed at various public centers for the same. AI has made it possible to detect the coronavirus by using special cameras that are well equipped with thermal sensors.

AI can also contribute to vaccine development by inspecting the components of the virus. This might support researchers and authorities to gain a piece of basic knowledge about the cause and ultimately, aid them in developing treatments that can be considered for preclinical trials.

Apart from the health effects, the COVID-19 situation has drastically affected the economy worldwide. Even in the economic sphere, AI and big data make it possible for us to investigate the influence and frame the responses (Kolozsvari et al. 2020). Where AI has found its effective contribution to the detection of the infected patient, the big data have solved a bigger problem, i.e., improving the observation systems for tracking the virus distribution.

5.4.2 Nanotechnology

Advancement in nanotechnology may significantly be helpful to battle COVID-19. Research and technological developments are the best weapons in the fight against COVID-19. To diagnose, cure, and prevention of this disease, nanotechnology methods can be employed and modified. Nanoparticles may be used for diagnostic techniques, medicinal carriers, and the development of vaccines (Chan 2020). Built on a variety of products that can be classified nanotechnology provides a variety of approaches to cope up during the emergency, due to its valuable physicochemical properties with flexible chemical functionalization. Taking into account what already knew about the virus lifecycle a few key steps to counteract the disease in nanotechnology can be considered. Next, by their intrinsic antipathogenic properties, nanoparticles may provide alternate approaches to conventional healthcare disinfection procedures. It could also be explorative in patients for SARS-CoV-2 inactivation. Nanomaterials can be used to supply pulmonary device drugs to avoid interactions between the receptors of angiotensin converting enzyme 2 (ACE2) and S viral protein. In fact, the "nano-immunity by design" concept will allow one to develop materials for immune control and applications for the production of vaccines against SARS-CoV-2 or cytokine surge counteraction. Nanomaterials can help to improve immune system upregulation and direct immune responses to antigens specifically in vaccine and immunization studies in several ways. Immune-specific nanotherapeutic products can be generated via appropriate nanoscale development that can stimulate the immune response, for example, as vaccine adjuvants. Nanotechnology has a significant role in diagnostics as well as in disease prevention and therapeutics that are easy and economically viable nanotechnologies to diagnose the presence of SARS-CoV-2 and associated biomarker (Weiss et al. 2020). Recently, pristine nanoparticles (NPs) of zinc ferrite (ZNF) have been developed by the auto-combustion sol-gel process. This model allows the viral RNA from many specimens to be extracted in the cycle of automation and may significantly decrease the level of discomfort regarding COVID-19, working time, and criteria for new molecular stage diagnosis. For the surface-functionalized ZNF NP by silica and polyvinyl carboxylate alcohol, it was suggested a basic but feasible RNA extraction protocol to active detection of the COVID-19. This approach is simple and cost-effective and will shortly be a replacement for traditional approaches to cope with tough times and the need for human capital, and is thus applicable for active identification of COVID-19 (B. Somvanshi et al. 2020). The plasma membranes of the human lung epithelial type II or human macrophages comprise two types of cell nanosponges. These nanosponges have the same known and unidentified protein receptors, which are needed for cell entry by

SARS-CoV-2. Essentially the architecture of nanosponge is agnostic for both viral and possible viral mutations. As long as the virus target remains, nanosponges can neutralize the virus, which has been established as a host cell (Zhang et al. 2020). The laboratory research findings using the Nanofibre Filter PVDF 6-layer primed have resulted in a filtration effectiveness of 92%, 94%, and 98% (qualified for "N98," respectively), using monodispersed NaCl aerosols between 50, 100, and 300 nm standard) with a 26 Pa decrease of equivalent pressure. The 2%–6% efficiency discrepancy in the lack of contact in laboratory aerosols and monodispersed NaCl aerosols of different dimensions was primarily due to NaCl aerosols. The disparity can be minimized further with the increasing number of modules in the filter and aerosol greater than 300 nm. A 6-layer nanofiber filter has been classified "N98 respirator." The water pressure fall of just 2.65 mm was 1/10 below normal N95 (98% capture capacity for 300 NaCl), with a water column of 25 mm (exhalation) to 35 mm (inhalation). A 6-layer polyvinylidene fluoride filter can be used effectively for prevention as it provides strong personal protection from pollutants based on the N98 model against airborne COVID-19 and nanoaerosol, but at least 10× more appealing than traditional N95 respirators (Zhu et al. 2020; Das et al. 2020). Considering the progress of the pandemic COVID-19, a life-threatening respiratory infectious disease coronavirus (SARS-CoV-2) where the quick, pick, and responsive identification is required; biosensors with nanosensing range are highly resourceful. When suitable for tailored health wellness management, biosensors are evolving as efficient (sensitive and selective) and inexpensive analytical diagnostic devices. Ever since biosensors have been established, efforts have been made to test their applicability in clinical applications with a demonstrated and streamlined sensing system. To determine disease development under treatment, low rates of identification of biomarker controlled diseases were highly helpful. The efficiency of prescription medication, maximizing therapy, and matching the degree of the biomarkers of pathogenesis diseases are expected in these bioinformatics and its multi-aspect-oriented analytics. Taking into account the technical advancement by 2022, a compound annual growth rate of 8.4% of biosensors is projected to amount to up to USD 28 trillion. There is a relentless initiative to encourage state-of-the-art biosensing technologies as a noninvasive diagnostics technique for next-generation diseases.

In light of this radical opinion essay discusses customized research instruments for access to better well-being for everyone and in general for the timely management of a healthy world. For the customized health management biosensors, nanosensors are efficient (sensitive and selective) and inexpensive methods to evaluate the early-stage disease detection; these inexpensive intelligent diagnostic tools are required urgently to manage the predictions and achievements. In one of the study biosensor-dependent Field Effect transistor (FET) tool was used in clinical samples to detect COVID-19. A highly sensitive COVID-19 diagnostic device and the promising SARS-CoV-2 FET biosensor, which does not need samples pretreatment or protocol (Mujawar et al. 2020; Seo et al. 2020). Figure 5.4 is showing the tools based on nanotechnology for detecting the effect of COVID-19. There are a lot of other scientific advances that are still in progress for decreasing the effect of COVID-19.

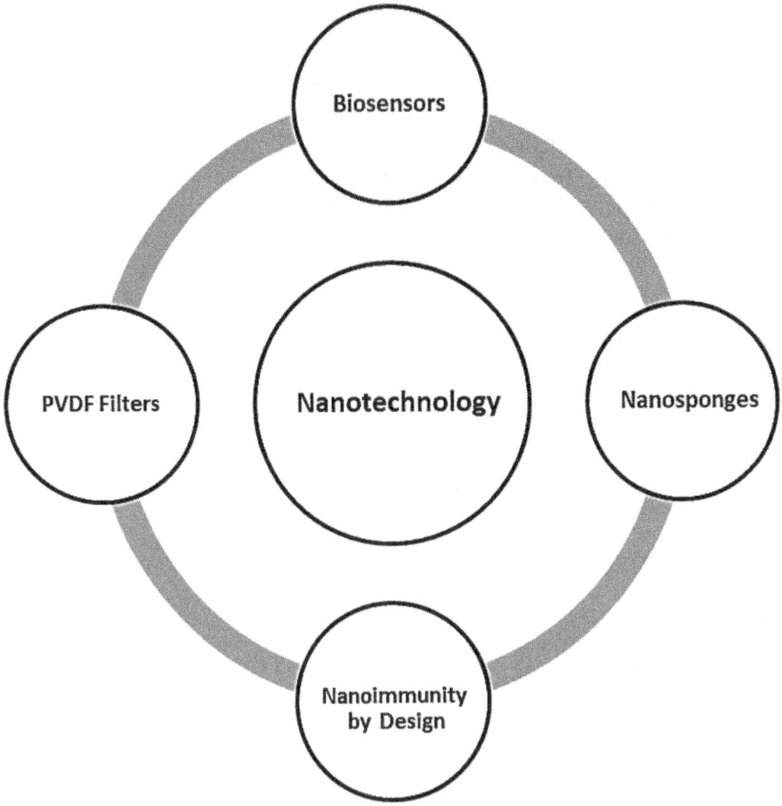

FIGURE 5.4 Nanotech tools for detecting COVID-19.

5.5 CONCLUSION

COVID-19 is declared the first pandemic of the 21st century and is known to affect humans drastically. Since the effective vaccines are yet to be developed for inhibiting the growth of this viral infection, the various methods can be employed to deal with its ill effects in daily life. In this context, awareness among human beings about the effect of such viruses, and how to decrease its ill effects of the hazardous virus is very important. Strict self-quarantine measures and social distancing play a significant role in properly dealing with coronavirus. We can decrease the effect of COVID-19 by avoiding the direct use of fomites and following the sanitization process strictly. The ultrasound waves, UV light, and the sunlight play a vital role in inhibiting/killing microorganisms/bacteria which is enough to combat the deadly virus too as in the case of packaging food, schools/colleges, and in public places. The Ayurveda is known to serve mankind for a longer time than we can imagine and it is a well-known fact by now that stronger the immunity, more easily you can fight the COVID-19 virus if infected. As it is said, "Prevention is better than cure," it is highly expected that various Ayurvedic herbs can be run through trials and hence employed for enhancing the immunity and can be used on preventive aspect against COVID-19. Moreover, alcohol-based sanitizers are being remarkably used to avoid getting in contact with the coronavirus through any of the fomites. Among the scientific and technological inventions, various techniques based on machine learning and AI as AI sensors-supported thermal scanners can be used to keep a track of the spread of the disease and may render a life-saving effort by reducing the spread of transmission. It has been found that biosensors, primarily the nanosensors, are a cost-efficient and effective way of detecting the virus at an early stage.

ACKNOWLEDGMENTS

The authors are gratified to Dr. M A Shaz, Department of Physics, BHU, for providing salient suggestions on the manuscript; Prof. Anand Chaudhary, IMS-BHU, and Dr. Rohit Sharma, IMS-BHU, for their feedback and suggestions. Author "H.A." would like to acknowledge the CSIR India for providing senior research fellowship (09/013(0763)/2018-EMR-I).

REFERENCES

Andersen, Kristian G., Andrew Rambaut, W. Ian Lipkin, Edward C. Holmes, and Robert F. Garry. 2020. "The Proximal Origin of SARS-CoV-2." *Nature Medicine*. doi: 10.1038/s41591-020-0820–9.

Ankri, Serge, and David Mirelman. 1999. "Antimicrobial Properties of Allicin from Garlic." *Microbes and Infection*. doi: 10.1016/S1286–4579(99)80003–3.

B. Somvanshi, Sandeep, Prashant B. Kharat, Tukaram S. Saraf, Saurabh B. Somwanshi, Sumit B. Shejul, and Kamalakar M. Jadhav. 2020. "Multifunctional Nano-Magnetic Particles Assisted Viral RNA-Extraction Protocol for Potential Detection of COVID-19." *Materials Research Innovations*: 1–6. doi: 10.1080/14328917.2020.1769350.

Boone, Stephanie A., and Charles P. Gerba. 2007. "Significance of Fomites in the Spread of Respiratory and Enteric Viral Disease." *Applied and Environmental Microbiology*. doi: 10.1128/AEM.02051–06.

Chan, Warren C.W. 2020. "Nano Research for COVID-19." *ACS Nano*. doi: 10.1021/acsnano.0c02540.

Chang, R. Shihman, lu Ding, Chen Gai-Qing, Pan Qi-Choa, Zhao Ze-Lin, and Kevin M. Smith. 1991. "Dehydroandrographolide Succinic Acid Monoester as an Inhibitor against the Human Immunodeficiency Virus." *Proceedings of the Society for Experimental Biology and Medicine* 197 (1): 59–66. doi: 10.3181/00379727–197–43225.

Chen, Jieliang. 2020. "Pathogenicity and Transmissibility of 2019-NCoV—A Quick Overview and Comparison with Other Emerging Viruses." *Microbes and Infection* 22 (2): 69–71. doi: 10.1016/j.micinf.2020.01.004.

Das, Oisik, Rasoul Esmaeely Neisiany, Antonio Jose Capezza, Mikael S. Hedenqvist, Michael Försth, Qiang Xu, Lin Jiang, Dongxiao Ji, and Seeram Ramakrishna. 2020. "The Need for Fully Bio-Based Facemasks to Counter Coronavirus Outbreaks: A Perspective." *Science of the Total Environment* 736. doi: 10.1016/j.scitotenv.2020.139611.

Egwari, L.O., O.O. Ayepola, B. Adekeye, T. Adegbayi, S. Olurunshola, and A. Kuye. 2016. "Evaluating the Effect of Hand Washing and Sanitization on the Microbial Burden of the Hand." *International Journal of Infectious Diseases* 45: 277. doi: 10.1016/j.ijid.2016.02.613.

Fino, Viviana R., and Kalmia E. Kniel. 2008. “UV Light Inactivation of Hepatitis A Virus, Aichi Virus, and Feline Calicivirus on Strawberries, Green Onions, and Lettuce.” *Journal of Food Protection* 71 (5): 908–913. doi: 10.4315/0362-028X-71.5.908.

Gautret, Philippe, Jean-Christophe Lagier, Philippe Parola, Van Thuan Hoang, Line Meddeb, Morgane Mailhe, Barbara Doudier, et al. 2020. “Hydroxychloroquine and Azithromycin as a Treatment of COVID-19: Results of an Open-Label Non-Randomized Clinical Trial.” *International Journal of Antimicrobial Agents*, 105949. doi: 10.1016/j.ijantimicag.2020.105949.

Hernández-Alvarez, Noemi, Juan Manuel Pascasio Acevedo, Enrique Quintero, Inmaculada Fernández Vázquez, María García-Eliz, Juan De La Revilla Negro, Javier Crespo García, and Manuel Hernández-Guerra. 2017. “Effect of Season and Sunlight on Viral Kinetics during Hepatitis C Virus Therapy.” *BMJ Open Gastroenterology* 4 (1). doi: 10.1136/bmjgast–2016–000115.

Ho, Dean. 2020. “Addressing COVID-19 Drug Development with Artificial Intelligence.” *Advanced Intelligent Systems* 2 (5): 2000070. doi: 10.1002/aisy.202000070.

Jing, Jane Lee Jia, Thong Pei Yi, Rajendran J.C. Bose, Jason R. McCarthy, Nagendran Tharmalingam, and Thiagarajan Madheswaran. 2020. “Hand Sanitizers: A Review on Formulation Aspects, Adverse Effects, and Regulations.” *International Journal of Environmental Research and Public Health*. doi: 10.3390/ijerph17093326.

Jordan, Michael I., and Tom M. Mitchell. 2015. “Machine Learning: Trends, Perspectives, and Prospects.” *Science*. doi: 10.1126/science.aaa8415.

Kolozsvari, Laszlo Robert, Tamas Berczes, Andras Hajdu, Rudolf Gesztelyi, Attila Tiba, Imre Varga, Gergo Jozsef Szollosi, Szilvia Harsanyi, Szabolcs Garboczy, and Judit Zsuga. 2020. “Predicting the Epidemic Curve of the Coronavirus (SARS-CoV-2) Disease (COVID-19) Using Artificial Intelligence.” *MedRxiv*. doi: 10.1101/2020.04.17.20069666.

Kraay, Alicia N.M., Michael A.L. Hayashi, Nancy Hernandez-Ceron, Ian H. Spicknall, Marisa C. Eisenberg, Rafael Meza, and Joseph N.S. Eisenberg. 2018. “Fomite-Mediated Transmission as a Sufficient Pathway: A Comparative Analysis across Three Viral Pathogens 11 Medical and Health Sciences 1117 Public Health and Health Services.” *BMC Infectious Diseases* 18 (1). doi: 10.1186/s12879-018-3425–x.

Kricka, Larry J, Sergei Polevikov, Jason Y Park, Paolo Fortina, Sergio Bernardini, Daniel Satchkov, Valentin Kolesov, and Maxim Grishkov. 2020. “Artificial Intelligence-Powered Search Tools and Resources in the Fight Against COVID-19.” *EJIFCC* 31 (2): 106–116. http://www.ncbi.nlm.nih.gov/pubmed/32549878%0Ahttp://www.pubmedcentral.nih.gov/articlerender.fcgi?artid=PMC7294813.

la Fleur, P, and Jones, S. 2017, March 16. *Non-Alcohol Based Hand Rubs: A Review of Clinical Effectiveness and Guidelines* [Internet]. Ottawa, ON: Canadian Agency for Drugs and Technologies in Health. PMID: 29266912.

Lan, Lan, Dan Xu, Guangming Ye, Chen Xia, Shaokang Wang, Yirong Li, and Haibo Xu. 2020. “Positive RT-PCR Test Results in Patients Recovered from COVID-19.” *JAMA - Journal of the American Medical Association*. doi: 10.1001/jama.2020.2783.

Liebert, Ann, Brian Bicknell, Daniel M. Johnstone, Luke C. Gordon, Hosen Kiat, and Michael R. Hamblin. 2019. “‘Photobiomics’: Can Light, Including Photobiomodulation, Alter the Microbiome?” *Photobiomodulation, Photomedicine, and Laser Surgery*. doi: 10.1089/photob.2019.4628.

Luengo-Oroz, Miguel, Katherine Hoffmann Pham, Joseph Bullock, Robert Kirkpatrick, Alexandra Luccioni, Sasha Rubel, Cedric Wachholz, et al. 2020. “Artificial Intelligence Cooperation to Support the Global Response to COVID-19.” *Nature Machine Intelligence* 2 (6): 295–297. doi: 10.1038/s42256-020-0184–3.

Mei, Xueyan, Hao Chih Lee, Kai yue Diao, Mingqian Huang, Bin Lin, Chenyu Liu, Zongyu Xie, et al. 2020. “Artificial Intelligence–Enabled Rapid Diagnosis of Patients with COVID-19.” *Nature Medicine*. doi: 10.1038/s41591-020-0931–3.

Mujawar, Mubarak A., Hardik Gohel, Sheetal K. Bhardwaj, Sesha Srinivasan, Nicolerta Hickman, and Ajeet Kaushik. 2020. "Nano-Enabled Biosensing Systems for Intelligent Healthcare: Towards COVID-19 Management." *Materials Today Chemistry*. doi: 10.1016/j.mtchem.2020.100306.

Norval Mary. 2006. "The Effect of Ultraviolet Radiation on Human Viral Infections." *Photochemistry and Photobiology* 82 (6): 1495–1504. doi: 10.1562/2006-07-28-IR–987.

Pan, Xingchen, David M. Ojcius, Tianyue Gao, Zhongsheng Li, Chunhua Pan, and Chungen Pan. 2020. "Lessons Learned from the 2019-NCoV Epidemic on Prevention of Future Infectious Diseases." *Microbes and Infection* 22 (2): 86–91. doi: 10.1016/j.micinf.2020.02.004.

Panyod, Suraphan, Chi Tang Ho, and Lee Yan Sheen. 2020. "Dietary Therapy and Herbal Medicine for COVID-19 Prevention: A Review and Perspective." *Journal of Traditional and Complementary Medicine*. doi: 10.1016/j.jtcme.2020.05.004.

Piyasena, P., Eugene Mohareb, and R. C. McKellar. 2003. "Inactivation of Microbes Using Ultrasound: A Review." *International Journal of Food Microbiology*. doi: 10.1016/S0168–1605(03)00075–8.

Puri, Anju, Ragini Saxena, R. P. Saxena, K. C. Saxena, Vandita Srivastava, and J. S. Tandon. 1993. "Immunostimulant Agents from Andrographis Paniculata." *Journal of Natural Products* 56 (7): 995–999. doi: 10.1021/np50097a002.

Sarvazyan, Armen P., Oleg V. Rudenko, and Wesley L. Nyborg. 2010. "Biomedical Applications of Radiation Force of Ultrasound: Historical Roots and Physical Basis." *Ultrasound in Medicine and Biology*. doi: 10.1016/j.ultrasmedbio.2010.05.015.

Savan, Dipti, Kumar, and Jaiswal Mohanlal. 2019. "Karira (Capparis Decidua Edgew.) As Food and Medicine: A Comprehensive Chronological Review." *International Journal of Research in Ayurveda and Pharmacy* 10 (2): 1–6. doi: 10.7897/2277–4343.100224.

Seo, Giwan, Geonhee Lee, Mi Jeong Kim, Seung Hwa Baek, Minsuk Choi, Keun Bon Ku, Chang Seop Lee, et al. 2020. "Rapid Detection of COVID-19 Causative Virus (SARS-CoV-2) in Human Nasopharyngeal Swab Specimens Using Field-Effect Transistor-Based Biosensor." *ACS Nano* 14 (4): 5135–5142. doi: 10.1021/acsnano.0c02823.

Shi, Feng, Jun Wang, Jun Shi, Ziyan Wu, Qian Wang, Zhenyu Tang, Kelei He, Yinghuan Shi, and Dinggang Shen. 2020. "Review of Artificial Intelligence Techniques in Imaging Data Acquisition, Segmentation and Diagnosis for COVID-19." *IEEE Reviews in Biomedical Engineering*. doi: 10.1109/RBME.2020.2987975.

Szeto, Wai, W. C. Yam, Haibao Huang, and Dennis Y.C. Leung. 2020. "The Efficacy of Vacuum-Ultraviolet Light Disinfection of Some Common Environmental Pathogens." *BMC Infectious Diseases* 20 (1). doi: 10.1186/s12879-020-4847–9.

Tan, W. S.Daniel, Wupeng Liao, Shuo Zhou, and W. S. Fred Wong. 2017. "Is There a Future for Andrographolide to Be an Anti-Inflammatory Drug? Deciphering Its Major Mechanisms of Action." *Biochemical Pharmacology*. doi: 10.1016/j.bcp.2017.03.024.

Tang, Yi Wei, Jonathan E. Schmitz, David H. Persing, and Charles W. Stratton. 2020. "Laboratory Diagnosis of COVID-19: Current Issues and Challenges." *Journal of Clinical Microbiology*. doi: 10.1128/JCM.00512–20.

Thomson, Martha, and Muslim Ali. 2005. "Garlic [Allium Sativum]: A Review of Its Potential Use as an Anti-Cancer Agent." *Current Cancer Drug Targets* 3 (1): 67–81. doi: 10.2174/1568009033333736.

Weber, N. D., D. O. Andersen, J. A. North, B. K. Murray, L. D. Lawson, and B. G. Hughes. 1992. "In Vitro Virucidal Effects of Allium Sativum (Garlic) Extract and Compounds." *Planta Medica* 58 (5): 417–423. doi: 10.1055/s–2006–961504.

Webmeter. 2020. "Coronavirus Age, Sex, Demographics (COVID-19) - Worldometer." Www.Worldometers.Info.

Weiss, Carsten, Marie Carriere, Laura Fusco, Laura Fusco, Ilaria Capua, Jose Angel Regla-Nava, Matteo Pasquali, et al. 2020. "Toward Nanotechnology-Enabled Approaches against the COVID-19 Pandemic." *ACS Nano*. doi: 10.1021/acsnano.0c03697.

Welch, David, Manuela Buonanno, Veljko Grilj, Igor Shuryak, Connor Crickmore, Alan W. Bigelow, Gerhard Randers-Pehrson, Gary W. Johnson, and David J. Brenner. 2018. "Far-UVC Light: A New Tool to Control the Spread of Airborne-Mediated Microbial Diseases." *Scientific Reports* 8 (1). doi: 10.1038/s41598-018-21058–w.

Xing, Jida, Shrishti Singh, Yupeng Zhao, Yan Duan, Huining Guo, Chenxia Hu, Allan Ma, et al. 2017. "Increasing Vaccine Production Using Pulsed Ultrasound Waves." *PLoS ONE* 12 (11). doi: 10.1371/journal.pone.0187048.

Yu, Fei, Lanying Du, David M. Ojcius, Chungen Pan, and Shibo Jiang. 2020. "Measures for Diagnosing and Treating Infections by a Novel Coronavirus Responsible for a Pneumonia Outbreak Originating in Wuhan, China." *Microbes and Infection* 22 (2): 74–79. doi: 10.1016/j.micinf.2020.01.003.

Zhang, Qiangzhe, Anna Honko, Jiarong Zhou, Hua Gong, Sierra N. Downs, Jhonatan Henao Vasquez, Ronnie H. Fang, Weiwei Gao, Anthony Griffiths, and Liangfang Zhang. 2020. "Cellular Nanosponges Inhibit SARS-CoV-2 Infectivity." *Nano Letters*. doi: 10.1021/acs.nanolett.0c02278.

Zhou, Peng, Xing Lou Yang, Xian Guang Wang, Ben Hu, Lei Zhang, Wei Zhang, Hao Rui Si, et al. 2020. "A Pneumonia Outbreak Associated with a New Coronavirus of Probable Bat Origin." *Nature* 579 (7798): 270–273. doi: 10.1038/s41586-020-2012–7.

Zhu, Hai Dong, Chu Hui Zeng, Jian Lu, and Gao Jun Teng. 2020. "COVID-19: What Should Interventional Radiologists Know and What Can They Do?" *Journal of Vascular and Interventional Radiology*. doi: 10.1016/j.jvir.2020.03.022.

Weiss, Carmen, Marc Carriere, Laura Fusco, Ivana Capua, Jose Angel Regla-Nava, Matteo Pasquali, et al. 2020. "Toward Nanotechnology-Enabled Approaches against the COVID-19 Pandemic." ACS Nano. doi:10.1021/acsnano.0c03697.
Welch, David, Manuela Buonanno, Veljko Grilj, Igor Shuryak, Connor Crickmore, Alan W. Bigelow, Gerhard Randers-Pehrson, Gary W. Johnson, and David J. Brenner. 2018. "Far-UVC Light: A New Tool to Control the Spread of Airborne-Mediated Microbial Diseases." Scientific Reports 8 (1). doi:10.1038/s41598-018-21058-w.
[illegible] 2017. "[illegible] Vaccine Protection Using [illegible]." [illegible]
[illegible] 2020. "[illegible] for [illegible] and Tracking Infections by [illegible] Pneumonia Outbreak Originating in Wuhan, China." [illegible]
[illegible] 2020. "[illegible] SARS-CoV-2 [illegible]." [illegible]
Zhou, Peng, Xing-Lou Yang, Xian-Guang Wang, Ben Hu, Lei Zhang, Wei Zhang, Hao-Rui Si, et al. 2020. "A Pneumonia Outbreak Associated with a New Coronavirus of Probable Bat Origin." Nature 579 (7798): 270–273. doi:10.1038/s41586-020-2012-7.
[illegible] 2020. "[illegible] COVID-19: What Should [illegible]." [illegible]

6 An Efficient Control Strategy for Prevention and Identification of COVID-19 Pandemic Disease

M. Parimaladevi, T. Sathya, V. Gowrishankar, and G. Boopathi Raja
Velalar College of Engineering and Technology

S. Nithya
SRMIST

CONTENTS

6.1 INTRODUCTION

Nowadays, a few virus infections arise with the pandemic outbreak and an assortment of ailments to human with nature of disease as yet present [1,2]. The origin of virus has been proposed for a large number of these sicknesses henceforth it is needed to concentrate on the significance of scanning for new infections [1]. Corona virus, also known as Corona Viridae, has a genome size of 27–32 kb RNA. Also, three numbers of unique gatherings of corona infections are explored. These infections are distinguished from various animal sources, for example, chickens, hounds, cattle, cat, ponies, bunnies, pig, turkeys and people too [3,4]. The resultant irresistible illnesses produced for the hepatic part, gastrointestinal organ and also for the respiratory tract. The respiratory illness are identified as the unavoidable contamination found in patients, while the inner contamination is brought about by porcine and ox-like corona infections which brought about an extreme financial misfortune [5–8].

All the corona infections affect the respiratory tract, that is, respiratory sicknesses called SARS-CoV (Severe Acute Respiratory Syndrome Corona Virus), where an aggregate of six animal groups were included for illness to patients [9]. In the mid-1960s, few species such as HCoV-OC43, HCV-229E, etc., were identified and recognized, which causes a typical cold. Later, other possibilities such as MERS-CoV (Middle East Respiratory Syndrome–Corona virus), SARS-CoV (Severe Acute Respiratory Syndrome–Corona Virus), HCoVNL63 virus was identified from a newborn child with bronchiolitis in the Netherlands, and HCoV-HKU1 virus was observed in a grown-up tolerant with extreme pneumonia in Hong Kong [10]. Among these corona virus families, SARS and MERS corona viruses were seen as profoundly more pathogenic. The SARS corona virus distinguished as the more harmful pathogenic human corona infection which causes perilous pneumonia [11–13]. It spreads to people through the unpredictable and exponential way, as this infection dwelled to peoples [14–16]. As of late, a more difficult and fatal one the Novel Corona Virus COVID-19 corona infection affect the respiratory tract which is considered for the study of disease transmission, etiology and clinical qualities [17–23].

COVID-19 a quickly dynamic infection announced initially with 29 pneumonia instances with obscure etiology at the office of World Health Organization (WHO) (Wuhan, Hubei territory, China) on 31 December 2019. This report stated that it is independent on hereditary arrangement and was immediately distinguished as novel β-corona infection on 12 January 2020. Later a month, WHO named the illness as Novel Corona infection Disease 2019 (COVID-19), brought about by the novel corona infection [1,23–29]. Fast progressive indicators of COVID-19 were observed as inadequacy and high infectivity of a compelling treatment. It is the most extreme significance to create effective methodologies to prevent the pathogenesis of novel corona virus. In spite of the fact that numerous remedial alternatives were at that point presented, the utilization of current and new age antiviral, steroids and distinctive restorative treatment, moreover, the ideal technique for extreme COVID-19 stays indistinct. In the recent few days, the affirmed instances of COVID-19 quickly gathered, yet because of the constrained data on novel corona infection and its pathogenesis, it turns into a difficult issue for every worldwide researcher. All things considered, no clear treatment has been recognized up until now, this makes it incredibly hard for medical examinations. This chapter gives consolidated data of corona infection sources, diseases, manifestations and its preventive, careful steps needed to apply and new logical difficulties for residents, researcher and social orders which may assist with discovering successful systems for COVID-19 treatment.

6.1.1 Clinical Symptoms of COVID-19 Patients

The basic symptoms of Novel Corona Virus COVID-19 include high fever, dry cough, severe breathing issue and weariness or myalgia. The early stage side effects were diarrhea, nausea and severe fever however may not be prime side effects of this disease. A few patients may experience the ill effects of migraines (headache) or hemoptysis and even generally asymptomatic [14,21] and [30]. In some cases, more established age patients have respiratory disappointment because of serious alveolar harm [18]. This disease spreads very rapidly before predicting the symptoms, that bringing about the abnormal function of the heart or kidney organs [21–24]. The COVID-19 infected patients indicated lower immune power by bringing down defensive white blood cells with expanded and initiated thromboplastin time and also with expanded C-reactive protein level [14,18,21,24]. So it requires checking the most widely recognized manifestation identified with this malady normally and numbness of this may bring about a national/worldwide harm.

6.1.2 Persistence and Precautions

The WHO and Center for Disease Control and Prevention released authority archives to give between time rules to the assortment, dealing with and diagnostics of medical examples that may include corona virus SARS-CoV-2. The rules demonstrated control measures under five unique measures, for example, cleanliness support, utilization of biocidal agents, social distancing, isolate individuals from others (quarantine each other) and mentality [1].

6.1.3 Hygiene Maintenance

Cleanliness and well-being successfully diminish the transmission of respiratory diseases. In this manner, personal defensive schemes including continuous washing of hands and utilization of face cover (mask) have been broadly sent danger of respiratory beads or droplets and aerosols [1,31].

The handling of several types of alcohol- or organic-based hand sanitizers and also medical disinfectants, for example, disinfectant bleach, has assumed a significant job in destroying SARS-CoV-2 virus. Also, the utilization of good quality FFP3 face covers or N95 masks forestalls severe respiratory tract diseases. Environmental disinfection, for example, surfaces and substances that can represent a hazard with suitable disinfectant is additionally suggested [32].

6.1.3.1 Utilization of Biocidal Agents

Different sorts of biocidal or medical agents, for example, alcohols, sodium hypochlorite (NaClO), hydrogen peroxide (H_2O_2) or benzalkonium chloride were utilized all over the world for sanitization. Mostly, the suggested biocidal agents such as 95% ethanol with 30 seconds treatment found to be an efficient factor for sanitization in

COVID-19 precaution. Because of the tremendous interest for biocidal necessities, WHO broadcasts a manual for nearby creation of organic-based handwash method for human security [33].

6.1.4 Social Distancing

Get-together or public gathering is the fundamental well-spring across the board network spread of COVID-19, and along these lines, regional mitigation or network alleviation provides control of the neighborhood spread, and it must be carefully thought of and applied at every possible opportunity. Also, this incorporates the temporarily closure of open community places for public gatherings, for example, schools, universities or privately owned businesses, transportations and so on. Its all-inclusive rendition remembers prohibiting for mass gathering, the compulsory wearing of veils, detachment of sick people and fitting sterilization/cleanliness measures could lessen mortality. Time limit and long-term nation lockdown will be a compelling method to diminish outbreak and spread the scourge over a more extended period of time. In this manner, the general effect of social distancing strategy controls the peak of the epidemic growth, by decreasing the cumulative sum of affected people and the general well-being as well [34].

6.1.5 Quarantine People

Isolate the individuals having fundamental manifestations and/or corona-positive patients in isolation wards or medical clinics. The nearby contacts of such quarantined or isolated people must be carefully monitored, followed and isolated at home or in assigned isolate offices. A large portion of the nations (Hong Kong, Singapore, Taiwan, and so forth) utilize these practices to maintain a strategic social distance from the transmission of this ailment. Until this point, these regulation measures seem to have had the option to forestall continued neighborhood transmission. However, isolated or quarantine measures can be an expensive testing to authorize and present area with explicit moral challenges [32].

6.1.6 Mentality and Diagnosis of COVID Patients

The individual behavior or community attitude with positive intuition brought an effective impact and firing soul against savage viral infections. Accordingly, self-assertion about the asymptomatic patient and its disconnection activity just as acknowledgment of well-being division and legislative choices is critical [32].

A Novel Corona virus has exponential growth and maximal infectivity and the inaccessibility of adequate drugs is the significant worry to handle the COVID-19 pathogenesis. Despite the fact that numerous restorative alternatives of Ayurvedic, antiviral or antibacterial steroids, allopathic, and conventional Chinese medication are in existence still are considered as the most positive and viable medication against COVID-19. Up until this point, no positive or proper treatment has been identified or distinguished, which causes it incredibly hard for hospital and worldwide administration [35].

The profound sequencing genomic procedures or polymerase chain reaction (PCR)-based approach is utilized at present for the recognition of infection and diagnosis [36,37]. In any case, for recognition with these procedures, an adequate amount of test assortment (nearness of viral genome) for intensification must be required. Correspondingly, a mistaken example assortment can constrain the convenience of qPCR-based examination [38,39]. In such a situation, an extra screening strategy is expected to guarantee the opportune determination of a contaminated individual.

6.2 CHALLENGES

The research challenges associated with novel corona virus causing COVID-19 are due to the following cases [1]:

a. In vitro replication confinement to Corona infections and also utilization in viral diagnostics.
b. In vitro replication confinement of those corona infections that provide a cytopathic impact get disappointment to infection recognizable proof techniques.
c. Explain wrong acknowledgments about antibodies and treatments.
d. Research on the pathogenicity.
e. Identify successful, simpler, less expensive helpful operator, treatment technique or immunization for patients.
f. To identify the utilization of virus disclosure dependent on complementary deoxyribonucleic acid–amplified fragment length polymorphism technique.
g. To train the individual's mentality and battling soul against COVID-19.

Therefore, COVID-19 treatment is not a simple task for researcher and need to investigate corner to corner by thinking about all prospects of science, bioengineering, expository apparatuses, just as psychology [1].

6.3 LITERATURE REVIEW

In this case [40], plasma treatment was provided to five patients critical of COVID-19. Their clinical conditions were monitored continuously and identified as improved, reflected with improved PAO2/FIO2, reduced body temperature and chest imaging within the next days after treatment with convalescent plasma. Four patients undergoing mechanical ventilation and Extra-Corporeal Membrane Oxygenation did not require breathing assistance by 9 days following the plasma transfusion.

Maghdid et al. [41] uses the smartphones that are fitted with sensors. The measurement of sensor's information was used for the symptoms of coronavirus diseases. The symptoms were observed for human-health purposes and algorithms are applied in each reading of those sensors. The outcomes of the symptoms reported by the application are then predicted and stored as a single record in the data set. Thus, the records collected from various patients were used as input to Machine Learning (ML) algorithm. In addition, a transfer learning method can also be used to further develop the ML model in the final stage, when the model operates on the cloud.

A prototype of an AI-based system, namely α-Satellite, is proposed in [42] to evaluate the risk of infection at the community level of the specific geographical region. The system collects various types of big data from heterogeneous sources in large scale. The social media data available for a given area may be limited so that they are enriched by the conditional generative adversarial nets [43] to learn the public awareness of COVID-19. A heterogeneous graph autoencoder model is then devised to aggregate information from neighborhood areas of the given area in order to estimate its risk indexes. This risk information enables residents to select appropriate actions to prevent them from the virus infection with minimum disruptions in their daily lives. It is also useful for authorities to implement appropriate mitigation strategies to combat the fast evolving pandemic.

Story et al. [44] use smartphone's videos for nausea prediction while Lawanont et al. [45] use camera images and inertial sensors' measurements for neck posture monitoring and human headache level prediction. Alternatively, audio data obtained from smartphone's microphone are used for cough type detection in [46,47].

Taking into consideration China's emerging condition [48], smart disease detection systems based on Internet of Things (IoT) may be a big breakthrough in the fight against current pandemic. The role that these technologies play to restrict the pandemic spread involves accumulating and interpreting of gathered information, given many infrastructures in place. If feasible measures are taken to increase and make use of the information, the joint part of IoT and emerging innovations may enhance the initial epidemic diagnosis. IoT-based intelligent disease monitoring systems would provide concurrent controlling and recording, intercepting and notifications, data range and interpretation as well as remote health solutions to be implemented for diagnosing and controlling the infectious disease.

Judson et al. [49] implement an efficient Corona Virus Symptom Checker. It is nothing but a digital patient self-planning as well as self-directed tool in a large academic healthcare system to the diagnosis of COVID-19 pandemic. Such applications thus help in preventing unnecessary direct consultations and utilizing the facilities of healthcare.

The Chinese Government has adapted a mobile application as described by Hua et al. [50] and is a successful example of contact tracking. In Wuhan territory and later in the entire Hubei province, the screening of people was implemented with Quick Response (QR) code-based scan. This QR code has been used to map the movement of people, in particular, in public transport entering public areas. Three color codings such as red, green and yellow were given to each citizen using big data and mobile phones.

Taiwan [51] has been able to track the people at high risk of COVID-19 infections due to the recent travel histories in the affected regions. In the quarantine, the subjects were electronically tracked on their mobile phones when identified as being high risk. Then, the Entry Quarantine System was launched: travelers could acquire a rapid immigration clearance by completing the form of health statement (need QR code scan leading to an online form, before or upon leaving to the Taiwan airport).

These two papers [52,53] indicate that online search engines tracking information from public health sector may be complementary to traditional systems for public health monitoring in forecasting the future COVID-19 waves.

Lin et al. [54] outlines a prospective information technology–surveillance program (i.e., using an electronic medical record surveillance algorithm) to identify the hospital in patients with pneumonia that has not progressed dramatically during antibiotic therapy and to alert the primary healthcare teams every day to these conditions.

Turer et al. [55] recommended the use of electronic Personal Protective Equipment (ePPE) to secure and sustain the staff to provide fast usage of medical services and complying with health emergency care. ePPE has been potentially applicable to emergency healthcare facilities, health clinics and intensive care units.

China is practicing surveillance with more than 100 drones (The Micro Multi Copter company) over many cities [56]. This measure is considered to be useful to prevent viral transmission by alarming people if the interpersonal distance between individuals becomes less than a "specific" value or if people are walking on public places without a mask.

In India, states like Delhi, Kerala and Assam are making announcements during surveillance across cities via drones. Maharashtra is a step ahead as it has generated data analysis reports of the area being covered via drones [57,58].

Adarsh Kumar et al. [59] proposed an unmanned aerial vehicle-based smart healthcare system for COVID-19 monitoring, sanitization, social distancing, data analysis and control room. The existing framework gathers data by either through wearable sensors, movement sensors deployed in the targeted areas or through thermal image processing. The data are processed through multilayered architecture for analysis and decision-making. In multilayered architecture, edge computing controls the proposed drones' collision-resistant strategies. Whereas, fog and cloud computing approaches build commuters and patient profiles before making decisions. It is found that within a short period a wide distance can be covered and the proposed drone-based healthcare system is effective for COVID-19 operations in Delhi/NCR regions.

Golchin et al. [60] proposed a mesenchymal stem cells (MSCs)-based treatment. In this approach, MSCs gain interest in clinical studies based on their immunomodulatory and regenerative properties. Large populations of stem cells are injected in the lung buildup to protect alveolar epithelial cells.

The Cloud-based Internet of Medical Things (CIoMT) can be used by the public and welfare departments to monitor and screen the facial identification dependent on thermal imaging [61,62] at different points of entry at airports, railway stations, hostels, etc.

The areas are rapidly clustered and categorized as containment zones, buffer zone, red zone, orange zone, green zone, etc., via CIoMT based on how many reports have been verified. If the medical and health services are linked through IOT, we can collect the sophisticated locationwise positive cases in real time. The data and health check-ups for the concerned area may be accessible by the government through AI framework [63,64].

Swati et al. [65] suggests that the disruptive technology from the CIoMT can be applied to intelligent healthcare and tackling COVID-19. CIoMT is a technology promising to quickly diagnose, monitor, track, manage and control the virus better, while not distributing the infection to others. Additionally, the emergency procedures can be carried out economically, the pressure of medical devices scarcity can be minimized, a systematic modeling of database is maintained and the disease activities can be predicted, preparedness and online consultation can be strengthened.

6.4 CONTROL STRATEGY

The essential measures or steps that can be implemented by any country to successfully overcome the outbreak of COVID-19 pandemic are discussed. The COVID cases growing exponentially all over the world, so the proposed measures can reduce the number of COVID-19 infections worldwide.

The strategies that are followed to mitigate the spread of COVID-19 are as follows:

1. Early detection of entry of any COVID-19 patient in the border of any country.
2. Regulating COVID-19 proactive case and allocating containment.
3. Controlling and allocation of proper resource.
4. Creating awareness among the public.

Initially, steps should be taken by every nation for screening people from high-risk areas where COVID-19 disease was spreading intensively [66]. The persons who are returning from the COVID-19 pandemic areas must be quarantined themselves at least for 14 days. Screening test such as temperature and blood (antibody) tests should be conducted at airports, railway station and individuals who tested positive for the COVID-19 must be quarantined either at isolation ward or at home. The COVID-19 infection database should be integrated with the immigration database which will help the hospitals.

In addition, also the responsibility of each government is to maintain strict control over the adequate production and supply of PPE to all healthcare professionals within the country. The government should also ensure masks that any person could purchase within the given period and there should not be any demand for mask. The daily production of masks should rise so that the production should meet the population of the country.

Daily press meets as well as travel risk advisory board should be set to spread the updates of COVID-19 infections to the public via social media. An independent COVID-19 care center should be established in COVID-19 high-risk areas or states with separate entrance and separate wards to completely eliminate COVID-19.

The government will also be able to track the quarantine compliance effectively of COVID-19-affected persons, since they are returning from high-risk areas and are to be identified by the usage of smartphone technology.

The technology-enabled environment is very much useful for the effective implementation of status of COVID-19 infection through public warning system so as to warn the public about the high-risk areas or containment zone of the COVID-19 patients who are tested positive. This warning system permits the overall population to preemptively keep away from these control regions.

6.5 COVID-19 MULTIPARAMETER SENSING AND MONITORING SYSTEM

The fundamental strategy of the public health is social distance measure at the time of COVID-19 pandemic [67].

The key measures or strategies that are followed during the COVID-19 pandemic are as follows:

1. isolating the sick people
2. contact tracing and creating the containment zone
3. isolating COVID-19 affected patients
4. closing the educational institutions
5. ensuring safe work environment
6. avoiding gathering of the unaffected public

6.6 AI-BASED SOCIAL DISTANCE MONITORING AND COVID-19 SYMPTOMS IDENTIFICATION DEVICE

The efficient control strategy for the prevention of COVID-19 pandemic disease is social distancing. When individual persons limit their interactions thereby chance of disease spread reduces. The new AI-based framework is proposed to detect the individual with COVID-19 symptoms and to also maintain the social distance in public places such as banks, ticket counters, offices, factories and stores. The proposed AI-based device is low in cost, compact and portable to place in the office or in factories. The proposed AI-based framework not only maintains the social distance and it also detects the individual with fever and coughing in public places. The proposed AI-based framework reads the camera sensor module and thermal imaging signals measurement to analyze and collect the grade of human distance and body temperature and provides real-time decision to prevent the disease transmission and movement. The AI-based device measures the number of people and social distance with each other's using computer vision, image processing and ML algorithms. The video-enabled observation and analysis of data are done using Raspberry Pi in the python programming language.

The microphone sensor measures the cough voice level in the working place, and using ML techniques adaptive threshold method, the framework predicts the grade of infections. The inertia sensor measures the orientation of the body and it helps to identify the individual headache and pain by their behavior. Combined camera and inertial sensors measure the neck position to detect fatigue and headache.

The data visualization of human's body temperature level, cough voice level and social distance are obtained using Smartphone apps of concern authorities of the office or workplace. The microphone record the voice signals of the workroom and analyzes the data, identifies the series of coughing that is three in a row and the alert message would be sent to the manager or head of the office.

6.6.1 Building Social Distance Alert Device

Using computer vision algorithms the device would count, track all the people in the live video stream and analyze the distance between all the people, if any two-person distance was less than N pixel apart, the alert notification will be sent to the official authorities and also sends the alert audio to the speaker; for example, a woman in the blue dress you are very close to another woman. Please keep the distance. If needed we could also use light bulbs for triggering the alert. The captured video is processed frame by frame using image processing algorithms. This tailor-made device suits both to indoor and outdoor conditions. The high-resolution camera with night mode has been utilized in the design. Figure 6.1 shows the flow chart to identify social distancing in public places.

The ML libraries utilized by the devices for social distance measurement are Numpy, Scipy, OpenCV and TensorFlow. The thermal infra image processing is done using Flirpy libraries. For converting data message to audio, the Espeak library is used.

Figure 6.2 shows the proposed framework for COVID-19. The ML algorithms are used to count and localize the staffs and customers in the room which are then mapped to a bird's eye view projection for analysis. The algorithms would find the

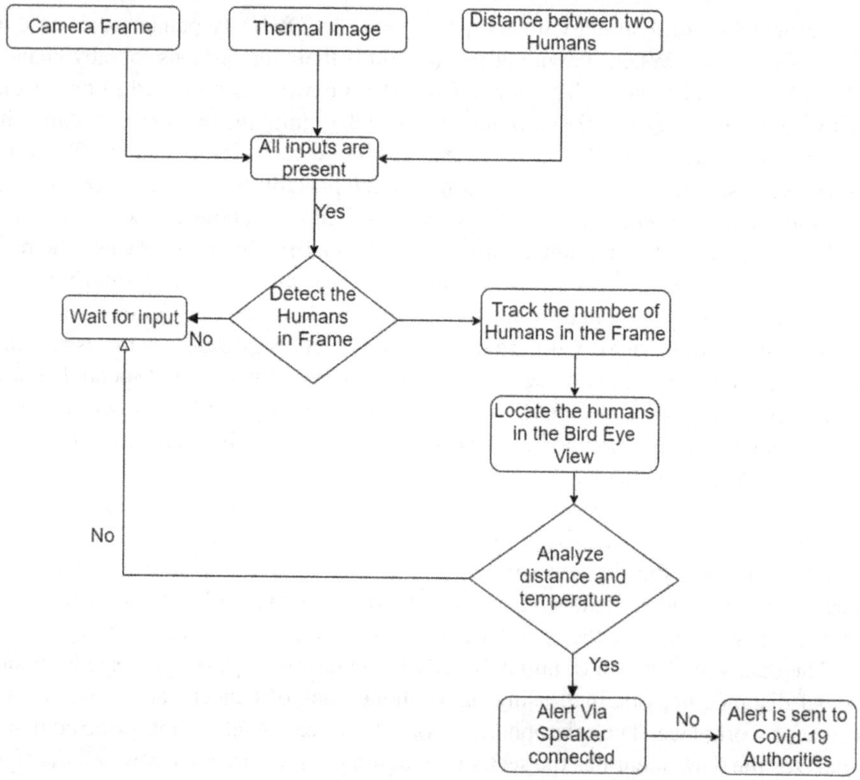

FIGURE 6.1 Flow chart to identify social distancing in public places.

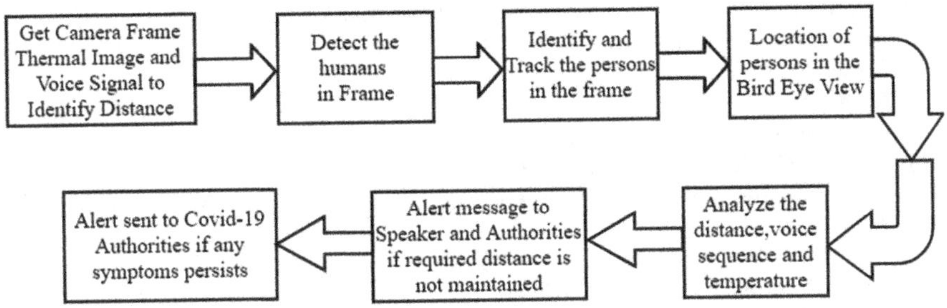

FIGURE 6.2 The flow chart of the proposed AI-based framework for COVID-19 prevention strategy.

coordinates of the individual persons in the bird's eye view and then physical distancing between nodes is easily measured.

6.7 CONCLUSION

The proposed AI-based framework provides the live video surveillance and social distance monitoring in the crowded rooms or places along with identifications of the individual with COVID-19 symptoms. This automated framework is highly suitable for COVID-19 hotspot areas that can be monitored from a remote station without risk for COVID-19 authorities. During the pandemic outbreak, contactless imaging using AI-based diagnosis device with flexible installation is proposed. Most importantly this AI-based device could easily incorporate into the existing closed circuit television surveillance systems.

REFERENCES

1. Bhusare BP., Zambare VP., & Naik AA. (2020). COVID-19: Persistence, Precautions, Diagnosis and Challenges. *J Pure Appl Microbiol.*, May 14, 823–829.
2. Otter JA., Donskey C., Yezli S., et al. (2016) Transmission of SARS and MERS Corona Viruses and Influenza Virus in Healthcare Settings: The Possible Role of Dry Surface Contamination. *J. Hosp Infection*, 92, 235–250.
3. Gallagher TM., & Buchmeier MJ. (2001). Coronavirus Spike Proteins in Viral Entry and Pathogenesis. *Virology*, 279, 371–374.
4. Hoek L., Pyrc K., Jebbink MF., et al. (2004). Identification of a New Human Coronavirus. *Nat. Med.*, 10, 368–373.
5. Bailey OT., Pappenheimer AM., Sargent F., et al. (1949). A Murine Virus Causing Disseminated Encephalomyelitis with Extensive Destruction of Myelin. II. Pathology. *J Exp Med.*, 90,195–212.
6. Cheever FS., Daniels JB., Pappenheimer AM., & Baily OT. (1949). A Murine Virus (JHM) Causing Disseminated Encephalomyelitis with Extensive Destruction of Myelin. I. Isolation and Biological Properties of the Virus. *J Exp. Med.*, 90,181–194.
7. Weiss SR., & Navas-Martin S. (2005). Corona Virus Pathogenesis and the Emerging Pathogen Severe Acute Respiratory Syndrome Corona Virus. *Microbiol. Mol. Biol. Rev.*, 69, 635–664.
8. Guy JS., Breslin JJ., Breuhaus B., et al. (2000) Characterization of a Corona Virus Isolated from a Diarrheic Foal. *Journal Clin. Microbiol.*, 38, 4523–4526.

9. Guo L., Ren L., Yang S., et al. Profiling. (2020). Early Humoral Response to Diagnose Novel Coronavirus Disease (COVID). Clinical Infectious Diseases. Infectious Diseases Society of America.
10. Su S., Wong G., Shi W., et al. (2016). Epidemiology, Genetic Recombination and Pathogenesis of Corona Viruses. *Trends Microbiol.*, 24(6), 490–502.
11. Drosten C., Günther S., Preiser W., et al. (2003). Identification of a Novel Corona Virus in Patients with the Severe Acute Respiratory Syndrome. *N. Engl. J. Med.*, 348, 1967–1976.
12. Lai MM., (2003). SARS Virus: The Beginning of the Unraveling of a New Corona Virus. *J. Biomed. Sci.*, 10, 664–675.
13. Peiris JS., Chu CM., Cheng VCC., et al. (2003). Clinical Progression and Viral Load in a Community Outbreak of Corona Virus Associated SARS Pneumonia: A Prospective Study. *Lancet*, 361, 1767–1772.
14. Guan Y., Zheng BJ., He Y., et al. (2003). Isolation and Characterization of Viruses Related to the SARS Corona Virus from Animals in Southern China. *Science*, 302, 276–278.
15. Martina BE., Haagmans BL., Kuiken T., et al. (2003).Virology: SARS Virus Infection of Cats and Ferrets. *Nature*, 425, 915.
16. Kanwar A., Suresh S., & Frank E. (2017). Human Corona Virus (HCoV) Infection among Adults in Cleveland, Ohio: An Increasingly Recognized Respiratory Pathogen. *Open Forum Infect. Dis.*, 4(1), S312.
17. Chang D., Lin M., Wei L., et al. (2020). Epidemiologic and Clinical Characteristics of Novel Corona Virus Infections Involving 13 Patients Outside Wuhan, China. *JAMA*, 323(11), 1092–1093.
18. Chen N., Zhou M., Dong X., et al. (2020) Epidemiological and Clinical Characteristics of 99 Cases of 2019 Novel Coronavirus Pneumonia in Wuhan, China: A Descriptive Study. *Lancet*, 395, 507–513.
19. Del RC., & Malani PN. (2020). 2019 Novel Corona Virus-Important Information for Clinicians. *JAMA*, 323(11), 1039–1040.
20. Li Q., Guan X., Wu P., et al, (2020). Early Transmission Dynamics in Wuhan, China, of Novel Corona Virus- Infected Pneumonia. *N. Engl. J. Med.*, 382, 1199–1207.
21. Wang D., Hu B., Hu C., et al. (2020). Clinical Characteristics of 138 Hospitalized Patients with 2019 Novel Corona Virus-Infected Pneumonia in Wuhan, China. *JAMA*, 323(11), 1061–1069.
22. Wu F., Zhao S., Yu B., et al. (2020). A New Corona Virus Associated with Human Respiratory Disease in China. *Nature*, 579, 265–269.
23. Zhu N., Zhang D., Wang W., et al. (2020). A Novel Corona Virus from Patients with Pneumonia in China, 2019. *N Engl J Med.*, 382, 727–733.
24. Huang C., Wang Y., Li X., et al. (2020). Clinical Features of Patients Infected with 2019 Novel Corona Virus in Wuhan, China. *Lancet.*, 395, 497–506.
25. Whitworth J. (2020). COVID-19: A Fast-Evolving Pandemic. *Trans R Soc Trop Med Hyg.*, 114(4), 241–248.
26. Bridges CB., Kuehnert MJ., & Hall CB. (2003). Transmission of Influenza: Implications for Control in Health Care Settings. *Clin. Infect Dis.*, 37(8), 1094–1101.
27. Brankston G., Gitterman L., Hirji Z., et al. (2007). Transmission of Influenza in Human Beings. *Lancet Infect Dis.*, 7(4), 257–265.
28. Boone SA., & Gerba CP. (2007). Significance of Fomites in the Spread of Respiratory and Enteric Viral Disease. *Appl Environ Microbiol.*, 73, 1687e1696.
29. Spicknall IH., Koopman JS., Nicas M., et al. (2010). Informing Optimal Environmental Influenza Interventions: How the Host, Agent, and Environment Alter Dominant Routes of Transmission. *PLoS Comput Biol.* doi: 10.1371/journal.pcbi.1000969

30. Chan JF., Yuan S., Kok KH., et al. (2020). A Familial Cluster of Pneumonia Associated with the 2019 Novel Corona Virus Indicating Person-To-Person Transmission: A Study of a Family Cluster. *Lancet.*, 395(10223), 514–523.
31. Xiao J., Shiu EYC., Gao H., et al. (2020). Non-pharmaceutical Measures for Pandemic Influenza in Nonhealthcare Settings-Personal Protective and Environmental Measures. *Emerg Infect Dis.*, 26(5), 967.
32. Cowling BJ., & Aiello A.(2020). Public Health Measures to Slow Community Spread of COVID-19. *J. Infect. Dis.*, 221(11), 1749–1751. doi: 10.1093/infdis/jiaa123.
33. Kampf G., Todt D., Pfaender S., et al. (2020). Persistence of Corona Viruses on in Animate Surfaces and Its Inactivation with Biocidal Agents. *J Hosp Infect.*, 104(3), 246–251.
34. Bootsma MC., & Ferguson NM. (2007). The Effect of Public Health Measures on the 1918 Influenza Pandemic in U.S. Cities. *Proc Natl Acad Sci*, 104, 7588–7593.
35. Cascella M., Cuomo A., Cuomo A., et al. (2020). Features, Evaluation and Treatment Coronavirus (COVID-19). Stat. Pearls Publishing, Treasure Island.
36. Corman VM., Landt O., Kaiser M., et al. (2020). Detection of 2019 Novel Corona Virus (2019-nCoV) By Real-Time RT-PCR. *Euro Surveill.*, 25(3), 2000045. doi: 10.2807/1560-7917.ES.2020.25.3.2000045.
37. Lu R., Zhao X, Li J., et al. (2020). Genomic Characterization and Epidemiology of 2019 Novel Corona Virus: Implications for Virus Origins and Receptor Binding. *Lancet*, 395(10224), 565–574.
38. Rothe C., Schunk M., Sothmann P., et al. (2020). Transmission of 2019-nCoV Infection from an Asymptomatic Contact in Germany. *N Engl J Med.*, 382(10), 970–971.
39. Zu ZY., Jiang MD., Xu PP., et al. (2020). Coronavirus Disease 2019 (COVID-19): A Perspective from China. *Radiology*. doi: 10.1148/radiol.2020200490.
40. Shen C., Wang Z., Zhao F., et al. (2020). Treatment of 5 Critically Ill Patients with COVID-19 with Convalescent Plasma. *JAMA*, 323(16), 1582–1589.
41. Maghdid HS., Ghafoor KZ., Sadiq AS., et al. (2020). A Novel AI-Enabled Framework to Diagnose Coronavirus COVID-19 Using Smartphone Embedded Sensors: Design Study. arXiv preprint arXiv:2003.07434.
42. Ye Y., Hou S., Fan Y., et al. (2020). α-Satellite: An AI-Driven System and Benchmark Datasets for Hierarchical Community-Level Risk Assessment to Help Combat COVID-19. arXiv preprint arXiv:2003.12232.
43. Mirza M., & Osindero S. (2014). Conditional Generative Adversarial Nets. ar Xiv preprint ar Xiv:1411.1784.
44. Story A., Aldridge RW., Smith CM., et al. (2019). Smartphone-Enabled Video-Observed Versus Directly Observed Treatment for Tuberculosis: A Multicentre, Analyst-Blinded, Randomised, Controlled Superiority Trial. *Lancet.*, 393(10177), 1216–1224.
45. Lawanont W., Inoue M., Mongkolnam P., & Nukoolkit C. (2018). Neck Posture Monitoring System Based on Image Detection and Smartphone Sensors Using the Prolonged Usage Classification Concept. *IEEJ Trans. Electr. Electron. Eng.*, 13(10), 1501–1510.
46. Nemati E., Rahman MM., Nathan V., et al. (2019). A Comprehensive Approach for Cough Type Detection. *IEEE/ACM International Conference on Connected Health: Applications, Systems and Engineering Technologies (CHASE)*, Arlington, VA, 15–16.
47. Vhaduri S., Van Kessel T., Ko B., et al. (2019). Nocturnal Cough and Snore Detection in Noisy Environments Using Smartphone-Microphones. *IEEE International Conference on Healthcare Informatics (ICHI)*, Xi'an, China, 1–7.
48. Rahman MS., Peeri NC., Shrestha N., et al. (2020). Defending against the Novel Corona Virus (COVID-19) Outbreak: How Can the Internet of Things (IoT) Help to Save the World? *Health Policy Technol.*, 9(2), 136–138. doi: 10.1016/j.hlpt.2020.04.005.

49. Judson TJ., Odisho AY., Neinstein AB., et al. (2020). Rapid Design and Implementation of an Integrated Patient Self-Triage and Self-Scheduling Tool for COVID-19. Published online ahead of print, Apr. 8, 2020. *J Am Med Inform Assoc.*, ocaa051. PMID: 32267928. doi: 10.1093/jamia/ocaa051.
50. Hua J., & Shaw R. (2020) Corona Virus (COVID-19). Infodemic and Emerging Issues through a Data Lens: The Case of China. *Int J Environ Res Public Health.*, 17(7), E2309. PMID: 32235433. doi: 10.3390/ijerph17072309.
51. Wang CJ., Ng CY., & Brook RH. (2020). Response to COVID-19 in Taiwan: Big Data Analytics, New Technology and Proactive Testing. Published online ahead of print, Mar. 3, 2020. *JAMA*. 2020. PMID: 32125371. doi: 10.1001/jama.2020.3151.
52. Wang W., Wang Y., Zhang X., et al. (2020). WeChat, a Chinese social media, may early detect the SARSCoV-2 outbreak in 2019. med Rxiv 2020.02.24. 2002 6682; doi: 10.1101/2020.02.24.20026682.
53. Ciaffi J., Meliconi R., Landini M., & Ursini F. (2020).Google Trends and COVID-19 in Italy. Could We Brace for Impact? (Preprint).
54. Lin CY., Cheng CH., Lu PL., et al. (2020). Active Surveillance for Suspected COVID-19 Cases in Inpatients with Information Technology. Published online ahead of print, Mar. 31, 2020. *J. Hosp. Infect.*, S0195–6701(20), 30126–2.
55. Turer RW., Jones I., Rosenbloom ST., et al. (2020). Electronic Personal Protective Equipment: A Strategy to Protect Emergency Department Providers in the Age of COVID-19 published online ahead of print, April. *J Am Med Inform Assoc.*, ocaa048. PMID: 32240303. doi: 10.1093/jamia/ocaa048.
57. Drone Technology: A New Ally in the Fight against COVID-19. https://www.mdlinx.com/internalmedicine/article/6767 (accessed Apr. 15, 2020).
58. Drone Association of Kerala: Latest News & Videos, Photos about Drone Association of Kerala. *The Economic Times.* https://economictimes.indiatimes.com/topic/Drone- (accessed Apr. 15, 2020).
59. Covid-19 Lockdown: Authorities Rely on Drone Eye to Maintain Vigil. *The Economic Times.* https://economictimes.indiatimes.com/news/politics-and-nation/covid-19-lockdown-authorities-rely-on-drone-eyeto-maintain-vigil/articleshow/75112745.cms (accessed Apr. 15, 2020).
60. Kumar A., Sharma K., Singh H., et al. A Drone-Based Networked System and Methods for Combating Corona Virus Disease (COVID-19) Pandemic. *arXiv preprint, arXiv:2006.06943.*
61. Golchin A., Seyedjafari E., & Ardeshirylajimi A. Mesenchymal Stem Cell Therapy for COVID-19: Present or Future. Stem Cell Reviews & Report, Apr. 2020, 1–7.
62. Ting DSW., Carin L., Dzau V., & Wong TY. (2020). Technology and COVID-19. *Nat Med.*, 26(4), 459–461.
63. Vaishya R., Haleem A., Vaish A., & Javaid M. (2020). Emerging Technologies to Combat COVID-19 Pandemic. *J Clin Exp Hepatol.* doi: 10.1016/j.jceh.2020.04.019.
64. Shi F., Wang J., Shi J., et al. (2020). Review of Artificial Intelligence Techniques in Imaging Data Acquisition, Segmentation and Diagnosis for COVID-19. *IEEE Rev Biomed Eng.*, 2(1), 220–235.
65. Rao ASS., & Vazquez JA. (2020). Identification of COVID-19 Can Be Quicker through Artificial Intelligence Framework Using a Mobile Phone-Based Survey When Cities and Towns Are Under Quarantine. *Infect Control Hosp Epidemiol.*, 41(7), 826–830.
66. Swati S., & Chandana M. (2020). Application of Cognitive Internet of Medical Things for COVID-19 Pandemic. *Diabetes Metab Syndr*, 14(5), 911–915.
67. Fong MW., Gao H., Wong JY., et al. (2020). Nonpharmaceutical Measures for Pandemic Influenza in Nonhealthcare Settings—Social Distancing Measures. *Emerg. Infect. Dis.*, 26(5), 976–984.

7 Strategies for Prevention of COVID-19

Satyendra Pratap Singh, Mahak Nischal, and Aditi Saxena

CONTENTS

7.1 INTRODUCTION

We have seen the conditions reaching catastrophe level and as the time flew on wings, we will have to find the right approach to deal with it. Since the year 2020 started the world is terrified with the name called COVID-19. Impact of coronavirus is getting worse day by day affecting the lives on emotional, social and economic level and various other physical and psychological levels. As we will go far in this chapter, the main aim is to focus on methodologies needed to manipulate and regulate the control of this same widely distributed COVID-19 at every standard and regulations by the World Health Organization (WHO) and various other countries [1,2].

As we approach the different aspects and perspective of several countries we came across that though several countries are trying to just get the life train back on the right track, and out of others some are trying to fight back some to prevent the spread and some to take cautions and these same impediments were also growing. It all started on December 31, 2019, the date when WHO and China got a hint about a virus infected disease in their country under the case of pneumonia patients [3]. Till that time the effect was limited to the city of China. Measures on the government level were taken to quarantine the whole city so as to prevent the spread. Since then the limited disease in China is now a pandemic for the world, as on June 19, 2020; there have been more than 8 million cases all over the world with the death rate of 41,34,210 death rate and still increasing with the rapid rate globally.

The most affected countries include US, Brazil, Russia with the highest rate of patients followed by UK, Spain, and India according to the analysis provided by the Centers for Disease Control and Prevention associated with WHO. Although on March 11, 2020, WHO declared COVID-19 the pandemic that needs to be controlled before it leads to its worst stage [4]. After this many countries took immediate step of blocking the ingoing and outgoing of people to restrict them to their particular countries and those who are urgently traveling back need to home isolate them for 14 days until they got their corona test negative so as to prevent the further spread [5].

Among those countries India managed to control its effect for a long span of time and restricted their patient numbers to the lowest as they can by announcing a lockdown and restricted the traveling from different parts of world. Prime Minister, Narendra Modi, declared a complete lockdown on March 22 till then India was the least affected country with around 500 people affected and this step was appreciated by WHO because it is better to prevent when you do not have the cure.

But since then the rate is rising with a total number of more than 2.5 million people infected and a death rate of 12,611 which is a huge number for such a populated country with more than 13 million people and having lesser medical facilities to deal with the worse impact. This is not only the condition of a particular country as we have known Italy for its one of the best medical statements and facilities and India still not in the list of top 50 countries with good medical support and healthcare centers but then Italy faced the largest loss in terms of lives. Now although Indian government decided to reopen the social world slowly with obeying the precaution required to prevent coronavirus but the risk of increasing rate is terrifying the whole country [6].

Since the breakout of COVID-19, there have been more than 200 days of existence of COVID-19. The impact of this on emotional, social and economic as well as on physical platforms is unbelievable. Now the question is how much of the people are aware of this COVID-19, what actually this disease is, how to prevent it and many more to be included? We know that COVID-19 is a respiratory disease that damages your lungs and other tissues to recover oneself on the critical stage. It is related to the SARS and MERS which was before in time caused by the virus itself, but due to the unpredictable nature of the coronavirus, we found that it has the ability to change its form which is unparalleled in such lifeforms. Data state that it is a transmission disease that travels from person to person and can be transmitted along both flora and fauna.

It is important to maintain a low level or no continuous transmission, as the pandemic is growing. The health and socio-economic impacts were profound and disproportionately affected the vulnerable. Many populations have already had lack of access to routine, essential healthcare. For migrants, refugees, displaced people and people living in high-density and informative settlements, interruptions of already limited health and social services are particularly important. Close schools increase the risk that certain students are neglected, abused or exploited and run the risk of breakage of basic services, such as school lunches. Although this impact was on the economic conditions too and to not make it even worse more, stern actions are to be taken at the earliest to curb the menace. The fear of being reintroduced and recovered from the condition shall continue, as the virus circulates between and within countries, to be sustainably controlled through the strict application of public health measures and preventing any immigrants and international traveling of their citizens. In the final analysis, the development and delivery of safe, effective vaccines or therapy can result in removal of certain measures needed to maintain this low level or lack of transmission.

7.1.1 Key Sights of COVID-19

Corona is initially a virus detected in Wuhan city in China in the month of December 2019 and till then the conditions are degraded as time is still counting. Its spread was such that whole city leads to the crises of lives in every way possible. Its cause was bat soup which Chinese people consume stated as per the research in China. From there we got to know that this is a respiratory disease that creates congestion in our lungs damaging the recovering tissue of the whole body and blocking the potential of the body to intake oxygen and makes it difficult to breathe on its own when leads to its critical stage. It is proved that this is a transmission disease that takes place from person to person interaction and also by the surface contact, and this also led us to the information that it can be transferred from animals to humans as well but not such cases are moreover discovered in animals [7].

When the air droplets from our sneeze, cough or even when we are talking develop a contact with the outside air, and air being a transmission medium transmits it to other surfaces and enters to the human body whenever got in contact with it. The cause of corona is somewhat related to the previous virus like severe acute respiratory syndrome (SARS) and Middle East respiratory syndrome (MERS) but has unpredictable nature [8,9].

Data received from different countries affected provide us the different percentage levels that lead to the stages of this COVID-19 from low, moderate to the major cases, according to which only 5% considering the population of a particular country respectively cases lead to the severer level of coronavirus infection. Especially to the people who are old, young children and those who are already suffering from cancer, any respiratory disease, diabetes and similar problems have more chances than other to get infected from this disease. Overall, 83,325 people infected with the death rate of 4,634 in China till now.

After this impact many strategies to prevent its spread were taken at government, health organization and local levels, and this has to be done keeping in mind the ongoing conditions of this virus. The major step as mentioned above of this lockdown and high testing rates in the whole city created a great impact on stopping the rise in the number of patients in Wuhan [1]. Although now they are bringing the life back on track with proper precautions to prevent any further step. Along with lockdown they also increased the rate of testing by making testing centers everywhere possible and also providing mobile application in each citizen's mobile phones which detect its surrounding is corona-free or not and what is its overall health status and as WHO suggested a complete plan of strategies and preventions but local use so that everyone can assess the information and help themselves on the local causes to prevent it globally.

7.2 GLOBAL STRATEGIES

As we all know that different minds lead to different decisions and in the same scenario each country follows multiple approaches based on its socioeconomic background, taking into consideration this infection as a major catastrophe. Each country was motivated primarily by influence of the transmitting than the further distribution. Everyone's first phase was to stop the nationwide interactions to significantly reduce the continent's standard and also to prevent possible reunification, using shutdowns across all of the countries. Studies indicate that the person that sends this virus can spread the virus to four persons within a two-day period and, in 15 days, can infect 3,000 persons. The most effected countries like US, Brazil, Russia and India have taken many measures [5,10].

Starting from Italy which was first country affected badly by this pandemic and as a consequence restricted the movement of citizen and declared a complete lockdown on 12 March 2020, and it further gets extended with strict implementation of rules and guidelines. Although with this now the increasing rate is almost in control in Italy by increasing the rate of detection and avoiding mass gatherings as much as possible [11,12].

Some countries like France and Spain have told people that they need authorization to move and tighten restrictions as virus cases continue to grow. On the other hand, United Kingdom has very restrictively joined other countries, although people can practice once a day, shop for essential needs, work as doctors and other necessities if any, go to work. Now considering in the United Kingdom and elsewhere there were questions about how the rules should be interpreted, and criticism about how the authorities apply them [13,14].

One of the toughest decisions taken by these countries like France, Spanish, Italian and UK is all these countries' authorities have imposed fines for those who are disregarding the rules. These are as high in one section of Italy, Lombardy as to control massive gathering so that being the most affected country the spread does not became uncontrollable. Well, talking about some countries like Germany, lockdown rules vary slightly between states and people but allow for fresh air in their homes. After observing the strategies of different countries there are major restrictions in the USA now. There are unusual ways in which countries manage lockdowns. Unless it is absolutely necessary, the state of California in the US stops people leaving home and forces firms to close off that is not essential. New York State has also introduced a strict lockdown with the highest number of coronaviral cases in the US. When the breakout of COVID-19 in China started, the authorities limited their national travel and told people to stay at home, only recently to relax it.

Travel restrictions are now eased in Hubei province, where the virus began, allowing people in China to enter and leave. And on 8th April, Wuhan city should leave the lockdown. India has locked its 1.3 billion inhabitants strictly after a sharp increase in the number of cases of coronavirus, and Russia has urged people to stay at home. Many other countries also have limited movements to different levels, although Sweden has less restriction than others in Europe. In order to find out who is infected, the WHO urged countries to test as much as possible and thus help cut down on the propagation of the virus (Figure 7.1).

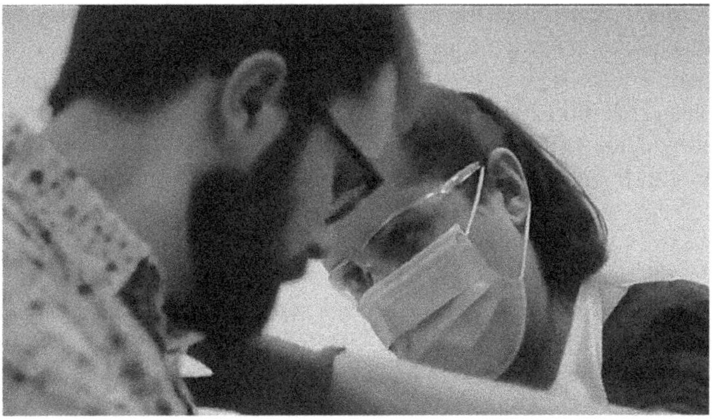

FIGURE 7.1 Testing the citizens.

South Korea tested the most people per head of population, while other testing was significantly lower, although they are now increasing in number. The United States, which was relatively slow when its testing program was launched, has now substantially ramped up this across the country. The UN estimates that approximately 87% of educators worldwide are affected by school and college closures. UNESCO's Education, Science and Culture body says that over 180 countries closed their schools on 30th March.

The pandemic of coronavirus also had a major effect on the sporting calendar as countries tried to restrict mass rallies. In addition to the shutdown, large-scale testing also is important because this virus in some conditions may not show symptoms in people but may make this person a large-scale transmitter of this virus. With this the countries were also focused on creating and transforming different places like banquet halls, stadium, school building and even the big ship especially in US into large quarantine centers to provide sufficient beds to each citizen. Because if the quarantine centers and hospitals run parallelly, the situation of increasing rates of patients infected will also increase as the patients who are already in the hospital or in other healthcare centers due to their health issues other than corona are already having weak immune system which will lead them an easy predator for coronavirus to attack on. So, there is a need to separate these in support of the other strategies.

Getting from the news provided by New Zealand being the first country to be corona-free taken all the above measures seriously and early than the other countries. Although the population of this country is less than the other major infected countries but becoming corona-free is incredible in these conditions. The people and government work strictly on the lockdown rules and also have sufficient testing capacity to overcome through this pandemic soon. People in New Zealand maintain their personal record of the day consisting the number of people they meet and the places they visited [15].

With this the most important focus of each country was to test more and more people as fast as they can. Starting from the people those who have any interaction with other countries and also the people they meet and the places they visit to the normal citizen who are staying in the country itself for many years. If after testing the person is COVID positive then they are sent to quarantine centers for under observation and recommended others to isolate themselves in their homes and keenly observe the symptoms [16]. Being the insurable disease that may not show the symptoms at the start but can harm you with the passage of time is a big challenge to the high-tech medical staff all over the country.

In this century when mobile phones are more reliable than human the application provided by many countries which help to detect the status of the particular users tells them the condition of corona in their surroundings and also helps government to maintain the record according to the infected people. Although after this many countries are getting back to their normal life with many restrictions of wearing masks and gloves necessary for all citizens and also maintain social distancing at every place [17].

7.3 COMMON STRATEGIES

As an individual can contribute there are various basic steps and guidelines to help in ways you can and henceforth we need an approach in which every entity and society, each corporation and philanthropic, that each governmental agency, each NGO, each international coalition, every domestic and national democratic accountability body along with the support of civilians and citizens can be unified to work together in order to make international cooperation possible against with the COVID-19. In preventing COVID-19, everyone should have an important part to play [18]:

Personal: People should preserve ourselves and some by incorporating decisions like handwashing, trying to avoid contact of face, causes great breathing relabels, individually trying to distance ourselves from one another, and using any object until it was kept untouched for at least a day, alienating ourselves in a communal environment. As the online precautions on WHO website mention the below discussed details [19].

To forestall the spread of COVID-19:

- Clean your hands regularly. Use cleanser and water or a liquor-based hand rub.
- Keep up a sheltered good way from any individual who is hacking or sniffling.
- Try not to contact your eyes, nose or mouth.
- Spread your nose and mouth with your twisted elbow or a tissue when you hack or sniffle.
- Remain at home on the off chance that you feel unwell.

On the off chance that you have a fever, hack and trouble breathing, look for clinical consideration. Bring ahead of time. Maintaining a strategic distance from unneeded visit to clinical offices permits human services frameworks to work more successfully, hence ensuring you and others. The laboratory or residence because once unwell identified as a confirmation case interaction where necessary and cooperated with physical removal measures and regulations of activism because once requested [20].

Social: Societies should be encouraged to help make sure that their communication and individual branches allow for the planning and adaptation of support and services. Essential operates such as educational programs, vulnerability protection, healthcare professional's assistance, case discovery, interaction tracking and collaboration.

Mostly with the assistance of each portion of the surrounding populations can physiologically trying to distance actions take place [21].

Authorities: Authorities should result but instead synchronize ruling coalition responses so that all people and communities can and do possess reactions through all the communication, education, participation, infrastructure development and assistance. Authorities should also reconstruct but instead involve well all capacities accessible to the public, authorities and private sectors to expand public healthcare systems to quickly identify, evaluate, segregate and treat documented cases [18] and recognize, trace, containment and assistance contact information, whether it is at residence but at healthcare center. Authorities should therefore simultaneously support the medical system.

Sufferers must be easily diagnosed with COVID-19 and several other vital health and development service providers must be maintained for everyone. Authorities may also need to apply wholesale, health risk-related physical separation regulations and physical restrictions, because more duration is needed to implement above that the initiatives.

Individual: This same consistency of vital services like the food supply chain, government infrastructure and the start manufacturing of humanitarian equipment must be ensured by private enterprises. The knowledge and experience and creativity can be supported and maintained by confidential companies, especially through to the manufacturing and consumption of diagnostic procedures, protective equipment facilities, ventilation systems as well as other critical clinical oxygen at reasonable rates and screening procedures, medications and influenza vaccine technology development.

7.4 STRATEGIES OF INDIA

Although countries those who are far better than India in many fields but still they break down in front of the crises faced by the COVID-19. From the day India got the information about the first patient from Delhi who came from Italy few weeks back the government started creating awareness among the citizen and just after few days of increasing number of infected cases, the lockdown is declared by the authorities from 22 March 2020. Every state according to their number of cases declares prevention and restrictions of their people. All closed their borders and also restrict the domestic flights interaction and also state to state interaction. But since then the rate has increased, with a total number of over 3 lakh infected persons and the death rate of 12,611, a huge increase of more than 13 million people in a populous country with healthcare facilities that have worse consequences.

Since the lockdown was declared giving relaxation to the essential services, the healthcare Indian government used that duration for the initial preparation of masks, sanitizers and other health gadgets as they were not expecting the outbreak of this disease at such a large scale. Many private and government laboratories supported the government by producing the advanced masks and testing kits further for more testing throughout the country. After some time they also created some small centers that air sanitized the clothes, outer body and objects of the individual before entering any centers and organization where they work or gone for some important issue. During these periods they also sanitized different places where the movement takes place so as to stop further spread [22,23].

But soon this badly affected the economic condition of India as many people led homeless due to the shutdown of the working sectors [24]. Considering this government decided to unlock from 8 June 2020 the country in different phases with slow and steady means so that with life on track, people should also follow the precautions associated with COVID-19. Meanwhile they developed an Aarogya Setu mobile application which also maintains the health and surrounding status of an individual and also helps the government to maintain the record of the same [25]. With all these lockdowns planning they increased the number of testing laboratories all over the country. All private and government laboratories are turned into corona testing centers. With all those the government divided the zones in India so as to work accordingly with that. The most infected area was declared red zone then the area with moderate rate of infected people as orange zone and the area with less than two infected people is considered the green zone. According to this the government decided its strategies to work upon the same and control the spread by increasing the testing in those particular areas at a rapid scale. Information according to the sources provides the following data related to the testing capacity of India. Mylab Discovery Solutions based in Pune has enhanced its own manufacturing capability from 20,000 testing methods each day in the primary portion of April to 2 lakhs evaluations still in connection with the Serum Institute of India (SII) Ltd. to ensure that certain yet another residential manufacturing company meets Indian recent test requirements anticipated at both the end of this month to reach further 1 lakh sample requirements [8]. The Center Drug Standard Control Organization endorsed Mylab because of its testing, which again approves six other international firms to furnish diagnostic kits which really satisfy the framework of that same Medical Research Council of India. India currently tests 75,000–80,000 specimens every day because of the latter's RT-PCR test, a reserve currency to validate COVID-19 instances, but have established a goal to assess at least 1 lakh specimens per day until late May. The sample targets for these tests are at least 1 lakh specimens. To date, Mylab Discovery alternatives have tested 65,000 throughout one-hundred and forty laboratories as well as healthcare facilities in nearly 20 countries. CSIR laboratory is also providing personal health protection suits for the protection of people getting in contact with the infected person (Figure 7.2) [16].

The unlock strategies that have three stages as mentioned above are discussed as follows. The phase one will be more concerned to the working sector of the society that is the business, company and other corporate agencies including malls and

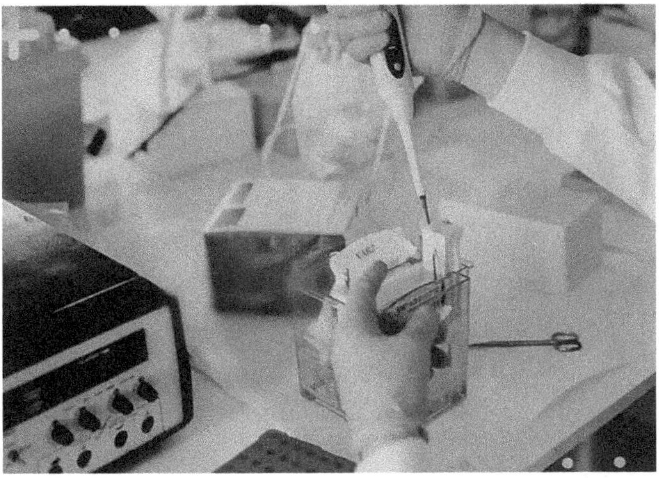

FIGURE 7.2 COVID-19 test samples.

spiritual places like temples, etc. providing the restriction that only 50% workers should be called for work.

They state interaction was also allowed. The phase two will focus on the education sectors, cinema halls and local transport. And last but not the least the third phase indeed focused on gyms and local life importance, that is, the metro services. All those things should be followed by keeping in mind the factor of social distancing and other precautions related to corona. The condition was that the next phase should enter after the success of the previous one.

Now the most affected states in India are Maharashtra and Delhi with the infected number of more than 1 lakh and more than 50,000 cases with death rate of 5,984 and 2,112 people, respectively. To dealing with this there is an instruction of the government to increase the amount of test in both states in which private and government sectors will work together. They also planned to increase the amount of quarantine centers like banquet halls, old schools' buildings, hotels and many places are converted so as to be prepared before it gets worse. Looking at the recent condition some states have again sealed their borders for better preventions. Being a tough time, the strategies are implemented day after day to prevent the spread as much possibly as they can and hoping better conditions in future [26,27].

7.5 MEDICAL RESEARCH AND INNOVATION

Immediate COVID-19 Global Research blueprint for guiding the union COVID-19 production and innovation strategy for 11 and 12 February 2020 was sponsored by the WHO in Geneva. The discussion board agreed that investigation and the development of medical measures to prevent, which include immunizations, pharmacogenomics and diagnostic equipment, were urgently required. Much initiatives have already been funded by significant investments as well as operations to tackle COVID-19 obstacle. A document upon it this same international clinical research environment is produced

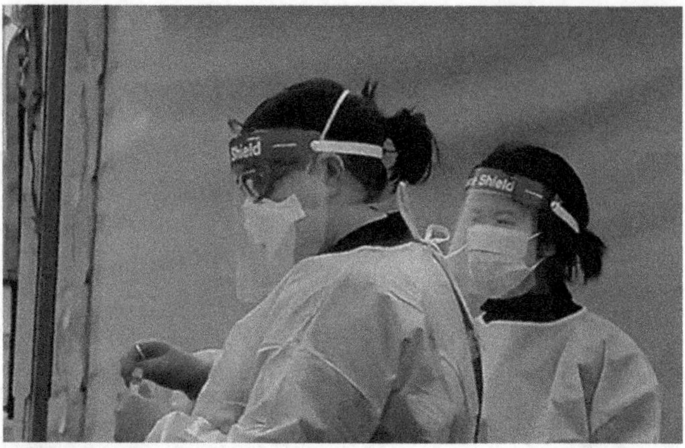

FIGURE 7.3 Testing by medical staff.

every week as well as notification alerts about the ongoing research and efforts for innovation including progress of front-runner immunizations available for clinical conditions, two of them are the steps of assessment (Figure 7.3) [28].

Several other initiatives are indeed structured and bankrolled individually. The international society needs a genuinely cohesive as well as a worldwide attempt to accomplish greatest exposure. Now, government and industry must unite to promote a consistent and integrated system and implement in pursuit of communal technology and development objectives.

The whole widely accepted danger has been addressed [29]. The international COVID-19 speeded up vaccine's enterprise was already developed with both the aim of aligning the environment toward a committed grand plan as well as of identifying every possibility for financial achievement of technology development and scale of delivery to organize the never-ending joint venture between interested parties in the WHO.

The whole specific proposal is indeed a distinctive as well as worldwide objective for aptitude test against COVID-19 at such a fast pace inside the framework of both the narrower implementation plan for technology and development. The WHO works with companies to achieve the structure besides synchronized technology and development as well as a detailed analysis including its magnitude of investments needed for underwriting by expanding as well as expanding mostly on international research and development plan [30]. Cohesion is required for the biggest human betterment as well as cooperation, the establishment, collaborative effort, bridge-agency and community-private joint ventures with adequate capital and facilitation of public access or the communication of knowledge. Assistance, expenditure, priority setting and appropriate management of resources would be needed along all communities, personal and charitable organizations and industries. Collaborative success will depend on cooperation as well as confluence of attempts. That being said devoted as well as calculated and personal as well as separated actions are not enough to meet the minimum COVID-19 obstacle.

We must group, construct as well as translate entrepreneurship from greater heights in order to be effective. The above requires proactive management action and the purpose to coordinate rather than just to measure and review operations extra passively. Considering these many countries also initiated the efforts for the same and India is no far in this case.

7.6 CONCLUSION

Leading to the scenario cases are being increased day by day and many countries are also focusing on vaccine and medicine related to it but observing the situations in the most affected countries like US, Brazil, India, etc., [5] the best way to plan strategies is keeping in the articular populations associated with different countries and also taking deaccession according to the welfare of society and the lives of their citizen. Because when you are facing an unknown rival and you are not aware how it will react and its weakness is then it is important to be prepared so as to reduce the chances of your loss. Same is the case with coronavirus; as its vaccine and any medicine are not known, it becomes important to have full-fledged strategies to prevent a large-scale damage. The strategies prepared by each and every country and also the suggestions by WHO proven to be remarkable and the best front foot forward which can play a vital role in bringing the prevention strategies to avoid coronavirus disease. As discussed in the above chapter about different strategies and protocols were built to control the spread and transmission portrait the alarming quality of the world. Although many countries are still facing the increasing rate but according to the situation strategies are modified as per the strategies mentioned about India. Considering the statistics of all countries and cases is a major and important initial step as conditions get clear on how to implement strategies. All are working upon it together as it is since 100 years when world faced this kind of pandemic after the Spanish flu which took the world in the same crisis which it is in now.

REFERENCES

1. "Strategy unveiled to prevent spread of COVID-19", *Udhagamandalam*, 25 June 2020.
2. "Asian scientist magazine, India's skilled health workforce falls short of WHO threshold", Retrieved from: https://www.asianscientist.com/ 2019/06/health/indiaskilled-health-workers-bmj-open/
3. Sabino-Silva, R., Jardim, A.C.G. and Siqueira, W.L. "Coronavirus COVID-19 impacts to dentistry and potential salivary diagnosis", Clinical Oral Investigations, 24, no. 4 (20 February 2020): 1619–1621.
4. "Coronavirus disease (COVID-2019) situation reports with special reference to INDIA", https://www.worldometers.info/coronavirus/country/india/
5. "World Bank sees FY21 India growth at 1.5-2.8%, slowest since economic reforms 30 years ago", *The Hindu*. PTI, 12 April 2020. ISSN 0971-751X. Retrieved 13 April 2020 from: https://www.thehindu.com/business/world-bank-sees-fy21-india-growth-at-15-28-slowest-since-economic-reforms-30-years-ago/article31322011.ece.
6. Shefali, D. "Impact of goods and service tax (GST) on Indian economy", *Business and Economics Journal*, 7, no. 4 (2016). doi: 10.4172/2151–6219.1000264.
7. Muthulakshmi, E.K. "Impacts of demonetisation on Indian economy- issues & challenges", *Journal of Humanities and Social Science*, 6, no. 10 (2017): 34–38.

8. Cascella, M., Rajnik, M., Cuomo, A., Dulebohn, S.C. and Di Napoli, R. "Features, evaluation and treatment coronavirus (COVID-19)", 18 May 2020.
9. Wisconsin Department of Health Services. "COVID-19 coronavirus disease", 15 March 2020. Retrieved from Wisconsin Department of Health Services: https://www.dhs.wisconsin.gov/covid-19/index.htm
10. "Interim guidance on scaling-up COVID-19 outbreak in readiness and response operations in camps and camp-like settings (jointly developed by IFRC, IOM, UNHCR and WHO)", Posted on Tuesday, 17 March 2020.
11. "Coronavirus disease (COVID-2019) situation reports", https://www.worldometers.info/coronavirus/
12. Tandon, A., Murray, C.J.L., et al. World Health Organization (WHO), measuring overall health system performance for 191 countries, GPE, Paper Series: No. 30. Retrieved from: https://www.who.int/healthinfo/paper30.pdf
13. Government of CANADA. "Community-based measures to mitigate the spread of coronavirus disease (COVID-19) in Canada", 15 March 2020. Retrieved from government of Canada: https://www.canada.ca/en/public-health/services/diseases/2019-novel-coronavirus-infection/health-professionals/public-health-measures-mitigate-covid-19.html
14. NY Times. "The corona outbreak", 15 Month 2020. Retrieved from *New York Times*: https://www.nytimes.com/live/2020/coronavirus-usa-03-12
15. "COVID-19 strategy update", Printed in Geneva, Switzerland, 14 April 2020.
16. "Laboratory testing strategy recommendations for COVID-19", *Interim Guidance*, 21 March 2020.
17. Ting, D.S.W., Carin, L., Dzau, V., and Wong, T.Y. "Digital technology and COVID-19", *Nature Medicine*, 26, no. 4 (27 March 2020): 459–461.
18. "Considerations for quarantine of individuals in the context of containment for coronavirus disease (COVID-19)", *Interim Guidance*, 19 March 2020, COVID-19: Infection prevention and control / WASH.
19. Mental Health foundation. "Looking after your mental health during the Coronavirus outbreak", 11 Month 2020. Retrieved from mental heal the foundation: https://www.mentalhealth.org.uk/publications/looking-after-your-mental-health-during-coronavirus-outbreak
20. "Home care for patients with COVID-19 presenting with mild symptoms and management of their contacts", *Interim Guidance*, 17 March 2020, COVID-19: Clinical care.
21. Lee, I-K., Wang, C-C., Lin, M-C., Kung, C-T., Lan, K-C. and Le, C-T. "Effective strategies to prevent coronavirus disease-2019 (COVID-19) outbreak in hospital", *Journal of Hospital Infection*, 105, no. 1 (3 March 2020): 102–103.
22. "Government's multi-step strategy to tackle coronavirus spread in India", *Hindustan Times*, 5 May 2020.
23. Lora Jones, D. B. "Coronavirus: Eight charts on how it has shaken economies", 14 March 2020. Retrieved from BBC News: https://www.bbc.com/news/business-51706225
24. WHO. "Naming the coronavirus disease (COVID-19) and the virus that causes it", 03 October 2020. Retrieved from WHO: https://www.who.int/emergencies/diseases/novel-coronavirus-2019/technical-guidance/naming-the-coronavirus-disease-(covid-2019)-and-the-virus-that-causes-it
25. Center for Disease Control and Prevention. "Manage anxiety & stress", 15 March 2020. Retrieved from Coronavirus Disease 2019 (COVID-19): https://www.cdc.gov/coronavirus/2019-ncov/prepare/managing-stress-anxiety.html
26. Singh, S., "Demonetization and its impact on Indian economy", *International Journal of Engineering, Science and Technology*, 6, no. 11 (2017): 141–144.
27. Bhika, L.J. "Impact of GST on Indian economy", *International Journal of Recent Scientific Research*, 8, no. 6 (2017): 17505–17508.

28. Adhikari, S.P., Meng, S., Wu, Y-J., Mao, Y-P., Ye, R-X., Wang, Q-Z., Sun, C., Sylvia, S., Rozelle, S., Raat, H. and Zhou, H. "Epidemiology, causes, clinical manifestation and diagnosis, prevention and control of coronavirus disease (COVID-19) during the early outbreak period: A scoping review", Infectious *Diseases* of *Poverty*, 9, no. 1 (17 March 2020): 1–12.
29. Freeman, S. "Systemic social issues reflected in coronavirus outbreak", 05 March 2020. Retrieved from I politics: https://ipolitics.ca/2020/03/05/systemic-social-issues-reflected-in-coronavirus-outbreak/
30. Gandhi, M., Yokoe, D.S., and Havlir, D.V. "Asymptomatic transmission, the achilles' heel of current strategies to control COVID-19", 24 April 2020.

8 Strategies for Prevention of Coronavirus Disease

M. Sudha, R. Subasri, and A. Dinesh Karthik
Shanmuga Industries Arts and Science College

A. Poongothai
Sacred Heart College (Autonomous)

CONTENTS

8.1 INTRODUCTION

Coronavirus sickness 2019 (COVID-19) is the desirable weakness transported nearby Severe Acute Respiratory Syndrome Coronavirus 2 (SARS-CoV-2). This new infection and the illness brought about by it were incomprehensible before the episode happening in Wuhan, China, in Dec 2019. Coronavirus is unique of the most infectious ailments to have hit us in decades. As governments' and well-being associations' journey to contain the spread of coronavirus, they need all the backing they can get, including from man-made reasoning [1].

The COVID-19 pandemic consumes significantly exaggerated population well-being and affluence. Exploration endeavors are in progress to distinguish antibodies, advance testing, comprehend transmission, create serologic tests, create treatments, anticipate hazard and create relief and avoidance approaches. Biomedical informatics is integral to every one of these examination endeavors and for the conveyance of social insurance for COVID-19 patient [2]. The pandemic has in a general sense adjusted numerous parts of our life and society with both present moment and likely long haul impacts on well-being frameworks specifically. Well-being informatics apparatuses can assist better with supporting the well-being framework's endeavors to plan for the current pandemic and the difficulties related with this errand [3].

8.2 THE *CORONAVIRIDAE* FAMILY

The family *Coronaviridae* includes two genera, Coronaviruses and Toroviruses, and is assembled with two different families, the Arteriviridae and the Roniviridae, into the request Nidovirales. The name Nidovirus (from the Latin "nido"—home) alludes to the capacity to translate an alleged "settled set" of subgenomic mRNAs. Coronaviruses have genomes of around 30,000 nt, a length that is phenomenal among RNA infections. They are composed of "crown-like spikes" and found to have α, β, γ and δ subgroupings. Figure 8.1 shows the coronavirus structure.

8.3 HUMAN CORONAVIRUS

The humanoid coronaviruses are two alpha coronaviruses (229E and NL63; examples of Middle East Respiratory Syndrome (MERS)-CoV and Severe Acute Respiratory Syndrome (SARS)-CoV) and two coronaviruses (OC43 and HKU1); these are known as human crown infections [4].

8.3.1 Alpha Coronavirus

This gathering of infection is partitioned into bunches 1a and 1b and dependent on phylogenetic grouping. The 1b coronaviruses are in certainty "nonbunch 1b" coronaviruses, instead of having regular highlights that make them an unmistakable family. Genomes of the considerable number of individuals from 1a contains one (NS7a) or

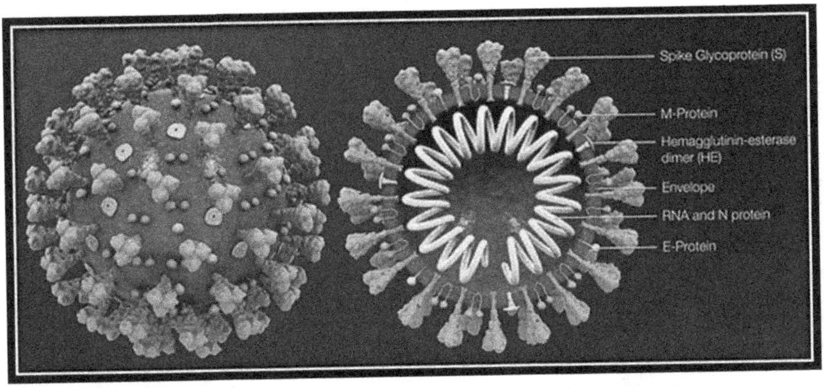

FIGURE 8.1 Coronavirus.

two (NS7z and 7b) open understanding edges downstream to N, in this manner the name Gesela virus, which means "quality seven last," the genomes of some gathering 1b coronavirus [5]. Figure 8.2 shows the structure of alpha coronavirus.

8.3.2 Beta Coronavirus

Before the disclosure of SARS-CoV, bunch 2 coronaviruses were considered to incorporate one parentage yet finally it was seen that 19 out of 20 cysteine build-ups were spatially monitored with those of the accord succession for bunch 2 coronaviruses. The structure of two subgroups, two novel genealogies, most firmly identified however in particular from bunch 2a and bunch 2b coronaviruses [6]. Figure 8.3 indicated the beta crown infection adjustment to human.

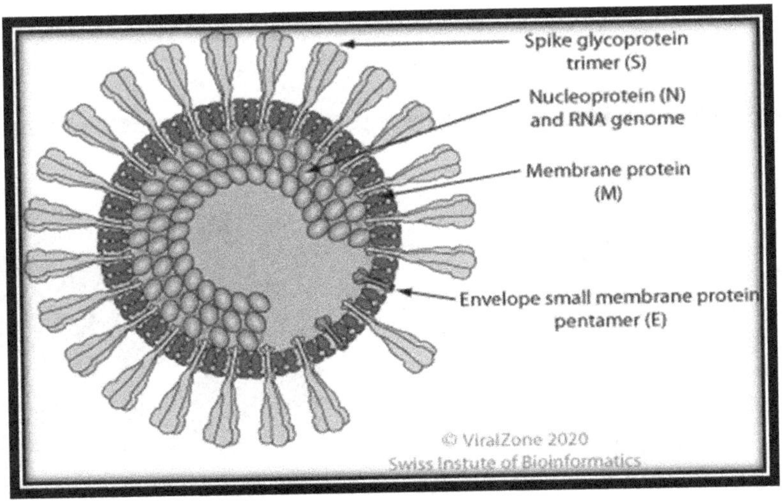

FIGURE 8.2 Structure of alpha coronavirus.

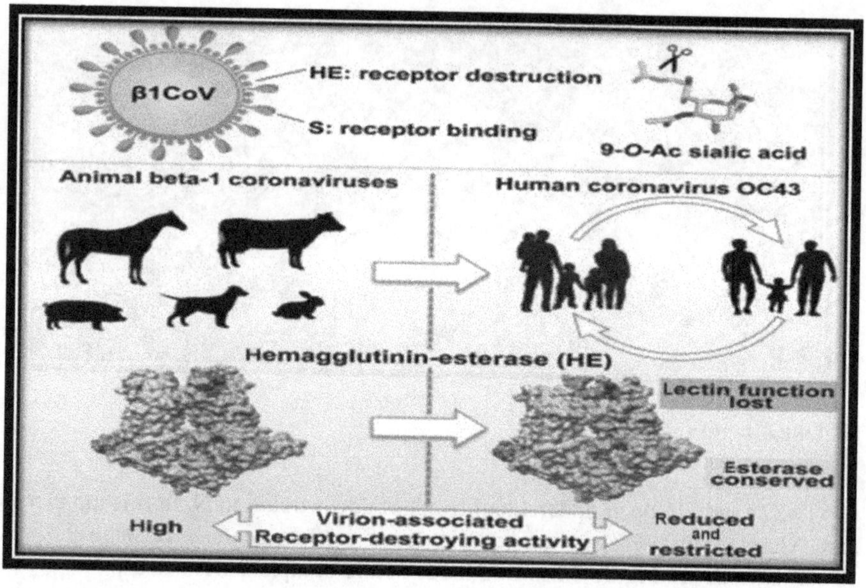

FIGURE 8.3 Beta coronavirus adaptation to human.

8.3.3 Gamma Coronavirus

Since 1937 infectious bronchitis virus has been the main type of gathering 3 coronavirus for more than 50 years. Presently a couple more added to the rundown including Turkey Coronavirus (TCoV) and they have been depicted in different types of flying creatures, with a portion of their genomes sequenced. The sizes, G+ C substance, and genome associations of their genomes were comparable, demonstrating that they most likely have wandered from a similar precursor as of late [7]. Figure 8.4 indicated the beta crown infection adjustment to human.

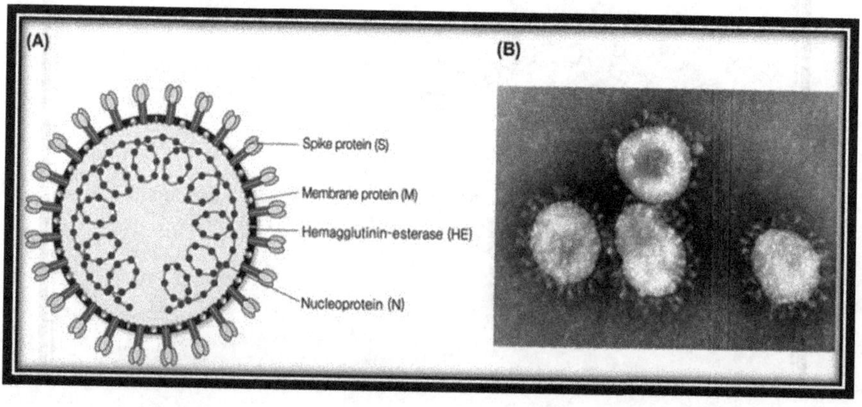

FIGURE 8.4 Gamma coronavirus: an overview.

8.4 ANATOMY OF A CORONAVIRUS

It is the crown-like projections on their surfaces that make us to call coronavirus. Thus the name coronavirus is enveloped with single-stranded RNA viruses and each viral particle is wrapped in a portion envelope. It is sphere-shaped or soberly discrete club-like projections made up by the S protein. Helically regular single-stranded and positive sRNA in the size of 26 to 32 kb [8]. Figures 8.5 and 8.6 show the anatomy of coronavirus.

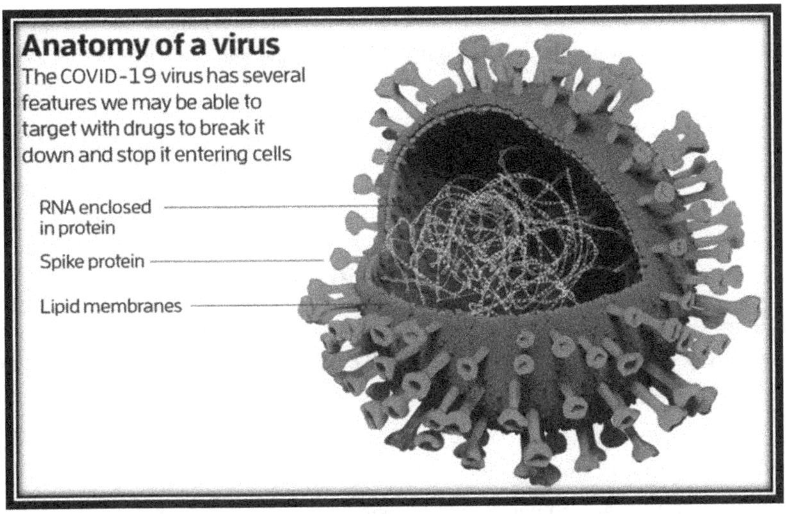

FIGURE 8.5 Anatomy of coronavirus.

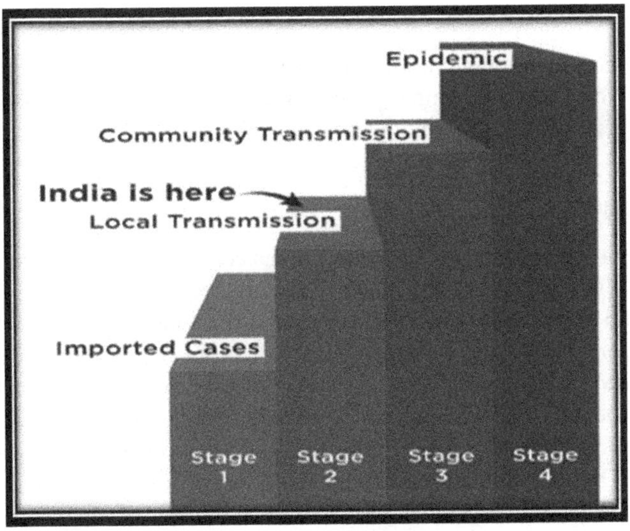

FIGURE 8.6 Stages of COVID-19.

8.5 DIFFERENT STAGES OF CORONAVIRUS

There are four stages: (1) Imported Cases; (2) Local Transmission; (3) Community Transmission; (4) Epidemic. **Stage 1:** These are the individuals who have headed out to infection hit outside nations and have returned to India. **Stage 2:** These are those cases who have interacted with patients who had movement in the past. **Stage 3:** Public spread is the point at which a patient not presented to any tainted individual or one who has made a trip to any of the influenced nation's that tests positive. Enormous zones get influenced when network transmission happens. **Stage 4:** This is the last and the most exceedingly awful stage where the ailment takes the state of a pandemic with no unmistakable end point as it did in China. Figure 8.5 shows the various phases of COVID-19.

8.6 COVID-19 EPIDEMIC

The four primary stages, well-being data innovation members, administration beneficiaries, advances and application situations, are discussed below.

8.6.1 Health Information Technology Participants

Well-being data innovation members who incorporate government organizations, innovation organizations, clinical offices and examination foundations and influence data advancements to react to the pandemic in this national crusade were recognized.

8.6.2 Recipients

These incorporate affirmed patients, distrusted patients, close contacts, patients with incessant ailments and the general population.

8.6.3 Technologies

An assortment of developing innovations, for example, distributed computing, huge information, the Internet of Things, Artificial Intelligence (AI)-portable web and 5G are being utilized for scourge anticipation and control.

8.6.4 Application Scenarios

Application situations basically incorporate data conveyance, case discovery, screening, online administrations, chance evaluation and insightful analysis. These situations are solid appearances of the coordinated utilization of different data advances for observation just as anticipation and control administrations. The health industry is unquestionably tested because of the general interruption to the social insurance framework; the pandemic without a doubt underscores the chance and significance of well-being informatics [9]. Figure 8.7 shows the proposed well-being data innovation system for COVID-19. In restoration informatics, physiological information should be investigated to get whether and how much patients with different level of COVID-19 diseases will get full recuperation of organ (lung and heart) capacities post recuperation. In irresistible ailment demonstrating, the study of disease transmission

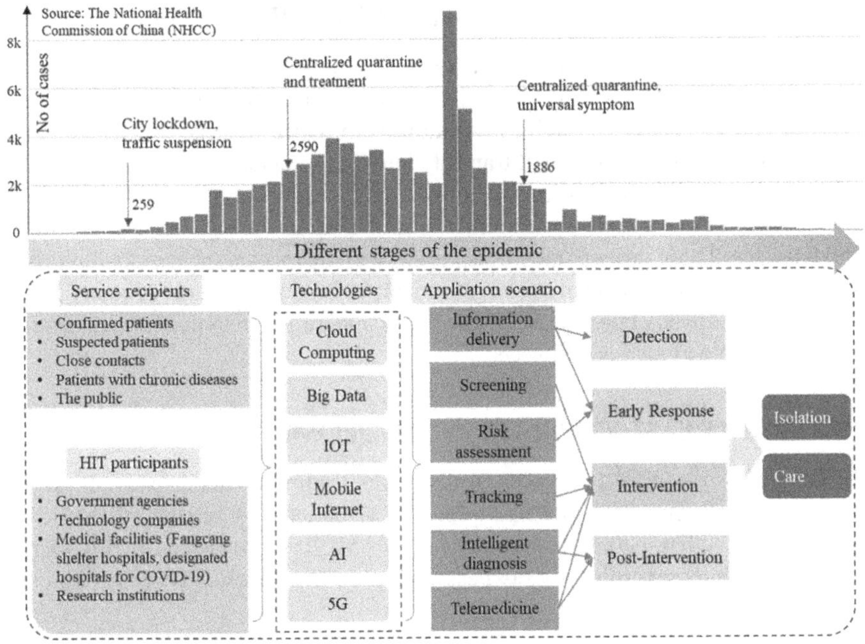

FIGURE 8.7 Health framework of COVID-19.

models should be utilized with field information to foresee COVID-19 spread speed so as to help strategy producers in taking legitimate activities.

In general well-being informatics, flare-up information should be dissected for populace well-being the executives and COVID-19 consideration asset flexibly chain the board. This exceptional issue points are (1) to urge the partners identifying with COVID-19 to share information source, information harmonization and instruments, which can accelerate COVID-19 examination for quite a long time to come; (2) to rouse new informatics strategy improvement for quick testing of infection in people; (3) to introduce propelled informatics arrangements that use AI and man-made consciousness strategies, for example, profound figuring out how to investigate COVID-19 information for conclusion, treatment and guess; (4) to create computational models and apparatuses to follow infection proliferation and repeat; (5) to display episodes for strategy producers for better dynamic. Informatics objectives incorporate information harmonization, information quality control, and multimethodology information coordination, and propelled investigation pipeline, for example, profound learning, and causal derivation, continuous dynamic and interpretable models.

The worldwide pandemic of novel COVID-19 transported about by thoughtful intense respiratory condition and since spread around the world. There have been more than 1.2 million announced cases and 69 000 passing's in excess of 200 nations. This beta coronavirus is like serious intense breathing disorder severe coronavirus and Middle East respiratory disorder coronavirus hereditary vicinity; it likely began from bat-inferred coronaviruses with passing through an obscure halfway well-evolved creature host to people [10].

8.7 SEVERE ACUTE RESPIRATORY SYNDROME

The SARS is an ssRNA-wrapped infection, target cells over the pathological basic S protein that ties to the ACE2 receptor and host cells type 2 trans film serine protease encourages cell passage through the spike protein. Once internal of the cell, viral polyproteins are incorporated that translate into RTC complex. The infection at that point orchestrates. Figure 8.8 shows the schematic introduction of SARS [11]. Basic proteins remain orchestrated prompting culmination of get together and arrival of viral particles. These virus-related lifespan steps give expected focuses to tranquilize treatment. Promising medication targets incorporate nonstructural proteins RNA-subordinate and RNA polymerase. Extra medication aims incorporate viral section and invulnerable guideline pathways [12].

8.8 MIDDLE EAST RESPIRATORY SYNDROME

MERS is a virus-related breathing ailment, was leading detailed in Saudi Arabia in the month of Sep, 2012, and has spread to 27 nations, as per the World Health Organization. A few people contaminated with MERS coronavirus (MERS-CoV) create serious intense respiratory sickness, including fever, hack and brevity of breath.

8.9 ANTIVIRAL AND ANTIMALARIAL DRUGS OF COVID-19

The therapeutic choices exist strained are based on the previous involvements with other viral epidemics like SARS/MERS as well as genomic arrangement of the fresh virus. Numerous prospective medicines including penciclovir, ribavirin, nafamostat, interferon, lopinavir-ritonavir, corticosteroids, nitazoxanide, remdesivir and old-fashioned drug.

Medications under this classification for the most part follow both of the accompanying three instruments in the infection viral repetition restraint, particle channel hindrance and serine protease restraint. Industrially accessible antiviral medications generally focus on the four significant gatherings of infections: human immunodeficiency infection, herpes, hepatitis and Prior flare-up scenes of such virus-related contaminations like SARS and MERS fair as hemorrhagic illness [13].

8.9.1 Oseltamivir

Oseltamivir a neuraminidase inhibitor endorsed for the handling of flu and the COVID-19 flare-up in China at first happened during top flu season so an enormous extent of patients got observational oseltamivir treatment pending the finding of sever coronavirus. Below Figure 8.9 demonstrated the Oseltamivir [14].

8.9.2 Umifenovir

Umifenovir is recognized as Arbidol more encouraging repurposed antiviral specialist by one of a kind system of activity focusing on the S protein/ACE2 communication and repressing film combination of the viral envelope. The operator is as of now

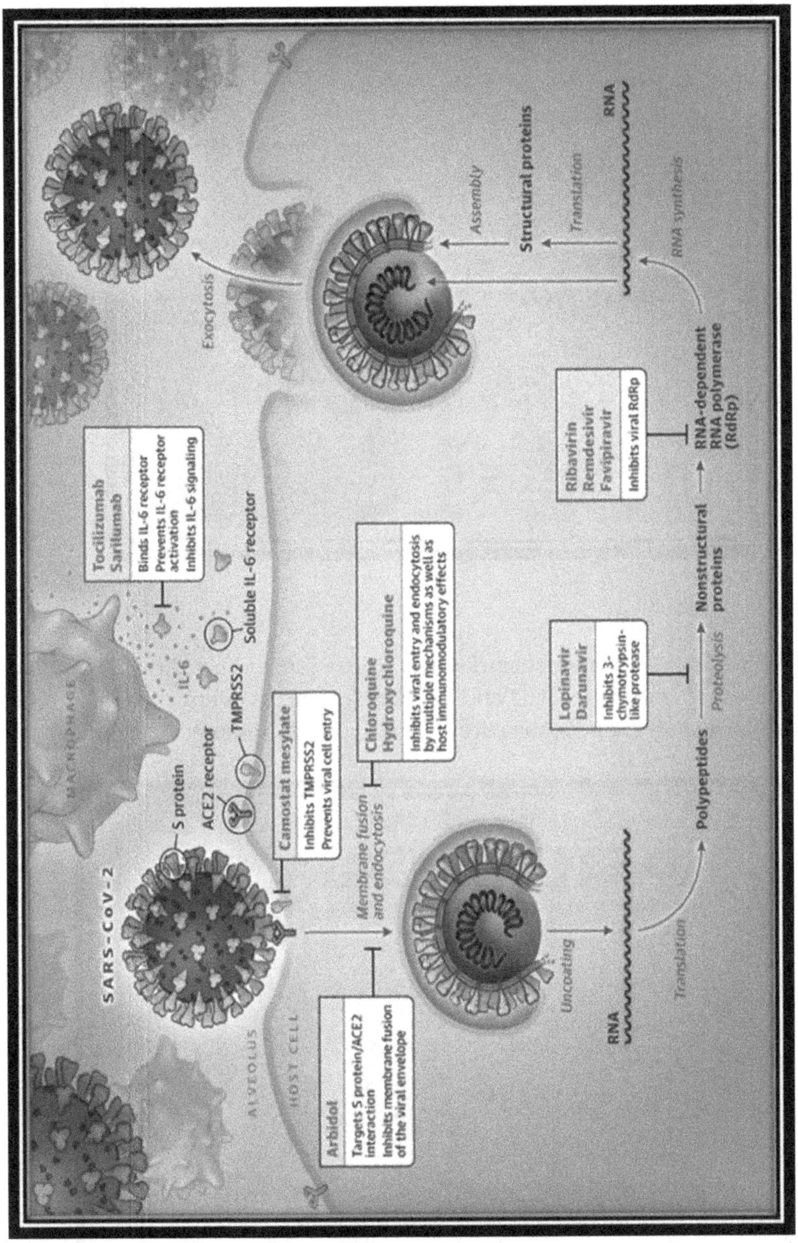

FIGURE 8.8 Schematic presentation of SARS.

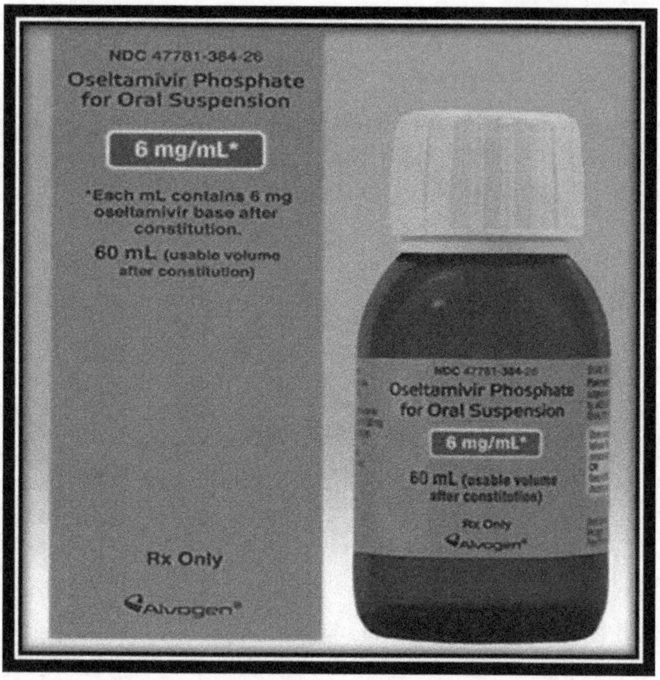

FIGURE 8.9 Oseltamivir.

confirmed in China and other countries like Russia for the cure flu and expanding enthusiasm aimed at rewarding COVID-19 dependent on the information in vitro suggesting action against severe coronavirus 2 [15]. Figure 8.10 shows the Umifenovir.

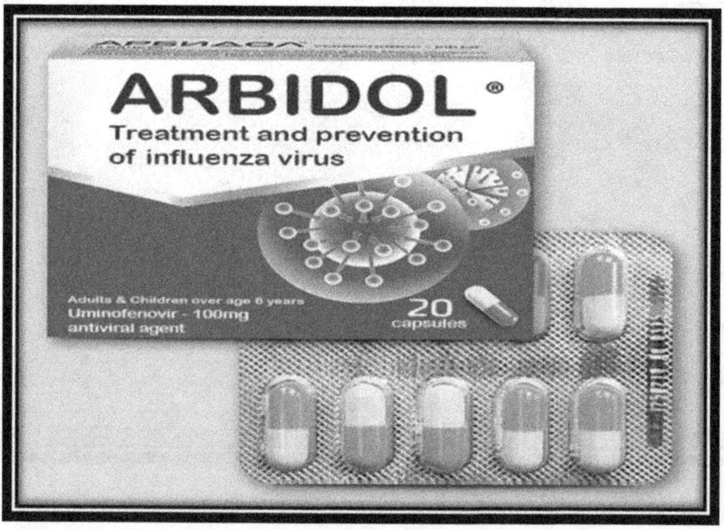

FIGURE 8.10 Umifenovir.

8.9.3 Chloroquine and Hydroxychloroquine

Chloroquine is chloroquine phosphate, sensitive malaria of *Plasmodium falciparum* and chief medication in *Plasmodium vivax* malaria in India. Its recommended quantity to cure malaria is 600 mg base orally followed by 300 mg by 6 hours, 24 hours and 48 hours. The total quantity is 1500 mg base per 3 days and also loading dosage for per day one due to the huge amount of transmission. Figure 8.11 shows the instrument and outline of antiviral impacts.

Chloroquine and hydroxychloroquine have not been demonstrated to be protected and powerful for rewarding or forestalling COVID-19. They are being read in clinical preliminaries for COVID-19, and FDA approved their brief use during the COVID-19 pandemic under restricted conditions through the EUA, and not through customary FDA endorsement. Used under the EUA when provided from the Strategic National Stockpile, the national vault of basic clinical supplies to be utilized during general well-being crises.

Chloroquine and hydroxychloroquine cause strange heart rhythms, for example, QT stretch prolongation and hazardously quick pulse called ventricular tachycardia. Risks that may increment when these meds are joined with different medications known to drag out the QT span, including the antitoxin azithromycin, which is additionally being utilized in some COVID-19 patients without FDA endorsement for this condition. Used with alert in patients who additionally have other medical problems, for example, heart and kidney sickness, who are probably going to be at expanded danger of these heart issues while accepting these prescriptions [16].

8.9.4 Lopinavir/Ritonavir

Ribavirin directed via venous brew at a dose of 500 mg for adult category persons, 2–3 times per day in mixture with interferon alpha or lopinavir/ritonavir. Lopinavir–ritonavir showed that the patients of COVID-19 showed no primary endpoint when compared to standard and some secondary endpoints. Presently, there is no evidence for antiviral cure against the new coronavirus and the origin of the respiratory problems. Researchers identified by *in vitro* studied the lopinavir aspartate protease

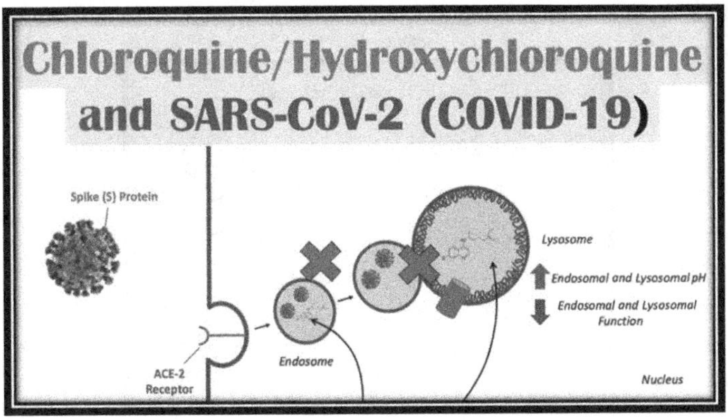

FIGURE 8.11 Mechanism and overview of antiviral effects.

inhibitor against this SARS-CoV-2 in the year 2003 [17]. Figure 8.12 shows the combination of lopinavir/ritonavir of COVID-19.

The accessibility and well-being outlines of lopinavir/ritonavir and interferon beta-1b reported that the expected viability for the curing of patients with MERS-CoV. Oral administration of lopinavir/ritonavir drugs in the animal model of MERS-CoV contamination brought about unobtrusive enhancements in MERS illness signs, including diminished aspiratory penetrates recognized by X-beam, pneumonia and weight loss [18].

8.9.5 Ribavirin

Ribavirin is a simple guanine viral RNA-subordinate RNA polymerase. The studies of *in vitro* movement compared to SARS-CoV constrained and vital high focuses to suppress on the viral-related imitation, requiring high-portion and blend treatment. Patients got whichever venous or enteral organization in previous studies. Figure 8.13 shows the structure of Ribavirin. In the treatment of MERS, ribavirin, for the most

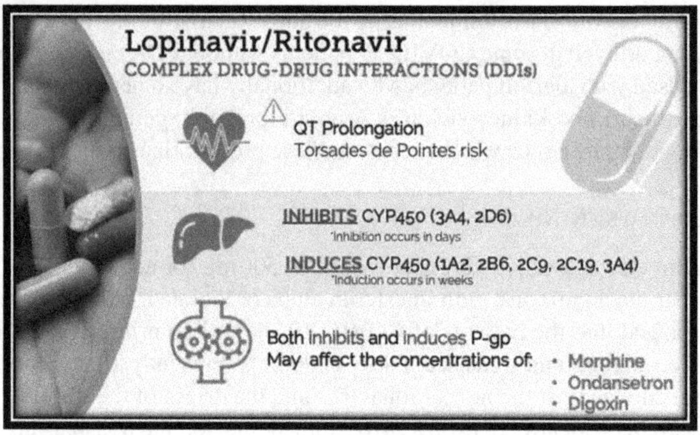

FIGURE 8.12 Lopinavir/ritonavir.

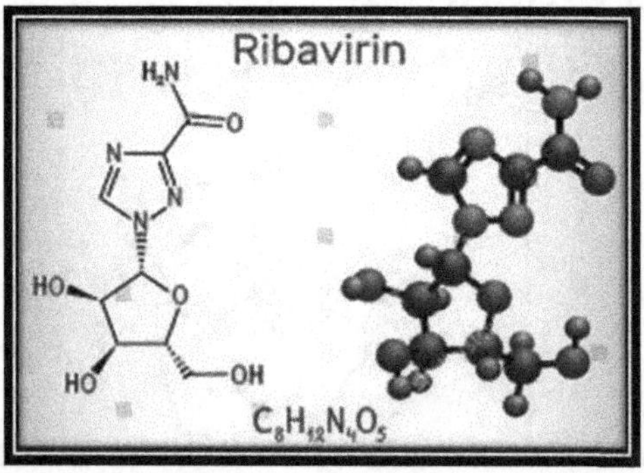

FIGURE 8.13 Structure of ribavirin.

part in mix with interferons, showed no detectable impact on clinical results or viral leeway. The clinical information with ribavirin for SARS-CoV-2 methods is its remedial job must be extrapolated from other nCoV information [19].

8.10 ANTI-HIV AND ANTI-INFLAMMATORY DRUGS

These medications are grouped into various classifications dependent on their objectives: switch translation, retro-interpretation, proteolytic preparing and fuse of proviral host genome of DNA. Calming medicines, particularly JAK-STAT inhibitors, utilized against rheumatoid joint inflammation and might be compelling against raised degrees of cytokines and valuable in restraining viral disease [20].

8.10.1 Remdesivir

Remdesivir commonly identified as monophosphate drug that experiences digestion and functioning C-adenosine nucleoside triphosphate. The specialist was found in the midst of a showing procedure for antimicrobials through movement against RNA infections, for example, Coronaviridae and Flaviviridae. Currently, remdesivir is a hopeful likely cure for COVID-19 [21].

8.10.2 Favipiravir and Corticosteroids

Favipiravir is a purine nucleotide and ribofuranosyl-5′-triphosphate. The RNA polymerases are able to stop viral replication. The SARS- and MERS-affected patients with detailed no relationship of corticosteroids with better endurance showed a relationship with postponed viral liberty from blood high paces of confusions and respiratory tract including hyperglycemia and avascular corruption [22].

8.11 MONOCLONAL ANTIBODIES AGENTS

The infection is identified to arrive the host cells by restricting the S protein to ACE2 receptors and creating killing antibodies in contradiction of the receptors; there is a great opportunity aimed at lessening the seriousness malady presently, just a bunch of medications have been affirmed for usage against SARS-CoV-2 [23]. Table 1 shows the financially accessible medications of COVID-19.

TABLE 8.1
The Commercially Existing Drugs in Cure of Coronavirus

Inhibitors	Name of Drugs	Illnesses Treated
Reverse transcriptase	Zidovudine, Didanosine	SARS
Nucleotide reverse transcriptase	Tenofovir, Disoproxil	SARS
Nonnucleoside reverse transcriptase inhibitors	Nevirapine, Delavirdine	SARS
Protease inhibitors	Saquinavir, Ritonavir	SARS
Fusion inhibitor	Enfuvirtide, Lamivudine	SARS

Indicates: SARS, Severe Acute Respiratory Syndrome

8.12 CURRENT CLINICAL TRIALS OF DRUGS IN COVID-19

At present, there are various organizations that have practiced for clinical preliminaries to repurpose prevailing medications just as to create antibodies and medications to battle contrary to the quick passing COVID-19. "National Institutes of Health (NIH)," USA, to recognize illness explicit medications. The significant medications experiencing clinical preliminaries that can possibly treat this viral disease [24].

8.13 SIGNIFICANCE OF HERBAL MEDICINE

Indian conventional restorative frameworks are considered as perhaps the most seasoned treatment in mankind's history and it assumes a significant job in experiencing worldwide medicinal service's needs. Conventional Indian traditional medicine like Unani, Ayurveda and Siddha which are effectively rehearsed for rewarding different ailments. Accustomed Indian medication mainly used the plant sources for relieving human infections. Around 25,000 herbs-based details have been utilized in society cures in Indian medication [25].

8.14 HERBAL MEDICINE FOR TREATING COVID-19

The inclusive methodology of AYUSH frameworks of medication centers around anticipation through way of life adjustment, dietary administration, prophylactic mediations for improving the insusceptibility and basic cures dependent on introduction of the indications. There are several plant sources that might constrain the coronavirus (Figures 8.14–8.18).

Indian protective and restorative herbs suggested by AYUSH for coronavirus 19. Similarly, different investigations on COVID-19 utilizing therapeutic plant materials are somewhat insignificant in India, an examination consumes indicated against mouse coronaviral movement (a proxy of SARS-CoV) by some herbal plants like *I. tinctoria, V. trifolia, G. sylvestre, A. indicum* and *L. aspera* [26].

8.15 COMPREHENSIVE CHALLENGES OF COVID-19

At present, researchers are well-known that SARS-CoV-2 has primarily two categories of strains like "L" and "S" strains. These "L" strains are further shared and also advanced rate of repetition within the host cell [27]. Many researchers declare the

FIGURE 8.14 *Andrographis paniculata.*

FIGURE 8.15 *Clerodendrum inerme.*

FIGURE 8.16 *Vitex trifolia.*

FIGURE 8.17 *Euphorbia hirta.*

more seasoned persons with preceding disorders like hypertension, coronary illness and lung ailment [28]. The following are the defensive estimates given by World Health Organization (WHO) [29]. Washing our hands totally utilizing a liquor-based hand antiseptic sanitizer will be killing the infection; avoid contacting the eyes, the nose and the mouth from others when going outside; be refreshed about the infection; avoid voyaging or assembling in packed spots; women especially with newborn children are urged to breastfeed their infants to improve their imperviousness.

FIGURE 8.18 *Strobilanthes cusia.*

8.16 CONCLUSION

It can be concluded that the coronavirus was dispersing from human to human through transmission by way of close contacts via airborne droplets through coughing, sneezing, kissing, etc. Coronavirus may be transmitted through even the pet animals including dogs, cats, pigs, cows and turkeys. So it is always better to avoid contact of these and separate them if they are observed with any infection symptoms like diarrhea, cold and fever. The WHO and European Centre for Disease Prevention and Control (ECDC) guidelines are also mentioned to avoid the contact with sick persons and also avoid the markets or public places as much as possible. There are no scientific evidences of vaccination for coronavirus to inhibit or treatment and also have some supportive therapy. Forthcoming investigation is desirable to fight with coronavirus. This novel coronavirus outbreak challenged the healthcare sectors, economic and public arrangements of China and also other countries in the world.

REFERENCES

1. Singhal T. (2020). A review of coronavirus disease-2019 (COVID-19). *Indian J Pediatr.* 87(4): 281–286.
2. Chen F., Liu Z.S., and Zhan F.R. (2020). First case of severe childhood novel coronavirus pneumonia in China. *Zhonghua Er Ke Za Zhi.* 58: E005.
3. Russel C.D., Millar J.E., and Baillie J.K. (2020). Clinical evidence does not support corticosteroid treatment for 2019-nCoV lung injury. *Lancet.* 395: 473–475.
4. Meng L., Hua F., and Bian Z. (2019). Coronavirus disease 2019 (COVID-19): Emerging and future challenges for dental and oral medicine. *J. Dent. Res.* 99(5): 481–487.
5. National Institute of Allergy and Infectious Diseases. (2020).
6. Hoffmann M., Kleine-Weber H., and Schroeder S. (2020). SARS-CoV-2 cell entry depends on ACE2 and TMPRSS2 and is blocked by a clinically proven protease inhibitor. *Cell.* 2(3): 22–28.
7. Woo M., Scott F., and Hudson Z. (2020). Human viruses: Discovery and emergence. *Phil Trans R Soc B.* 367: 2864e2871.
8. Habibzade P., and Stoneman E.K. (2020). The novel coronavirus: A bird's eye view. *Int. J. Occup. Environ. Med.* 11: 65.
9. Ienca M., and Vayena E. (2020). On the responsible use of digital data to tackle the COVID-19 pandemic. *Nat Med.* 26(4): 463–464.

10. Kang L., Li Y., Hu S., Chen M., Yang C., and Yang B.X. (2020). The mental health of medical workers in Wuhan, China dealing with the 2019 novel coronavirus. *Lancet Psychiat.* 7(3): e14.
11. National Institutes of Health. (2020). National library of medicine. United States.
12. Adam O., Frid-Adar M., Greenspan H., Browning P.D., Zhan H., and Ji W. (2020). Rapid AI development cycle for the coronavirus (COVID-19) pandemic: Initial results for automated detection & patient monitoring using deep learning CT image analysis, 2020–04–16.
13. De Lang A., Osterhaus A.D., and Haagmans B.L. (2006). Interferon-gamma and interleukin-4 downregulate expression of the SARS coronavirus receptor ACE2. *Virology.* 353: 474–481.
14. De Clercq E. (2009). Anti-HIV drugs: 25 compounds approved within 25 years after the discovery of HIV. *Int. J. Antimicrob. Agents.* 33, 307–320.
15. Razonable R.R. (2011). Antiviral drugs for viruses other than human immunodeficiency. *Mayo Clin Pro.* 86(10): 1009–1026.
16. Wang L., Wang Y., Ye D., and Liu Q. (2020). A review of the 2019 novel coronavirus (COVID- based on current evidence. *Int. J. Antimicrob. Agents.* 6(4): 105–118.
17. Elfiky A.A. (2020). Anti-HCV, nucleotide inhibitors, repurposing against COVID-19. *Life Sci.* 248(1): 117–477.
18. Kim U.J., Won E.J. Kee S.J., Jung S.I., and Jang H.C. (2016). Combination therapy with lopinavir/ritonavir, ribavirin and interferon-α for Middle East respiratory syndrome. *Antivir.* 21: 455–459.
19. Zhou P., Yang X.L., and Wang X.G. (2020). A pneumonia outbreak associated with a new coronavirus of probable bat origin. *Nature.* 579: 270–273.
20. Stebbing J., Phelan A., Griffin I., Tucker C., Oechsle O., Smith D., and Richardson, P. (2020). COVID-19: Combining antiviral and anti-inflammatory treatments. *Lancet Infect.* 20(4): 400–402.
21. Wu C., Liu Y., Yang Y., Zhang P., Zhong W., Wan Y., Wang Q., Xu Y., Li M., and Li X. (2020). Analysis of therapeutic targets for SARS-CoV-2 and discovery of potential drugs by computational methods. *Acta Pharm. Sin. B* 10(5): 201–207.
22. Wu J.T., Leung K., and Leung G.M. (2020 Feb). Nowcasting and forecasting the potential domestic and international spread of the 2019-nCoV outbreak originating in Wuhan, China: A modelling study. *Lancet.* 395(10225): 689–697.
23. Zheng M., and Song L. (2020). Novel antibody epitopes dominate the antigenicity of spike glycoprotein in SARS-CoV-2 compared to SARS-CoV. *Cell.Mol. Immunol.* 17: 1–3.
24. Rudra S., Kalra A., Kumar A., and Joe W. (2017). Utilization of alternative systems of medicine as health care services in India: Evidence on AYUSH care fromNSS 2014. *PLoS One* 12: e0176916.
25. Gomathi M., Padmapriya S., and Balachandar V. (2020). Drug studies on Rett syndrome: From bench to bedside. *J. Autism Dev. Disord.* 50(8): 1–25.
26. Srivastava R.A.K., Mistry S., and Sharma S. (2015). A novel anti-inflammatory natural product from *Sphaeranthus indicus* inhibits expression of VCAM1 and ICAM1, and slows atherosclerosis progression independent of lipid changes. *Nutr. Metab.* 12: 20.
27. Habibzadeh P., and Stoneman E.K. (2020). The novel coronavirus: A bird's eye view. *Int. J. Occup. Environ. Med.* 11: 65.
28. Wu Z., and McGoogan J.M. (2020). Characteristics of and important lessons from the coronavirus disease 2019 (COVID-19) outbreak in China: Summary of a report of 72 314 cases from the Chinese center for disease control and prevention. *JAMA* 323(13): 1239–1242.
29. World Health Organization (2020). Surveillance case definitions for human infection with novel coronavirus (nCoV). https://apps.who.int/iris/handle/10665/330376.

9 An Exploratory Study to Analyze the Impact of COVID-19 on the Daily Lives of People

A Focus Group Discussion

Surbhi, Nirupuma Yadav, and Ashwini Kumar
University of Delhi

CONTENTS

9.1 INTRODUCTION

COVID-19 (Coronavirus) has affected day to day lives and is slowing down the international economy. This pandemic has affected lots of peoples, who are both ill or are being killed due to the spread of this disease. The most common symptoms of this viral contamination are fever, cold, cough, bone ache and respiratory problems, and sooner or later main to pneumonia. This, being a new viral disorder affecting people for the first time, vaccines are not available. Thus, the emphasis is on taking big precautions such as extensive hygiene protocol (e.g., frequently washing of hands, avoidance of face to face interplay etc.), social distancing, and carrying of masks, and so on. This virus is spreading exponentially area wise. Countries are banning gatherings of human beings to unfold and break the exponential curve. Many countries are locking their populace and implementing strict quarantine to manipulate and unfold the havoc of this rather communicable disease. COVID-19 has unexpectedly affected our day to day life, businesses and disrupted the world trade and movements. Identification of the disease at an early stage is indispensable to manage the unfolding of the virus due to the fact it very hastily spreads from man or woman to person. Most of the international locations have slowed down their manufacturing of the products. Quite a number industries and sectors are affected with the aid of the motive of this disease; these encompass the prescription drugs industry, photovoltaic energy sector, tourism, information and electronics industry. This virus creates significant knock-on effects on the daily life of citizens, as nicely as about the international economy. The coronavirus has changed how we work, play and learn: Schools are closing, sports leagues have been canceled, and many people have been asked to work from home.

9.1.1 Life Will Not Be Same after COVID-19 Crisis

a. Home as the New Office:

During quarantine, most are forced to work from home. There will be people who will, on the first day after the quarantine, race to meet colleagues and drink that office coffee. But there will be those who will not want to return to the office. More attention will be given to the arrangement of the workplace at home. Spatial organization will change, with the place to work at home no longer a desk with a parody of an office chair and a lamp, slotted somewhere in the corner of the living room or under the stairs. Now it will be a completely separate room with large windows, blackout curtains and comfortable furniture. It will be technically equipped and sound-insulated. In response, offices will make more effort to win us back. Everything that the top companies have will become commonplace.

b. Mental Health of Students:

The outbreak has highlighted the fragility of psychological and mental resilience and the want for the provision of coordinated psychological intervention to the nation. It is advised to undertake and enhance the cutting-edge intervention system. Only via strengthening the psychological protection can they proceed to combat this long-drawn war and tightly closed success for the future. Another study by Coa et al. (2020), states that about 24.9% of university college students have skilled nervousness due to the fact of this COVID-19 outbreak. Living in city areas, dwelling with parents, having constant household earnings have been shielding elements for university college students and may result in anxiousness throughout the COVID-19 outbreak. However, having a relative or an acquaintance contaminated with COVID-19 was once an unbiased chance thing for skilled anxiety. The COVID-19 related stressors that blanketed monetary stressors, effects on daily-life and tutorial delays had been positively related with the stage of nervousness signs of Chinese university and college students for the duration of the epidemic, whereas social guide was once negatively correlated with their anxiety (Brooks et al., 2020). The intellectual fitness of university college students is significantly affected when confronted with public fitness emergencies, and they require attention, help, and support of the society, families and colleges. It is cautioned that the authorities and colleges have to collaborate to unravel this trouble in order to supply high-quality, well timed crisis-oriented psychological offerings to university students.

c. Psychological impact on the caregivers of COVID-19 patients:

The other topic of concerns are the psychological health of the Frontline workers who are in direct contact with the patients, doctors, nurses and other health sectors workers are tremendously affected and impacted with this pandemic as they are the one who have been in more pain working several shifts without rest. In a study it was found that in-depth appreciation of the psychological ride of caregivers of sufferers with COVID-19 through a phenomenological approach. It is located at some point of the epidemic; tremendous and terrible feelings of frontline nurses in opposition to the epidemic interweave and coexist. In the early days, poor feelings have been dominant and high-quality thoughts seemed concurrently or gradually. Self-coping fashion and psychological increase are essential for nurses to keep intellectual health. This finds out about supplied essential facts for in addition psychological intervention (Sun et al., 2020).

d. Mental health of children during quarantine and the impact of COVID-19:

According to a study conducted by Einspieler and Marschik (2020), communicating with youthful teens has to no longer fully count on simplification of the language or standards used, however should additionally take into account children's comprehension of sickness and causality. Between about 4 and 7 years, appreciation is notably influenced through magical thinking, an idea that describes a child's trust that thoughts, wishes, or unrelated moves can purpose external events, and sickness can be precipitated via a specific thinking or behavior (UNICEF, 2020). The emergence of magical wandering takes place around the same time adolescents are growing a feel of conscience, while nevertheless having a bad grasp of how sickness is spread. Adults want to be vigilant that kids are no longer inappropriately blaming themselves or feeling that the sickness is a punishment for preceding awful behavior (Edwards & Davis, 1997). Therefore, listening to what teens agree with about COVID-19 transmission is essential; imparting youngsters with a correct clarification that is significant to them will make certain that they do not now experience unnecessarily worried or guilty.

e. Impact on Older adults with chronic illnesses:

When it comes to COVID-19, the sickness brought on by using the new coronavirus, older human beings are in particular inclined to extreme illness. Research is displaying that adults 60 and older, in particular those with preexisting scientific conditions, specifically coronary heart disease, lung disease, diabetes or most cancers are extra probably to have extreme even lethal and coronavirus contamination than different age groups (Hopkins medicine, 2020). According to medical news today, 2020, The COVID-19 disease, in itself, has hit older adults more difficult than different age groups. Older adults are greater probable to already have underlying prerequisites such as cardiovascular disease, diabetes, or respiratory sickness comorbidities that we now comprehend elevate the threat of extreme COVID-19 and COVID-19-related death. In addition, a possibly weaker immune gadget makes it more difficult for older adults to battle off infection. As a result, it has an impact on older adults. According to World Health Organization information from April 2020, extra than 95% of COVID-19 deaths had been among humans over 60 years of age, and extra than 1/2 of all deaths passed off in humans greater than 80 years.

9.1.1.1 Method

9.1.1.1.1 Design

This was an exploratory qualitative study seeking to comprehend the impact of the current pandemic on the lives of college going students. A single focus group discussion (FGD) virtually was employed to understand the impact and experiences of COVID-19 amongst college going students.

9.1.1.1.2 Sample

There were a total of 12 participants who were recruited through purposive sampling (it is also known as judgmental sampling). It is a non-probability technique that involves conscious selection by the researcher of certain people to include in the study. Consent mail was sent to 15 participants and out of that confirmation mail has been received from 12 participants. In the current study, all the participants were in the age group of 18-24 years and were unmarried. As in this pandemic it is not feasible to make everyone sit in a room for discussion so zoom was chosen to have a discussion there. Everyone was present at the given time and instructions were clearly given and individuals were asked to clear their doubts if any.

9.1.1.1.3 Procedure

- College going college students who fulfilled the inclusion standards for the study had been contacted and invited for the FGD via mail.
- The objectives of the research were shared beforehand. The participants have been told that they had been being invited to take part in a crew dialog in order to share their experiences.
- The members participated actively in order to make the dialogue open.
- It was once ensured that all individuals had been requested for their opinions. If a participant tended to stay quiet, researchers in particular requested for their opinion.
- Confidentiality was ensured to everyone.

9.1.1.2 Analysis

Everyone participated enthusiastically and extensive rich data were collected. It was great to see so many experiences participants shared. Many themes and subthemes emerged. To start with two very relevant themes emerged: "Thought Process" and "Feelings Experienced". It has been divided into positive and negative branches.

In the "Thought Process" theme—in Positive Subthemes being "Self Dependent" and "Hygiene Conscious" emerged, while in Negative theme—"Uncertainty" related to (Career, Work, Normal lives) evolved, "Family Adjustments", "Discrimination" emerged

Moving on to the next theme emerged—"Feelings Experienced"—Positive sub-theme emerged was "Gratitude". Negative subthemes emerged are "Anxious", "Stressed" and "Fear"

Main themes emerged during analysis were in relation to Issues and Challenges of Coronavirus on Daily life and living faced during "Working", "Shopping", "Transport" and "Restaurants"

In the first theme Issues and Challenges in Working, following sub-themes emerged—Digitalized World, Disrupted Routine, No Escape, Lots of Pressure

In the Second theme Issues and Challenges during Shopping following sub-themes emerged—"Expenditure", "Realization", "Shopkeepers Perspective"

Third theme emerged was Issues and Challenges related to Transport following sub-themes emerged: "Trust", 'Avoidance of Public Transport" and "No Vacations"

Last sub-theme emerged was Issues and Challenges with Restaurants: Plight of local vendors, Hygiene issue, Cash Flow

9.1.1.3 Results

Results have been analyzed using thematic analysis major themes emerged were "Thoughts during lockdown", "Feelings during lockdown", Issues and Challenges related to "Working", "Shopping", "Transport" and "Restaurants".

First theme emerged was **Thoughts during lockdown**

It was found during discussion that everyone is experiencing thoughts at different levels. Some are able to balance it out and try to stay sane while some are having negative thoughts irrespective of trying to be positive and coping. Hence the researcher cannot miss upon this dimension and came out with both positive and negative sub-themes.

9.1.1.3.1 Positive Subthemes

Self-dependency: It was found in many narratives that participants were trying to be positive and they are coping really well by rationalizing everything that they are doing. Statements in support of this theme are:

> Quarantine has made us more self-dependent and is giving us a chance to be better humans and now people are finally learning what is important, realizing very small and little details, taking care of each other etc.
>
> It is important to know everything sometime you don't have choices

Hygiene Consciousness: This is another important theme that got reflected in many of the views shared by students that hygiene has always been very important but we are realizing its importance today. Statements in support of this theme are as follows:

> I have realized that suddenly all of us are so hygiene conscious, it took a pandemic to make us realize this but ya that's how it is and it's a good start
>
> Now it's a habit of everyone at my home, to wash hands and feet every time they are coming home from outside
>
> It's great to see that even kids are learning this and our upcoming generation hopefully will make this part of their daily routines and we are making them understand the Whys of this

9.1.1.3.2 Negative Subthemes

Uncertainty: This is one of the sub-themes which became very important because each and every participant relates to this and they have shared lots of instances and experiences from their life that in their every sphere of life especially in career, work and other sphere of lives, they are feeling the biggest impact. Some statements supporting this sub-theme are:

> We are in third year right now and we don't know where we will be going in our masters or what kind of exams we will have in few days, everything is very bleak
>
> Medical chain is disrupted and it's affecting so many other people who don't have coronavirus, my grandmother is in last stage of cancer and so many hospitals denied admitting her, and it was so much uncertain what to do, also uncertainty about that if she is getting treatment for cancer but what if she gets infected with coronavirus
>
> I left everything in my hostel when I came back to my place, I have no idea when I will go back now, whether I will be able to find a place to live in Delhi, where am I going to get admission issues like this are very common in my head nowadays
>
> Lots of uncertainty that whether we will be able to go back to the job or not
>
> I attended my cousins wedding on zoom, it was so different and a challenge that we have to sacrifice that part of my life too
>
> Testing is a major challenge lots of discrepancies are happening, no idea and no idea whom to trust nowadays

Family Adjustment: As participants are now living with their family for a long time different experiences and thoughts are being shared and it was a mix of both but a bent has been found toward more negative thoughts because of various reasons. Statements which are given in support of this sub-theme are as follows:

People who are living together or living apart have explored a new aspect of relationships, some are good some are not and people who are living far from their loved ones it's a big challenge to maintaining relationships and keeping them healthy

There are lots of restrictions now what you can eat and what you can't, what you want to wear what you should not and other things about daily routine

Biggest challenge for me is living with my family in this pandemic; it's a very big adjustment problem

Sudden privacy beach from parents' side shows that they don't trust us

Discrimination: Participants have observed and felt that this pandemic has made us more strict and rude with our behavior as we need to be cautious all the time and we cannot just let go if anyone is behaving in a manner which is harmful to the society and hence discriminatory behavior is done and faced too. Statements supporting this sub-theme are as follows:

We are living with a stigma now that when someone coughs or sneezes the first thought that comes to our mind is that this person has coronavirus and it automatically leads to discriminatory behavior

Now there is a very thin line between stigmatizing discrimination and being careful and you don't know where to step when

I feel so bad that now we have to stand far from people and if anybody is not following hygiene rules we need to be strict in conveying our thoughts which makes people feel offended and I feel bad too but I can't help it, I need to save me and my family too

I saw in news few days back that people are behaving in a very derogatory manner with sweepers, maids, etc., it's really hurtful

Second theme emerged was **"Feelings Experienced"**

9.1.1.3.3 Positive Subtheme

Gratitude: Participants responded by having this thought in mind too that in spite of whatever we are facing in this pandemic we cannot ignore that we are privileged as compared to others and instead of complaining every time we should practice gratitude too. Here are some of the statements in support of this subtheme:

In everything that is going on I am very grateful that I got the time to do my art work and it has no restriction from anyone

Quarantine made me realize that it really doesn't matter from where you are coming or from which strata you belong our primary focus should be on that our basic necessities are fulfilled and health of our family members stay well

I have my bed, food on my table and things to do I am grateful for that

9.1.1.3.4 Negative Subthemes

Anxiety: Anxiety is a feeling which we all are facing nowadays and participants unanimously agreed on whether mild or severe because of this pandemic they are facing lots of anxiety issues which is creating issues for them. Some of the statements supporting this sub-theme are as follows:

> I have fear which creates anxiety because of so overwhelming news all the time, its taking a lot of time to process it, it took a lot of time dealing with it with avoidance but now I have realized that avoidance won't work
>
> Because of this pandemic we are so much vulnerable to have anxiety in these times due to various events that are happening right now
>
> I haven't still been able to come in terms with the fact that we are going through a pandemic yet because for some reason I still feel that this is part of a dream, it's very disorienting to see that now people don't come out of their home and it makes me feel so anxious
>
> One day I need to step out to take something from outside it teared me up that I was not able to step out again since then

Stress: In one way or another everyone is facing stress nowadays. Participants who reported feeling stress because of so many reasons which are theirs as well and some are which people from other than they are facing. Statements supporting this sub-theme are as follows:

> I am staying alone in Delhi and my mom dad are stuck in abroad, now I have lot of stress and all this is forcing me to think that in back when I was with them I shouldn't have fought, I should have listen, I should have spent some quality time, now this is creating so much stress that neither this time is going forward nor I can go back in time
>
> All my books and my laptop is in Delhi I can't do anything, I feel so bad that this time is getting wasted and I can't help it even
>
> Farmers are so stressed that they are ready to bring vegetables to cities but no means are available which leads to wastage of products

Fear: This is again one of the sub-themes on which everyone agreed and is commonly experienced by them and by people who are in their known too are experiencing it. Statements supporting this sub-theme are as follows:

> I am having lot of fear when everything will be fine
>
> Lot of fear of isolation and limitations that we are having right now because of various things
>
> Lot of fear that god knows when we will be thrown out of our jobs and when we will get another one
>
> We were taking lot of our relationships for granted earlier but now we understand its value, keep on talking to people and keep checking people out, it should be done

Third major theme emerged was **Issues and Challenges related to "Working."**

As everyone knows that our life has come to a standstill because of being at home and quarantined one cannot go out and do any work, whether it is a job or internship or any course they were pursuing and lots of stress, anxiety and frustration have been seen. This has emerged as a separate theme because of its importance and relevance during pandemic times. Following are the emerged subthemes:

Issue with technology: Participants reported this frequently ever since everyone agreed that understanding technology is not everyone's cup of tea to be precise secondly we are living in different parts of the country now and it is very difficult to assume that everyone has good internet connectivity so we cannot be on the same page. Some of the statements supporting this sub-theme are as follows:

> Everything is online whether it's an internship or lectures to be attended. Last semester I worked with an organization to gain practical experience there was more of hands on approach or practical approach now it's more learning oriented this is one of the issue I am facing and it's a challenge since I won't be able to able to gain practical knowledge this time
>
> Not everyone has access to good internet connectivity this is a major issue with school and college students
>
> It's not easy for everyone to understand and work with technology how one can think that everyone will learn at one pace

Disrupted Routine: Every participant agreed to this that routine has disrupted tremendously and it is not easy to go back to earlier ones because of which they are facing issues with family members too. Statements supporting this sub-theme are as follows:

> Earlier I was very working in a psychiatric clinic, it was a great experience interacting with new people and learning about new cases and within a week its shut, it's like suddenly all my routine is gone, now I have become much of a procrastinator, all day I am in bed and it's so much exhausting to get out of bed because there is no routine

> Earlier there was a routine but now routine just begins when you wake up, I have friends who have become completely nocturnal now, there cycles have shifted completely
>
> It has also affected people who are workaholics, they used to have a drive to go to work feeling needed and wanted that work can't happen without them and people now who have lost their jobs are feeling such a big void in their lives because of no routine
>
> Earlier after coming from college I used to do 5–6 hours of study and now since lockdown started I have not studied anything my study routine has completely deteriorated and I have become completely nocturnal, my mom scolds me everyday
>
> Whole schedule is disturbed, I sleep in day and work at night, it has affected my efficiency of working a lot

9.1.1.3.5 No Escape

It has been found from the narratives that coming to college or going to work has been an escape from being at home all day and that is completely shut now which is leading to negative thoughts and feelings for the participants. Statements supporting this sub-theme are as follows:

> You can't go anywhere, you are stuck at home and can't do anything about it
>
> College used to be my escape zone, I used to meet my friends and it is a very different place for me altogether since lockdown it's really frustrating for me and I keep calling my friends

Lots of Pressure: Participants are agreeing to this fact that yes they are facing a lot of pressure because they can see on social media and through their friend circle that everyone is doing something or the other to make it productive and hence they are feeling its impact top. Some of the statements supporting this sub-theme are as follows:

> I feel a lot of pressure to work somewhere or other because everyone is working and I used to overburden myself, now I am working on it
>
> Not everyone has the privilege of working in this pandemic so it should be understood and one should not feel pressured as it is not good for one's mental health
>
> When we will go to college, first day we will be asked is "What you did" "How you made this lockdown productive", someone might has not done it but you feel pressurized you know
>
> We should realize that the definition of productivity has changed now, it's not the same and hence we should not pressurize ourselves. Earlier definitions of being productive is not existing now

Fourth major theme emerged was **Issues and Challenges related to "Shopping"**

Expenditure: Participants agree on the fact that now expenditure and perception towards shopping have changed drastically and realization has been from a completely different angle that they never thought existed. Statements supporting this sub-theme are as follows:

> Earlier I used to go to markets and used to do online shopping too but as compared to earlier now expenditure is less
>
> We are spending more on groceries than we used to do before and it's a weird experience. I have never seen my parents panicking about something
>
> At my place expenditure on medications along with groceries has increased too
>
> Earlier in a month I used to put two three outfits for me, now I just scroll through online websites and then I close time is so uncertain we can't afford to do unnecessary shopping
>
> Everyone will stay same behavior wise but hand gloves and masks are just added expenditure

Realization: Participants collectively shared this that it is not only them but everyone they know has gone through a reality check in this pandemic and realization has been huge, and these issues and challenges are making everyone learn a lesson. Statements supporting this sub-theme are as follows:

> Earlier kids used to spend a lot but now kids are realizing the intensity of the situation and they are making sacrifices too, I can see my grandmother teaching my cousin brother this and it's learning for all of us
>
> I am an avid online shopper, when it stopped suddenly it was very very difficult for me, I have totally stopped it now other than grocery nothing else we are buying now
>
> Even online groceries are not cheap and most importantly not everything is available so something we buy by going outside some online it is another challenge
>
> Things online are much more expensive than usual and my family is big time involved in panic buying, my staircase is covered with grocery bags from big basket
>
> No butter no cheese no dairy products are ordered because you can't keep them outside with other vegetables
>
> Earlier I used to go to malls for buying stuff but in this pandemic I have realized that that those are not essentials, I can live without them

Shopkeepers Perspective: This sub-theme emerged in a different way when participants reported that shopping is not only related to buyers but also to the sellers hence when we are talking about we cannot ignore the other side and how this pandemic has posed issues and challenges for them. Statements supporting this sub-theme are as follows:

> From the shopkeepers side condition is not good too people prefer to go to only essentials shops now hence rest of the shopkeepers are suffering
>
> My dad is in educational business and even in this pandemic he has to go to universities to showcase his products and he is not even getting money because no admissions have taken place and he has to pay to the employees as well, so inflow of money is not there but outflow is happening
>
> My dad has a shop in karol bagh, people are honestly following the rules of odd even no. of shops to open, because of this dishonesty other shopkeepers are facing loss too
>
> Lot of fear and anxiety from shopkeepers perspective too that no one knows when suddenly lockdown will happen again and we need to sit back at home
>
> Mentality is this that shopkeepers are ready to pay fine but want to come on both days

Fourth major theme emerged was **Issues and Challenges related to "Transport"**

Trust: Major issue reported by everyone that now trust is not there on any public transport and at the same time not everyone has a privilege to take their private vehicle daily to work or college. Participants reported feeling anxious and stressed about it. Statements supporting this sub-theme are as follows:

> Even Though I am taking all the precautions I don't really know that whether the other person is doing it too or not, just because of someone else's fault I can't risk myself
>
> First thing that comes to my mind is whether I will be able to trust either railways or flights to travel because irrespective of whichever mode you are using you are at high risk of getting infected
>
> After travelling and going home I need to quarantine myself for 14 days then after reaching back I need to quarantine myself there is no way I can meet anyone as all time will go in quarantine only
>
> It's scary to be all alone in Delhi, I want to go back to my family but they are so far. I miss my family and I can't do anything

Avoidance of Public Transport: Common opinion by participants and it is reported that earlier when we boarded metro, bus or flight we were not that much concerned about safety and hygiene and now that is the first thing that comes to mind. Statements supporting this sub-theme are as follows:

> I can't think in my life now to board Delhi metro it's very unsafe
>
> There has never been a concept of space existing in public transport, now just the thought of travelling in that makes me anxious
>
> It was in news that in Delhi metro they will be managing by leaving one seat in between people, the lines are so long how it will be followed
>
> When it comes to Uber or Ola even though the side from driver will be covered but you don't know whether the car is sanitized properly or not and who sat at your place earlier

No Vacations: Participants reported that earlier in vacations family travels were a common thing and get-togethers used to take place now it seems like a dream to everyone. Statements supporting the following sub-theme are as follows:

> Me and my parents used to travel during summer holidays I don't think so it's gonna happen now
>
> I travel only when my dad comes but now because of Corona it's not gonna happen. This will be the biggest change

Fourth major subtheme emerged was **Issues and Challenges related to "Restaurants"**

Plight of local vendors: According to participants this is the worst hit business right now and these people have no clue what to do and how to sustain living. Statements supporting this sub-theme are as follows:

> Street food owners, their shop have been closed totally
>
> Jo saving tha wo tha, now they don't have any work
>
> Here people usually perceive ki yaha to ganda khana hi hota hoga no hygiene it's affecting local businesses

Hygiene Issue: This has come up again as one of the common sub-themes that people are heavily concerned about and it is actually affecting the restaurant business in a big manner and it will continue to do so for some period of time. Statements supporting this sub-theme are as follows:

> People because of that markaz thing are not taking food from muslim delivery persons
>
> During precovid times we were adventurous in exploring places but now we are shifting too much safer options like McDonalds and Dominos which are considered by us as much safer places
>
> So many delivery people are not taking care of hygiene issues: smoking, not using masks, talking closely with each other

Cash Flow: Another important theme pointed out here is not every restaurant owner is paying digitally and as in today's times people are not preferring to pay through cash, it has become another challenge for them to do their transactions for paying to laborers and paying for raw materials

Most restaurants were used to of cash flow so online payment is new, when they pay staff they pay them in cash, it's a big challenge for them to deal with it

Online payment is major issue since people are not used to it

Restaurants were not ready for lockdown and whatever raw materials were there everything got wasted, some small shops closed because of this

9.2 DISCUSSION

COVID-19 has heavily impacted human lives of every life stage be it children, adults, older people or professions be it frontline workers, daily wage earners and the general public (Chen et al, 2020; Yang et al, 2020). COVID-19 has swept economies not in terms of economic fall but also in terms of physical and mental health as well. Every day in media we hear or see figures relating to how many people lost their lives because of this pandemic and it is very overwhelming to see and hence everyone is concerned about the physical health that is one should not come in contact with a person who is infected or one should practice social distancing as much as they can and avoid stepping out of home. But none of the researches until now has focused on the other side of the phenomenon which is equally important but not easily accepted and that is how being in this pandemic is stressful, is creating a lot of anxiety, uncertainty towards future and besides this lot of fear; in other words if this can be put down it is that mental health is going through a massive hit which was never anticipated. Some research which are done in this area are talking about the anxiety, stress faced by college students and concern about future employment (Mei et al., 2011; Cornine et al., 2020; Wang et al., 2020) but none of it has catered to pay attention to the experiences one is having this research has filled the gap by catering to subjective experiences which are very important to understand.

FGD method was used in collecting data and it was found that participants are majorly facing issues and challenges and it has been found very concretely in their thoughts and feelings as well as different major themes that emerged which are working, shopping, travelling and restaurants—lots of uncertainty, anxiety and fear. Feeling overwhelmed and stressed because of these factors. According to Limcaoco et al. (2020), researchers describe a bigger affective sign due to COVID-19. This pandemic is elevating the anxiousness levels. The findings of study exhibit the affective and cognitive transformations human beings are going through. This survey is the first try to measure the psychological penalties this pandemic is having, in order to be capable to grant the aid to confront this international issue, addressing the intellectual fitness care that will be needed where one of the future goals is to aid fitness of employees and the customary public in managing emotional stress and associated personal, expert and household problems, all through the COVID-19 pandemic. While there is presently no encouraged treatment, vaccine- or antiviral medicinal drug for COVID-19, pharmaceutical businesses are racing for solutions. Our response to mitigate the affective and cognitive penalties on the quarantine can be based totally on Brooks et al.'s (2020), recommendations: giving humans as a great deal statistics as possible, offering ample supplies, decreasing the boredom and enhancing communication. According to a study by Roy et al. (2020), during this coronavirus pandemic, most of the educated humans and authorities are conscious of this infection, viable preventive measures, the significance of social distancing and authorities initiatives had been taken to restriction the unfold of infection. However, there are extended concerns and apprehensions among the public involving obtaining the COVID-19 infection. People have greater perceived wishes to deal with their intellectual fitness difficulties. There is a want to intensify the recognition application and tackle the intellectual fitness troubles of people all through this COVID-19 pandemic. There is no learning about to date that evaluated the intellectual fitness views of human beings all through the COVID-19 pandemic. It is vital to learn about the intellectual fitness influences in more than a few populations (general populations, instances of COVID-19, and shut contacts of COVID-19 and healthcare workers) for planning high-quality intervention techniques for them. Cûosiü et al. (2020) suggested that while coronavirus pandemic is truly in the focal point of international attention, precise influences on the magnitude of poor psycho-socio-economic fallout of COVID-19, which relies upon on a constellation of pandemic, pre-pandemic and post-pandemic elements in every United States and region, deserve unique attention. For illustration, the Zagreb earthquake in Croatia in March 2020 in the course of the ongoing pandemic represents an extra hazard element that may amplify the post-pandemic incidence of intellectual fitness issues in Zagreb past coronavirus outbreak alone. Since Croatia already has an excessive range of pre-pandemic instances of persistent post-traumatic stress disorder (PTSD) due to the 1990s' Homeland War, it is vital from the countrywide viewpoint to understand high-risk people for intellectual fitness problems in order to supply them well timed and environment friendly assistance and stop a new buildup of PTSD in Croatian populace brought on by means of coronavirus pandemic 2020.

Limitations of the present study which can serve as future implications are that mixed method approach could have been used so that results would have been more strengthened and concrete. Secondly FGD with more groups on the same topic could have been done to see similarity or differences in opinions with the larger group. Lastly future researchers can focus on the intergenerational study to get a more holistic picture.

This study can also work as a reference for mental health professionals to come forward and help people who are facing anxiety, stress and fear severely and it is impacting their daily lives negatively and also for the institutions involved in the area of mental health to start calling services to help people because people need help during this time.

REFERENCES

Chen, Q. et al. (2020). Mental health care for medical staff in China during the COVID-19 outbreak. *The Lancet Psychiatry*.

Brooks, S.K. et al. (2020). The psychological impact of quarantine and how to reduce it: Rapid review of the evidence. *Lancet*, 395: 912–920.

Coa, W. et al. (2020). The psychological impact of the COVID-19 epidemic on college students in China. *Psychiatry Research*, 287: 112934.

Cornine, A. (2020). Reducing nursing student anxiety in the clinical setting: An integrative review. *Nursing Education Perspectives*, 41(4): 229–234.

Cûosiü, K. et al. (2020). Impact of human disasters and COVID-19 pandemic on mental health: Potential of digital psychiatry. *Psychiatry Danubina*, 32(1): 25–31. doi: 10.24869/psyd.2020.25.

Edwards, M., and Davis, H. (1997) The child's experience. In: *Counselling children with chronic medical conditions*. Leicester: British Psychological Society: 28–48.

Einspieler, C., and Marschik, P. B. (2020). Protecting the psychological health of children through effective communication about COVID-19. May 2020. https://www.thelancet.com/action/showPdf?pii=S2352-4642%2820%2930097-3

https://hub.jhu.edu/2020/05/05/impact-of-covid-19-on-the-elderly/

https://www.hopkinsmedicine.org/health/conditions-and-diseases/coronavirus/coronavirus-caregiving-for-the-elderly

Medical new letter. https://www.medicalnewstoday.com/articles/the-impact-of-the-covid-19-pandemic-on-older-adults#Mental-health-and-elder-abuse

Health advisory for elderly population of India during COVID19. https://www.mohfw.gov.in/pdf/AdvisoryforElderlyPopulation.pdf

Caring for the elderly during the COVID-19 pandemic. https://www.unicef.org/india/stories/caring-elderly-during-covid-19-pandemic

Limcaoco, R.S.G. et al. (2020). Anxiety, worry and perceived stress in the world due to the COVID-19 pandemic, March 2020. Preliminary results. *MedRxiv*.

Mei, S.L. et al. (2011). Psychological investigation of university students in a university in Jilin province. *Med Society (Berkeley)*, 24(5): 84–86.

Roy, D. et al. (2020). Study of knowledge, attitude, anxiety & perceived mental healthcare need in Indian population during COVID-19 pandemic. *Asian Journal of Psychiatry*, 51: 102083.

Sun, N. et al. (2020). A qualitative study on the psychological experience of caregivers of COVID-19 patients. *American Journal of Infection Control*, 48: 592–598.

The COVID-19 pandemic and emotional wellbeing: Tips for healthy routines and rhythms during unpredictable times by international society of bipolar disorders (ISBD) Task force on chronobiology and chrono-therapy and the society for light treatment and biologic rhythms (SLTBR).

UNICEF. How to talk to your child about coronavirus disease 2019 (COVID-19). https://www.unicef.org/coronavirus/how-talk-your-childabout-coronavirus-covid-19 (accessed March 11, 2020).

Wang, C., et al. (2020a). A novel coronavirus outbreak of global health concern. *Lancet*, 395(10223): 470–473.

Yang, Y. et al. (2020). Mental health services for older adults in China during the COVID-19 outbreak. *Lancet Psychiatry*, 20: 30079–30081.

10 Role of Modern Technologies in Treating of COVID-19

V. Dhinakaran, R. Surendran, and M. Varsha Shree
Chennai Institute of Technology

Suman Lata Tripathi
Lovely Professional University

CONTENTS

10.1 INTRODUCTION

Coronavirus is one of the most important human pathogenic respiratory systems. Extreme acute respiratory syndrome (SARS), coronavirus (CoV), respiratory syndrome in the Middle East (MERS) and other recent CoV outbreaks were historically considered to be a serious threat to the public health. By the end of December 2019, a cluster of patients with initial pneumonic diagnoses were admitted to hospitals with unknown etiologic. The wholesale market of wet and marine feed in Wuhan, Hubei province, China [1], was the epidemiological link between these patients. The Director General of the World Health Organization, Tedros Adhanom Ghebreyesus, has announced an outbreak of COVID-19, the acronym for 'Coronavirus 2019,' [2] on 11 February 2020. COVID-19 is a disease that is generally acutely resolved but also 2% death. Serious alveolar damage and progressive breathing failure could lead to poor disease deaths. Around 7,668,683 cases and 453 deaths, including massive, human, but also animal-infected one-stroke RNA viruses, were reported on June 12 [3]. Tyrell and Bynoe, who developed viruses in common cold patients, first identified coronaviruses in 1966. They are called coronaviruses based on their anatomy (Latin: corona = corona). These spherical virions are identical to a Solar Corona, with core coats and surface projections. Coronaviruses are of four subtypes: alpha, beta, gamma and delta. Primates are present in alpha and beta coronaviruses, in particular monkeys, pigs and birds, and in gamma and delta viruses. The dimensions of the genome range from 26 to 32 kb. Beta coronavirus can lead to serious illness and fatality, and alpha coronavirus may cause asymptomatic or moderately symptomatic infections for seven subtypes of human-infected coronavirus. SARS-CoV-2 is a family of SARS-CoV-linked B beta-coronavirus. The four main structural genes that predominate in the hCoB-Oc43 and HCU1-beta-coronavirus are nucleocapside (N), spike protein, small membrane protein and glycoprotein membrane 5. At the complete genome stage, the SARS-CoV-2 is 96% similar to a bat coronavirus. Pneumonia SARS-CoV-2-related COVID-19 disease was the key clinical indication that allowed diagnosis of the case. Recent reports also describe gastrointestinal signs and infections, in particular in young children. Like other viruses, SARS-CoV-2 infects lung alveolar epithelial cells with angiotensin converting enzyme II (ACE-2) as a receiving receptor. Artificial intelligence (AI) predicts the possibility of inhibiting the entry into the viral target cells of AP2 protein kinases 1 (AAK1). There has been talk of the role of technologies to deal with COVID-19's pandemic problems.

10.2 SMART TECHNOLOGY FOR EARLY STAGE DIAGNOSIS

10.2.1 IoT-Based Smart Helmet Detection and Diagnostic System

The smart helmet was equipped with two different types of cameras, allowing the gathering of detailed information of the face detection details and temperature measurements. Optical camera and infrared thermal camera which provided information about the temperature at which the different focuses of interest were found. A thermographic camera, which sometimes named by thermal imager, thermal imaging, or infrared cameras, is a device using infrared radiation to create image similarly to traditional camera that utilizes visible light to produce image. This module regards segmentation approach of an image according to the recorded temperature and captured color images by both thermal and optical cameras. Thermal camera is utilized for hot body detection and recognizing by adopting the variability of high temperature compared with other objects within the scanned zone. If thermal camera visualizes high temperature body, then it creates high intensity levels of infrared spectra. This allows persons with increased body temperature to be identified quickly and reliably, and to be isolated for more exact testing. Beyond checking body temperature, AI is being used to diagnose COVID-19. Infervision, a software that automatically detects symptoms via screening images, can make diagnoses quicker and reduce the risk of human error. The figure illustrates how smart helmet works generally as telemedicine and portable medical care and management of health. With the Internet of Things (IoT) technology that has widely implemented in the health care sector, this study aims to design of system that has the capability to detect the coronavirus automatically from the thermal image with less human interactions using smart helmet with Mounted Thermal Imaging System. The thermal camera technology is integrated to the smart helmet and combined with IoT technology for monitoring the screening process to get the real-time data. The interfacing of a modular system that is based on IoT communication link and Global System for Mobile communications (GSM) is done. This system delivers a notification if detecting temperature higher than normal temperature. The Global Positioning System (GPS) module determines the position coordinates after tagging it and a notification is sent to assigned smart mobile through a GSM. The officer will get the data of people's face and temperature to identify someone who is indicating as infected of COVID-19. In three segments, the proposed intelligent helmet is integrated. The first part of the system comprises the mechanism's input source consisting of thermal cameras, optical camera applications and mobile phones. The second segment of system development was the development of the processor. This segment has been integrated with the Arduino IDE software by the microcontroller processor to code the source code. The program facilitates the compilation in the NODEMCU V2 processor of the necessary commands and source code. Two different kinds of cameras were installed to collect detailed information on the face detection and temperature measurements in the intelligent helmet. In order to tentatively authorize the program, the validation stage centered on rational technical interims to ensure that all claims were checked and that the operational intervals were conducted in experiments to identify errors. The defined input also ensures that real results are coordinated with the required results. Each research was implemented and conducted in programs and models. The circuit simulation mimics the actual electronic device performance and the circuit of the cellular software (Figure 10.1).

10.2.2 Intelligent Diagnostics and Treatment Assistant Program for COVID-19

The Request for Information (RFI) and Internet technology were initially known to the IoT on the basis of an established communication protocol for intelligent management of information. The Improvement Development Agency (IDA) (nCapp) program is based on the Cloud of Things to detect COVID-19 faster and boost care by means of digital technology. The IDA program is the IDA program. The selection key is used to implement Terminal 8 features in real time with "Edge." The current data, questionnaires and control results are automatically diagnosed, confirmed, suspected or suspected by the new coronavirus disease (2019-nCoV) [5]. Medical classifications are mild, moderate, extreme or serious. You can also install the COVID-19 online diagnostic model and update it in real time using the latest real-life case data to improve diagnostic accuracy. NCapp will guide you as well. Assessment and treatment are connected with doctors, consultants and directors. NCapp also offers long-term care for COVID-19 patients. The final objective of the program is to update

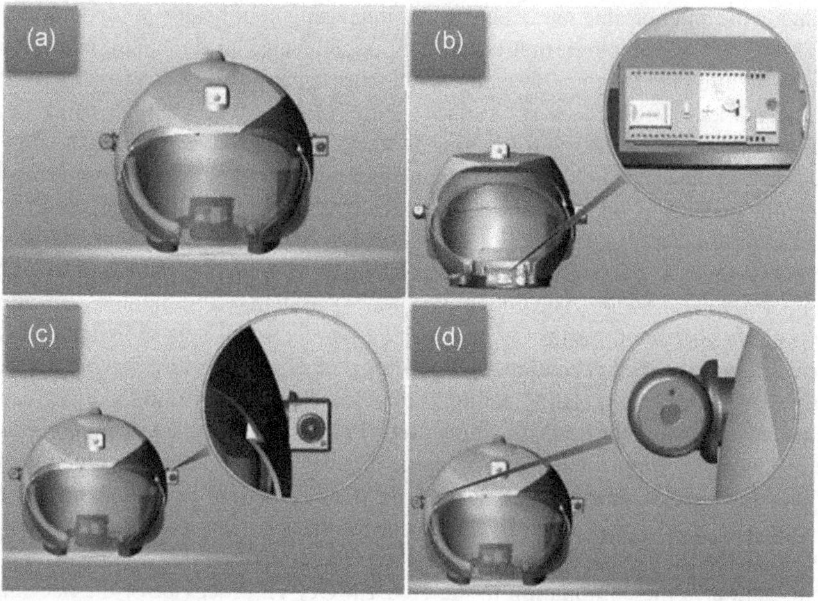

FIGURE 10.1 Configuration design of the system: (a) overall system, (b) system controller, (c) thermal camera and (d) optic camera [4].

diagnoses and treatment rates for COVID-19 in different hospitals nationally and internationally with NCapp's intelligent assistance. It helps them deter and control diseases of circulation, avoid human complications and eliminate epidemics as soon as possible. The COVID-19 control can also be used by MIoT. We will build a 3-stufe nCapp program based on medical theory and IoT technologies to diagnose and treat COVID-19. The virtual medical system's IoT health application contains key IoT and GPU functions. Increased support for deep mining and Smart diagnosis can be provided by cloud system links to the current electronic health records, image archiving, image archival and communication [6].

10.3 ANALYSIS OF COVID-19 USING MACHINE LEARNING AND DEEP LEARNING TECHNIQUES

Studying the spread of a disease is an old subject in statistical epidemiology and the actual disease is no exception. Many models of epidemiology monitoring rely upon the renowned Susceptible, Infected and Recovered (SIR) model, where R is the main interest parameter, as $R<1$ indicates that the infections will disappear naturally. The good thing about SIR is R's, a basic SIR parameter function. The SIR model in the basic version of the model is based on the presumption that the model parameters are constant along with time and space. However, modeling has evolved, including the epidemiology of more advanced models, since the application of spatial-temporal statistical analytics began. These are derived in various domains from statistical applications. We find COVID-19s development at the areal level (unlike dot data) in a given time domain from a spatial analysis. In the light of the insecurity of 1 arXiv:2005.10335v1 [stat. AP] 20 May 2020, we feel that Bayesian modeling is more than simply another way to build models with complicated structures, particularly with noisy data, must bear in mind any statements about the development of COVID-19 [7]. The issue with the normal Space-Time Model is the fact that the Bayesians consider such parametric models for evolution (e.g., Auto Regressive Moving Average (ARMA) process) across time and are distributed around areas often by a neighborhood matrix, which is meant to be sensitive to the observed data in the first place, utilizing a Conditional Autoregressive model. The latter approach is used to create such a super expert (of course data) as modern machine learning techniques will generate priors on a Bayesian model for the counts. A theorem by Bayes, which finally allows for the prediction of counts and associated uncertainty, correctly takes the discrepancy between observation and expert. In recent years, profound learning techniques have shown promising results in carrying out radiological tasks through the automatic analysis of multimodal medical images. Deep convolutional neural networks (DCNN) are one of the powerful architectures of deep learning and have been used intuitively in many practical applications such as pattern identification and image classification. DCNNs can handle four modes: (1) trainer of neural networks weighs on very big available data sets; (2) finalization of pertained DCNN network weights on the basis of small data sets; (3) pretraining unchecked to initialize network weights prior to application of DCNN models; (4) pretrained dynamic circuit network (DCSN) is also called a CNN off-shelf. DCNs were used to diagnose common thrust diseases like tuberculosis in previous studies in the X-ray image classification. The Oxford Robotics Institute of Karen Simonyan and Andrew Zisserman's VGG19: Visual Geometry Group Network (VGG) was developed using the convolution architecture of neural networks. This was addressed at the 2014 ILSVRC2014 Broad Scale Object Recognition Competition. On the imageNet data set, the VGGNet worked very well [8].

10.4 AI FOR COVID-19 DISEASES MANAGEMENT

Healthcare needs assistance from emerging technology like IoT, AI, big data and machines to combat and predict the current diseases. We are looking at AI's role in research, prevention and combat preparedness and COVID-19 (coronavirus) and other pandemics as a vital technology. It makes decision-making faster and more cost-effective. In COVID-19 cases it supports the construction of a new diagnostic and control system using useful algorithms [9]. AI facilitates the treatment of contaminated patients by medical imaging techniques such as magnetic resonance imaging and computed tomography (CT). AI can build an intelligent platform to automatically monitor individuals. It is going to predict the possible course of this disease. Analysis of information COVID-19 calls for the use of AI in drug studies. This is useful in designing and developing drug delivery. This technology is used to accelerate drug tests in real time when standard tests take a great deal of time and contribute substantially to the acceleration of this process. AI supports the development of vaccines and therapies much faster than usual and is helpful in the development of vaccines in clinical trials as well [10]. AI can provide updated information to prevent this disease by conducting real-time data analysis. AI is helpful. It can be used to predict possible transmission locations, virus growth and hospital requirements and health personnel during this outbreak. AI is important for the potential identification and reuse of past knowledge for viruses and diseases present at various times. This shows the characteristics, causes and causes of infection spread. This would be a big tool in the future to combat other outbreaks and pandemics. It can act in a way that prevents many other diseases. In future predictive and preventive healthcare AI should play a crucial role [11].

10.5 DIAGNOSIS OF COVID-19 USING SMARTPHONE-EMBEDDED SENSORS

While there are many approaches to get the disease test result, the proposal proposes a low cost and friendly solution. Radiologists or individuals who have smartphones at anytime and anywhere could use the solution. Therefore, in emergency situations, such a system is necessary useful. The proposed approach elaborates the process of creating a framework based on smartphones that includes smartphones, algorithms and embedded sensors [12]. In addition, a range of sensor technologies, including cameras, inertial sensors, microphone and temperature sensors, are mounted on the smartphones. Additionally, these sensor development readings were used for certain symptoms that occur in coronavirus diseases. As a state of the art, algorithms are applied separately on each of the readings of these sensors to detect the level of symptoms for the purposes of human health. For example, the temperature-fingerprint sensor as it is located under the touch-screen of the smartphone could be used to predict the level of fever in a new combination convolutional neural network (CNN) and recurrent neural networks (RNN) machine learning technique for the system [13]. Smartphones are pneumonia-probable case CT scanning images. CT scan images are certainly a key method for detecting COVID-19. An algorithm may be developed to diagnose and analyze the size and density of patients' lesions caused by COVID-19. The algorithm can compare several CT images of lesions in the lungs. Thus, the system data set can expand and create a broad data set by using the frame from various users or patients. In addition, such a process would provide the transition of learning to the latest smartphones from multiple smartphones and various on-board sensors. The rear camera sensor of smartphone is used to measure body temperature based on putting the user's index finger on it, so that our built algorithm captures the temperature of the body instantly. Finally, our developed function of our smartphone app captures the cough sound samples and each time a patient uses this function it stores the cough's voice sample three times [14]. The voice is captured through the microphone of the smartphone, and its voice signal readings are stored in a single file. The introduction of smartphones has made possible many of the e-health initiatives. The embedded software and hardware has enabled the emergence of numerous healthcare monitoring systems based on smartphones. In particular, the COVID-19 self-testing breathing software will be particularly useful for those with elevated risks of serious illness (e.g., older adults or patients with significant underlying medical conditions) as well as for high-risk patient clinical cohorts, where fast, regular daily self-testing can be done [15].

10.6 RAPID DETECTION OF COVID-19 USING FET AND METAL–OXIDE–SEMICONDUCTOR FIELD-EFFECT TRANSISTOR (MOSFET) BIOSENSOR

Biosensing technologies such as real-time monitoring, ultrasensitive detection, low cost and the convenience of extreme device miniaturization due to convenience dedicated to nanoscale materials are among others distinguishable by their unique characteristics. Nanodevices open up the door to detection of small biomolecules and minute tests as they are extremely susceptible to surface load modulation, which allows the increased screening at the point of diagnosis of different life-threatening infectious diseases. For the integration of the field effect transistor (FET) channel, the semi-conducting carbon nanotubes (SC-CNTs) are particularly promising, replacing voluminous silicon technology beyond the short-channel dimensions by its ultrathin 1D structure, superior electronic characteristics and biocompatibility [16]. Thus the inhomogeneous interface of SC-CNTs with metal and drain electrodes affects the biosensor output of CNT-FET. The electrical transduction biodevices are the most widely used so far, owing to their simple calculation definition and their lower costs. For the detection of different biological targets in multiple bioapplications, biosensors focused on field effects (BioFETs). Typical thresholds of this kind are essential biomarkers for the diagnosis of clinical diseases including cardiac diseases, kidney failure, diabetes, cancers, breathing diseases and contagion. Additional potential applications include viruses or bacterial identification for diagnosing infectious diseases, such as AIDS and hepatitis B, and other essential bacterial analyses as metabolites [17]. These developments have identified field effect transistor (FET)-based biosensors for next-generation point-of-care testing (POCT) testing as a good candidate. Integration of charged molecules with nanoobjects and inorganic/organic nanohybrid biosensors with BioFETs with semicircles is certainly a high label-free and real-time biosensing possibility [18]. The most common biochemical interactions used in transistor-based field-effect biosensors are DNA probe-target oligo's hybridization, antibody-antigen binding and enzyme substratum complexing. The geometry of the COVID-20 FET sensor was designed with the graph channel combined with the SARS-CoV-2 spike antibody. The FET was coated with phosphate saline tampon. The COVID-19 FET sensor in which the SARS-CoV-2 spike antibody is combined with the graphic sheet is used as a sensor area. The sensor detected SARS-CoV-2 virus in clinical samples. In addition, there was no major cross-reactivity to the MERS-CoV antigen. The functional graphic-based sensor platform therefore makes the detection of SARS-CoV-2 virus easy, fast and highly sensitive in clinical samples. This technology may also be adapted to assist in the diagnosis of other emerging viral diseases [19].

10.7 ENGINEERED NANOMATERIAL BIOSENSORS

Engineered nanomaterial biosensors critically summarize the latest drug development strategies, including the use of new and recurrent drug potentials. A Sirnaomics Polymer Nano System STP702 (FluquitTM) is being studied in preclinical terms. Systems include siRNA that can target retained influenza regions to significantly respond to H1N1 (swine flu, H5N1 (avian flu)) and the recent antiviral H7N9. Heat stimulation hydrogels, known as the nanotraps, capture live viral cells, RNAs and proteins [20]. Antiviral activity has also been shown through a reduction in DNA fragmentation, chromatin condensation and caspase-3. Nanocarriers Transmission Electron Microscopy, cell viability testing and cytopathic effect, and further oseltamivir control medicine have been detected in cell Madin–Darby canine kidney (MDCK) lines for oseltamivir-modified silver nanoconstructions. Nanoparticle (NP)-based biosensors are major tools to identify people with pathogens. Sensory systems for other microorganisms have been successfully recorded in literature [21]. Unfortunately, due to their long incubation time and extremely high infectivity, such quick and responsive diagnostic systems for COVID-19 have little or no evidence. Conventional methods of detection have special sensibilities to demerit I and laborious processing techniques, (1) long time taken among sample collection and findings control, (2) high fake negative rates and (3) a poor diagnosis of patients with a different infection with a virus. Nucleic acid detection is the basis of the approaches used. In a unique tube reaction path, Zhu et al. documented a biosensor system based on COVID-19 NPs. This sensor uses an isothermal amplification combined of a single-stage reverse transcription, reverse transcription loop-mediated isothermal amplification (RT-LAMP and NP) coupled with nanoparticles-based biosensor (RT-LAMP-NBS). Furthermore, isothermal supplying equipment (63°C) is necessary for 40 minutes and the time to interpret the sample collection is only about 1 hour. For test patients with promising sensitivities of 12 copies (each target) for COVID-19, the biosensor was shown successfully in non-SARS-CoV-2 models [22]. The non-SARS-CoV-2 models showed this biosensor. Traditional RT-PCR-based

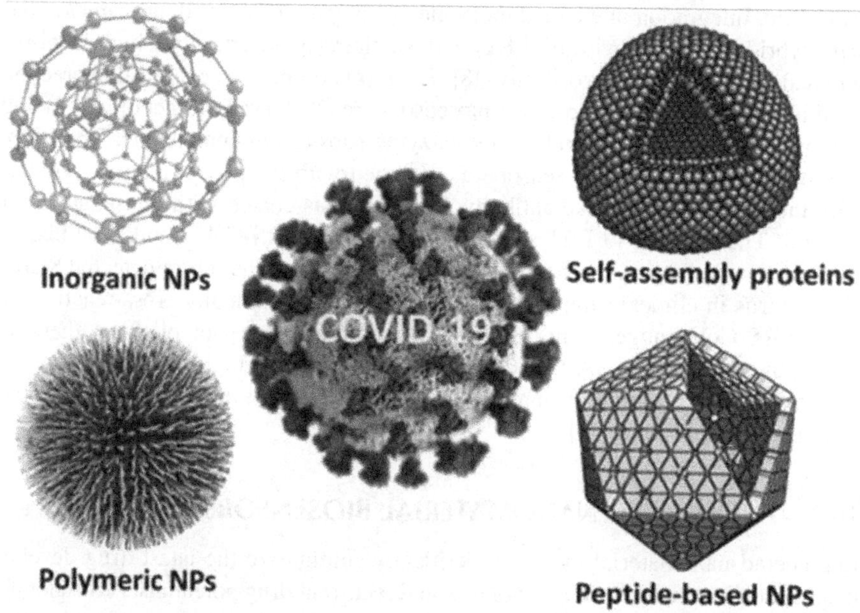

FIGURE 10.2 Applications of nanoscience to face COVID-19 [25].

sensor strategies are time-consuming and labor-intensive and not suitable if detected easily and successfully. A promising alternative to the procedure is the application of polyl(amino-ester) magnetic NP (pcMNP) by carboxyl groups (PCs). James et al. in a single phase of the combination of lysis with the binding point, the authors successfully developed and injected pcMNPs directly into the reaction RT-PCR. MNPs have been formerly detailed for effective use of sensory systems. This way is capable of purifying viral RNA from a large number of samples within 20 minutes and 10 minutes of copying sensitivity and has a strong linear correlation between SARS-CoV-2 and 105 PVP copies. The results are very positive since operational requirements are kept to a minimum in today's COVID-19 molecular diagnosis, which helps early diagnosed people [23]. While NP delivery approaches have shown significant potential applications, studies demonstrate that they can severely damage respiratory areas and even affect lung function [24].

10.8 BIOMEDICAL IMAGING AND SENSOR FUSION

Medical imaging such as X-ray and CT is supporting the global fight against COVID-19 while technology, which is evolving recently, further increases the power of imaging equipment and assisted medical experts. The improvement in AI image acquisition will make it much easier to scan and workflow with minimal interaction with patients will also be reworked and image technical experts will be best protected [26]. AI can also improve the quality of work by properly defining pathogens in X-ray and CT images and therefore promote further quantification. Computer-supported platforms also help radiologists take clinical decisions such as diagnosis, follow-up and the prognosis of diseases [27]. This research paper covers the whole COVID-19 process, including the collection of image, segmentation, identification and monitoring of medical imaging and diagnostic strategies. Physicians are greatly assisted in clinical practice by readily available imaging devices like chest X-rays and thoracic CT [28]. Added major parameters like ISO centering may be deduced by AI. In order to match the focus region with ISO centers and to improve overall image consistency, ISO centering refers to the alignment of the target body area of the subject [29]. Studies show that radiation dosage can be reduced while preserving comparable image quality with better ISO centralization. Radiographic images are generally considered to be less reliable than 3D chest-CT images, but they are standard first-line imaging in patients with COVID-19. Shanghai Unified Imaging Intelligence researchers have been working on developing patient-infection regions through machine learning and visualizing approaches in order to demonstrate changes in height, density and other clinically related variables [30]. AI was tested and tested for the entire COVID-19 diagnostic imaging system. In order to allow effective use of the correct amount of radiation, particularly for low-dose imaging, the X-ray exposure parameters can be automatically measured and adjusted using the patient thickness of the body area as defined by the AI [31]. Deep learning is the prevailing approach in combating COVID-19. Incomplete, improper and unreliable labels may, however, be the pictorial data of COVID-19 applications, posing challenges in the education of efficient segmentation and diagnostics networks [32]. Standard image features that can be useful to screen highly suspected cases and assess how severe and extinct their disease is in patients reported with COVID-19 pneumonia. For the diagnosis and treatment of COVID-19 patients the standard CT imaging tool is considered. In particular, in patients with an initial negative RT-PCR screening test [33], lung abnormalities can quickly be detected for suspect diseases. Typical picture features were included as ground-glass opacity (GGO) (86.1%), mixed GGOs (64.4%) and convergence and reticulations (48.5%). These results are similar to other pulmonary infections, such as SARS and MERS, with CT-like effects. Ironically speaking, we found that most patients had an acute inflammation reaction caused by a vascular enlargement of the lesion (71.3%) [34]. However, the change in the vascular system does not resemble changes caused in malignant lases such as pulmonary adenocarcinoma that have distorted or irregular vascular dilation and convergence by chronic progression and tumor infiltration. Concerning lesion distribution, COVID-19 is far more likely to have peripheral (87%), biological (82.2%, weak lung (54.5%) and a multifocal (54.5%) distribution than previous studies. In terms of lesion distribution in the emergency group, the patients were older than in the nonemergency group. In both classes the overall occurrence of disease was not significantly different, suggesting that other factors (e.g., viral load) would primarily reflect the incidence and severity of pneumonia COVID-19 [35]. Architectural distortion and bronchiectasis of traction and pleural effusion, which can be the viral and virulent charges of COVID-19, were statistically distinct from one group of two. In the intrathoracic lymph node, which is an unusual characteristic that significantly changes our cohorts (only a patient), further confirmation is necessary. In addition, diffused lesions in emergency patients have been shown to be higher than in the nonemergency group (78.6% vs. 24.1%). Not surprisingly, the score in the emergency category was even higher at the time than in the emergency group [36].

10.9 CONCLUSION

After the first outbreak of Coronavirus (COVID-19) in Wuhan, China, the disease has spread across the world. The biggest concern is for the people with high ages and individuals with immunological issues. The scientists will recognize disease, function and education. Health professionals should also be aware of the necessary precautions to prevent contractions and the spread of disease. Comprehensive steps should be taken to minimize transmission among individuals via COVID-19 in order to control the current outbreak. Careful consideration and steps to avoid or remove infection should be taken by vulnerable populations, including infants, healthcare providers and the elderly. Routine cleaners should be provided with decontaminating reactivity by public services and facility. In the handling of the virus, physical contact with wet and infected substances, in particular with agents like fecal and urine samples, that can serve as an alternative transmission route.

REFERENCES

1. Rothan, Hussin A., and Siddappa N. Byrareddy. "The epidemiology and pathogenesis of coronavirus disease (COVID-19) outbreak." *Journal of Autoimmunity* 109 (2020): 102433.

2. Cascella, Marco, Michael Rajnik, Arturo Cuomo, Scott C. Dulebohn, and Raffaela Di Napoli. "Features, evaluation and treatment coronavirus (COVID-19)." In Statpearls [internet]. StatPearls Publishing, 2020.
3. Xu, Zhe, Lei Shi, Yijin Wang, Jiyuan Zhang, Lei Huang, Chao Zhang, Shuhong Liu et al. "Pathological findings of COVID-19 associated with acute respiratory distress syndrome." *The Lancet Respiratory Medicine* 8, no. 4 (2020): 420–422.
4. Maghdid, Halgurd S., Kayhan Zrar Ghafoor, Ali Safaa Sadiq, Kevin Curran, and Khaled Rabie. "A novel ai-enabled framework to diagnose coronavirus covid 19 using smartphone embedded sensors: Design study." arXiv preprint arXiv:2003.07434 (2020).
5. Covid CDC., and Response Team. "Severe outcomes among patients with coronavirus disease 2019 (COVID-19)—United States, February 12–March 16, 2020." *Morbidity and Mortality Weekly Report* 69, no. 12 (2020): 343–346.
6. Mehta, Puja, Daniel F. McAuley, Michael Brown, Emilie Sanchez, Rachel S. Tattersall, Jessica J. Manson, and HLH Across Speciality Collaboration. "COVID-19: Consider cytokine storm syndromes and immunosuppression." *Lancet (London, England)* 395, no. 10229 (2020): 1033.
7. World Health Organization. "Coronavirus disease 2019 (COVID-19): Situation report, 72." (2020).
8. Zheng, Ying-Ying, Yi-Tong Ma, Jin-Ying Zhang, and Xiang Xie. "COVID-19 and the cardiovascular system." *Nature Reviews Cardiology* 17, no. 5 (2020): 259–260.
9. Remuzzi, Andrea, and Giuseppe Remuzzi. "COVID-19 and Italy: What next?" *The Lancet* 395, no. 10231 (2020): 1225–1228.
10. Cascella, Marco, Michael Rajnik, Arturo Cuomo, Scott C. Dulebohn, and Raffaela Di Napoli. "Features, evaluation and treatment coronavirus (COVID-19)." In Statpearls [internet]. StatPearls Publishing, 2020.
11. Dong, Ensheng, Hongru Du, and Lauren Gardner. "An interactive web-based dashboard to track COVID-19 in real time." *The Lancet Infectious Diseases* 20, no. 5 (2020): 533–534.
12. Velavan, Thirumalaisamy P., and Christian G. Meyer. "The COVID-19 epidemic." *Tropical Medicine & International Health* 25, no. 3 (2020): 278.
13. Rothan, Hussin A., and Siddappa N. Byrareddy. "The epidemiology and pathogenesis of coronavirus disease (COVID-19) outbreak." *Journal of Autoimmunity* 109 (2020): 102433.
14. Hollander, Judd E., and Brendan G. Carr. "Virtually perfect? Telemedicine for COVID-19." *New England Journal of Medicine* 382, no. 18 (2020): 1679–1681.
15. Wölfel, Roman, Victor M. Corman, Wolfgang Guggemos, Michael Seilmaier, Sabine Zange, Marcel A. Müller, Daniela Niemeyer et al. "Virological assessment of hospitalized patients with COVID-2019." *Nature* 581, no. 7809 (2020): 465–469.
16. Coronavirus, Novel. "Situation report, 22." World Health Organization. (2019). https://www.who.int/docs/default-source/coronaviruse/situation-reports/20200211-sitrep-22-ncov.pdf.
17. Dong, Liying, Shasha Hu, and Jianjun Gao. "Discovering drugs to treat coronavirus disease 2019 (COVID-19)." *Drug Discoveries & Therapeutics* 14, no. 1 (2020): 58–60.
18. Fang, Yicheng, Huangqi Zhang, Jicheng Xie, Minjie Lin, Lingjun Ying, Peipei Pang, and Wenbin Ji. "Sensitivity of chest CT for COVID-19: Comparison to RT-PCR." *Radiology* 296, no. 2: (2020): E115–E117.
19. Fang, Lei, George Karakiulakis, and Michael Roth. "Are patients with hypertension and diabetes mellitus at increased risk for COVID-19 infection?" *The Lancet Respiratory Medicine* 8, no. 4 (2020): e21.
20. Keesara, Sirina, Andrea Jonas, and Kevin Schulman. "Covid-19 and health care's digital revolution." *New England Journal of Medicine* 382, no. 23 (2020): e82.

21. Vaishya, Raju, Mohd Javaid, Ibrahim Haleem Khan, and Abid Haleem. "Artificial Intelligence (AI) applications for COVID-19 pandemic." *Diabetes & Metabolic Syndrome: Clinical Research & Reviews* 14, no. 4 (2020): 337–339.
22. Yang, Zifeng, Zhiqi Zeng, Ke Wang, Sook-San Wong, Wenhua Liang, Mark Zanin, Peng Liu et al. "Modified SEIR and AI prediction of the epidemics trend of COVID-19 in China under public health interventions." *Journal of Thoracic Disease* 12, no. 3 (2020): 165.
23. Ai, Tao, Zhenlu Yang, Hongyan Hou, Chenao Zhan, Chong Chen, Wenzhi Lv, Qian Tao, Ziyong Sun, and Liming Xia. "Correlation of chest CT and RT-PCR testing in coronavirus disease 2019 (COVID-19) in China: A report of 1014 cases." *Radiology* 296, no. 2 (2020): E32–E40.
24. Gozes, Ophir, Maayan Frid-Adar, Hayit Greenspan, Patrick D. Browning, Huangqi Zhang, Wenbin Ji, Adam Bernheim, and Eliot Siegel. "Rapid ai development cycle for the coronavirus (covid-19) pandemic: Initial results for automated detection & patient monitoring using deep learning ct image analysis." *arXiv preprint arXiv:2003.05037* (2020).
25. Santiago, Ibon. "Trends and innovations in biosensors for COVID-19 mass testing." *ChemBioChem* 21 (2020): 1–11.
26. Seo, Giwan, Geonhee Lee, Mi Jeong Kim, Seung-Hwa Baek, Minsuk Choi, Keun Bon Ku, Chang-Seop Lee et al. "Rapid detection of COVID-19 causative virus (SARS-CoV-2) in human nasopharyngeal swab specimens using field-effect transistor-based biosensor." *ACS Nano* 14, no. 4 (2020): 5135–5142.
27. Javaid, Mohd, Abid Haleem, Raju Vaishya, Shashi Bahl, Rajiv Suman, and Abhishek Vaish. "Industry 4.0 technologies and their applications in fighting COVID-19 pandemic." *Diabetes & Metabolic Syndrome: Clinical Research & Reviews* 14, no. 4 (2020): 419–422.
28. Zhu, Xiong, Xiaoxia Wang, Limei Han, Ting Chen, Licheng Wang, Huan Li, Sha Li et al. "Reverse transcription loop-mediated isothermal amplification combined with nanoparticles-based biosensor for diagnosis of COVID-19." MedRxiv (2020).
29. Russell, Steven M., Alejandra Alba-Patiño, Enrique Baron, Marcio Borges, Marta González-Freire, and Roberto de la Rica. "Biosensors for managing the COVID-19 cytokine storm: challenges ahead." *ACS Sensors* 5, no. 6 (2020): 1506–1513.
30. Chang, Hsiao Kang, Fumiaki Ishikawa, Po Chiang Chen, and Chongwu Zhou. "Label-free, electrical detection of the SARS virus n-protein with nanowire biosensors utilizing antibody mimics as capture probes." In Nanotechnology 2010: Bio Sensors, Instruments, Medical, Environment and Energy-Technical Proceedings of the 2010 NSTI Nanotechnology Conference and Expo, NSTI-Nanotech 3 (2010).
31. Singh, Ravi Pratap, Mohd Javaid, Abid Haleem, and Rajiv Suman. "Internet of things (IoT) applications to fight against COVID-19 pandemic." *Diabetes & Metabolic Syndrome: Clinical Research & Reviews* 14, no. 4 (2020): 521–524.
32. Peeri, Noah C., Nistha Shrestha, Md Siddikur Rahman, Rafdzah Zaki, Zhengqi Tan, Saana Bibi, Mahdi Baghbanzadeh, Nasrin Aghamohammadi, Wenyi Zhang, and Ubydul Haque. "The SARS, MERS and novel coronavirus (COVID-19) epidemics, the newest and biggest global health threats: What lessons have we learned?" *International Journal of Epidemiology* 49, no. 3 (2020): 717–726.
33. Mohammed, M. N., Halim Syamsudin, S. Al-Zubaidi, A.K. Sairah, and Eddy Yusuf. "Novel COVID-19 detection and diagnosis system using IOT based smart helmet." *International Journal of Psychosocial Rehabilitation* 24, no. 7 (2020): 2296–2303.
34. Ting, Daniel Shu Wei, Lawrence Carin, Victor Dzau, and Tien Y. Wong. "Digital technology and COVID-19." *Nature Medicine* 26, no. 4 (2020): 459–461.

35. Sivasankarapillai, Vishnu Sankar, Akhilash M. Pillai, Abbas Rahdar, Anumol P. Sobha, Sabya Sachi Das, Athanasios C. Mitropoulos, Mahboobeh Heidari Mokarrar, and George Z. Kyzas. "On facing the SARS-CoV-2 (COVID-19) with combination of nanomaterials and medicine: Possible strategies and first challenges." *Nanomaterials* 10, no. 5 (2020): 852.
36. Zhao, Wei, Zheng Zhong, Xingzhi Xie, Qizhi Yu, and Jun Liu. "Relation between chest CT findings and clinical conditions of coronavirus disease (COVID-19) pneumonia: A multicenter study." *American Journal of Roentgenology* 214, no. 5 (2020): 1072–1077.

11 Coronavirus Statistics
With Special Reference to Tamil Nadu

K. Vijayakumar, A. Dinesh Karthik, and A. Sivasankari
Shanmuga Industries Arts and Science College

CONTENTS

11.1 INTRODUCTION

A corona virus, assigned as 2019-nCoV, developed in Wuhan, China, toward the finish of 2019. As on January 24, 2020, in any event 830 cases had been analyzed in nine nations: China, Thailand, Japan, South Korea, Singapore, Vietnam, Taiwan, Nepal, and the United States. Twenty-six fatalities had happened, for the most part in patients who had genuine fundamental sickness [1]. Albeit numerous subtleties of the development of this infection, for example, its cause and its capacity to spread among people stay obscure, an expanding number of cases seem to have come about because of human-to-human transmission [2]. Given the extreme intense respiratory condition corona virus (SARS-CoV) [3] episode in 2002 and the Middle East respiratory disorder corona virus (MERS-CoV) flare-up in 2012, 2019-nCoV is the third crown infection to develop in the human populace in the previous two decades; a rise that has put worldwide general well-being establishments on high caution [4].

Crown infections make up a huge group of infections that can taint fowls and warm blooded creatures, including people, as indicated by world well-being association (World Health Organization) [5]. These infections have been answerable for a few flare-ups far and wide, including the extreme intense respiratory condition (SARS) [6] pandemic of 2002–2003 and the Middle East respiratory disorder (MERS) flare-up in South Korea in 2015. Most as of late, a novel crown infection (SARS-CoV-2, otherwise called COVID-19) [7] set off an episode in China in December 2019, starting as a global concern [8]. While some crown infections have caused destroying scourges, others cause gentle to direct respiratory contaminations, similar to the regular virus [9]. Under these conditions, it is important to make a precise report about the current circumstance on the corona infection cases in Tamil Nadu, Therefore in this examination endeavor has been made to recognize the corona infection cases in Tamil Nadu up to June 23, 2020.

11.2 OBJECTIVES OF THE STUDY

The following are the objectives of this study:

- To find out the corona virus impact and spread in the Tamil Nadu
- To find out the corona virus statistics up to June 23, 2020
- To identify the corona virus affected, recovered, and death rates in Tamil Nadu

11.3 METHODOLOGY

The corona virus data have been extracted from Health and Family Welfare Department, Government of Tamil Nadu website, during the last week of June 21, 2020. The study considered only the corona virus affected, recovered, and death in Tamil Nadu; data have been analyzed and interpreted to test and to fulfill the stated statistics. For this purpose Statistical Package for the Social Sciences and Microsoft Excel have been used for simple calculation.

11.4 DATA ANALYSIS

The analysis and interpretation of data extracted from the Health and Family Welfare Department of Tamil Nadu are presented. The data have been analyzed about the corona viruses affected, recovered and deaths in Tamil Nadu; further it classified 37 districts into eight zones and the data were presented in table and graph formats.

11.4.1 Corona Virus Affected, Recovered and Death in World, India, Tamil Nadu

The number of affected, recovered and deaths of COVID-19 in world, India, and Tamil Nadu outbreaks according to www.worldometers.info last updated: June 25, 2020, 09:33 GMT [10] and www.mygov.in last updated: June 25, 2020, 08:00 GMT as shown in Table 11.1 and Figure 11.1.

11.4.2 Corona Virus Affected, Recovered, and Death in States of India

The number of affected, recovered, and deaths of COVID-19 in India; state-wise outbreaks according to www.mygov.in COVID India as on: 25 June 2020, 08:00 IST (GMT+5:30) [11] as shown in Table 11.2.

According to the data released by the Ministry of Health and Family Welfare, it brings the total reported cases of corona virus in Tamil Nadu to 67,468. Among the total people infected as on date, 37763 (56%) have been recovered and 866 (1.28%)

TABLE 11.1
Corona Virus Affected, Recovered, and Deaths in World, India, Tamil Nadu

S.No.	Description	Affected	Recovered	Deaths
1	World	9,551,026	5,191,344	485,419
2	India	474,272	271,934	14,914
3	Tamil Nadu	67,468	37,763	866

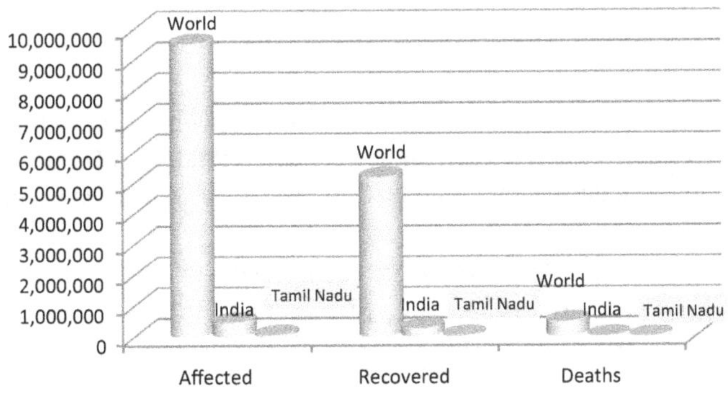

FIGURE 11.1 Affected, recovered, and deaths in world, India, and Tamil Nadu.

TABLE 11.2
Corona Virus Affected, Recovered, and Death in States of India

S.No.	State	Affected	Recovered	Deaths
1	Andhra Pradesh	10331	4779	124
2	Arunachal Pradesh	158	38	0
3	Assam	6198	3958	9
4	Bihar	8209	6113	57
5	Chandigarh	420	323	6
6	Chhattisgarh	2419	1627	12
7	Delhi	70390	41437	2365
8	Goa	951	289	2
9	Gujarat	28943	21088	1735
10	Haryana	12010	6925	188
11	Himachal Pradesh	806	466	8
12	Jammu and Kashmir	6422	3818	88
13	Jharkhand	2207	1570	11
14	Karnataka	10118	6151	164
15	Kerala	3603	1888	22
16	Latah	941	274	1
17	Madhya Pradesh	12448	9473	534
18	Maharashtra	142900	73792	6739
19	Manipur	970	328	0
20	Meghalaya	46	42	1
21	Mizoram	142	19	0
22	Nagaland	347	148	0
23	Odessa	5752	4123	17
24	Punjab	4627	3099	113
25	Rajasthan	16009	13611	375
26	Sikkim	84	39	0
27	Tamil Nadu	67468	37763	866
28	Telangana	10444	4361	225
29	Tripura	1259	897	1
30	Uttar Pradesh	19557	12586	596
31	Uttarakhand	2623	1721	35
32	West Bengal	15173	9702	591

have been passed away. Maharashtra is the highest corona virus affected state in India, followed by Delhi, Tamil Nadu, Gujarat, and Uttar Pradesh. Meghalaya (46) and Sikkim (84) are the states that are less affected by corona virus which is below 100.

11.4.3 Corona Virus Affected, Recovered, and Deaths in Districts of Tamil Nadu

The number of affected, recovered, and deaths of COVID-19 in Tamil Nadu; districtwise outbreaks according to www.covidindia.org as on: 22 June 2020, 08:42 pm IST (GMT+5:30) as shown in Table 11.3.

TABLE 11.3
Corona Virus Affected, Recovered, and Death in Districts of Tamil Nadu

S.No.	Districts	Affected	Recovered	Deaths
1	Ariyalur	432	387	0
2	Chengalpattu	3872	1968	53
3	Chennai	42752	23756	623
4	Coimbatore	280	164	1
5	Cuddalore	823	492	3
6	Dharmapuri	32	17	0
7	Dindigul	312	213	4
8	Erode	83	72	1
9	Kallakuruchi	395	301	1
10	Kancheepuram	1215	600	12
11	Kanyakumari	178	99	1
12	Karur	119	86	0
13	Krishnagiri	72	32	2
14	Madurai	849	389	8
15	Nagapattinam	219	67	0
16	Namakkal	89	85	1
17	Nilgris	31	14	0
18	Perambalur	151	144	0
19	Pudukottai	86	36	1
20	Ramanathapuram	317	131	2
21	Ranipet	525	391	2
22	Salem	352	204	0
23	Sivaganga	13	53	1
24	Tenkasi	261	108	0
25	Thanjavur	308	130	1
26	Theni	236	129	2
27	Thiruvallur	2645	1427	42
28	Thiruvarur	231	105	0
29	Thoothukudi	636	391	4
30	Tiruchirapalli	310	162	1
31	Tirunelveli	644	428	4
32	Tirupathur	83	40	0
33	Tiruppur	122	116	0
34	Tiruvannamalai	1199	465	7
35	Vellore	491	111	3
36	Villupuram	606	391	12
37	Virudhunagar	208	139	1

11.4.4 Zonewise in Tamil Nadu

Tamil Nadu government has announced an extension of the lockdown to combat the spread of COVID-19 until June 30 in containment zones [12]. The fifth phase of the lockdown being termed as Unlock 5.0 will see resumption of certain activities; the government has classified 37 districts into eight zones and the same has been shown in Table 11.4.

TABLE 11.4
Zones in Tamil Nadu

S.No	Zone	Districts
1	Zone 1	Coimbatore, Nilgiris, Tirupur, Erode, Karur, Salem, and Namakkal
2	Zone 2	Dharmapuri, Vellore, Tirupathur, Ranipet, and Krishnagiri
3	Zone 3	Villupuram, Tiruvannamalai, Cuddalore, and Kallakurichi
4	Zone 4	Nagapattinam, Tiruvarur, Thanjavur, Trichy, Ariyalur, Perambalur, and Pudukottai
5	Zone 5	Dindigul, Madurai, Theni, Virudhunagar, Sivagangai, and Ramanathapuram
6	Zone 6	Tuticorin, Tirunelveli, Kanyakumari, and Tenkasi
7	Zone 7	Kancheepuram, Tiruvallur and Chengalpet
8	Zone 8	Greater Chennai

TABLE 11.5
Affected, Recovereds and Deaths in Zone 1

		Zone 1		
S.No.	District	Affected	Recovered	Deaths
1	Coimbatore	280	164	1
2	Erode	83	72	1
3	Karur	119	86	0
4	Namakkal	89	85	1
5	Nilgris	31	14	0
6	Salem	352	204	0
7	Tiruppur	122	116	0

It can be seen from Table 11.5 that the districts such as Coimbatore, Erode, Karur, Namakkal, Nilgris, Salem, and Tiruppur are announced as zone 1 by the government according to Ministry of Health and Family Welfare data released and the same has been shown in Table 11.5.

11.4.5 Affected, Recovered, and Deaths in Zone 1

The data have been analyzed to identify the corona virus affected, recovered, and deaths in zone 1 and the same has been shown in Table 11.5.

The above table reveals that Salem district accounted for high number of affected cases (352) followed by Coimbatore (280) and Tiruppur (122), whereas in the recovered status 204 (57%) recovered in Salem district followed by Coimbatore 164 (58%) and Tiruppur116 (95%). In Coimbatore, Erode, and Namakkal districts only one death was cross-notified.

11.4.6 Affected, Recovered, and Deaths in Zone 2

The study has further extended to five districts in Zone 2 and the same has been shown in Table 11.6.

TABLE 11.6
Affected, Recovered, and Deaths in Zone 2

Zone 2				
S.No.	District	Affected	Recovered	Deaths
1	Dharmapuri	32	17	0
2	Krishnagiri	72	32	2
3	Ranipet	525	391	2
4	Tirupathur	83	40	0
5	Vellore	491	111	3

TABLE 11.7
Affected, Recovered, and Deaths in Zone 3

Zone 3				
S.No.	District	Affected	Recovered	Deaths
1	Cuddalore	823	492	3
2	Kallakuruchi	395	301	1
3	Tiruvannamalai	1199	465	7
4	Villupuram	606	391	12

Districtwise break-up is available for zone 2 of the total cases reported in the district. Ranipet had the highest number of COVID-19 cases at 525 confirmed infections, followed by Vellore (491 cases) and Tirupathur (83 cases).

11.4.7 Affected, Recovered, and Deaths in Zone 3

In zone 3, four districts such as Cuddalore, Kallakurichi, Tiruvannamalai, and Villupuram clearly explain affected, recovered, and deaths as shown in Table 11.7.

It is seen from the above table that Tiruvannamalai district reported highest in zone 3: 1199 was affected; of 1199 cases, 465 was recovered and only 7 (5.83%) deaths were found in that district; it is then followed by Cuddalore: 823 were affected, 492 were recovered, and 3 (3.64%) deaths in Cuddalore.

11.4.8 Affected, Recovered, and Deaths in Zone 4

The data have been analyzed to identify the corona virus affected, recovered, and deaths in zone 4 and the same has been shown in Table 11.8.

The number of active COVID cases has seen 432 affected cases rise in Ariyalur, while the district had reported that only 387 were recovered as on date; however, the number of deaths is not seen in Ariyalur, Nagapattinam, Perambalur, Thiruvarur whereas other districts such as Pudukottai, Thanjavur, Tiruchirapalli were identified with one death in the respective districts. Nearly 90% of the cases were recovered

TABLE 11.8
Affected, Recovered, and Deaths in Zone 4

		Zone 4		
S.No.	District	Affected	Recovered	Deaths
1	Ariyalur	432	387	0
2	Nagapattinam	219	67	0
3	Perambalur	151	144	0
4	Pudukottai	86	36	1
5	Thanjavur	308	130	1
6	Thiruvarur	231	105	0
7	Tiruchirapalli	310	162	1

TABLE 11.9
Affected, Recovered, and Deaths in Zone 5

		Zone 5		
S.No.	District	Affected	Recovered	Deaths
1	Dindigul	312	213	4
2	Madurai	849	389	8
3	Ramanathapuram	317	131	2
4	Sivaganga	13	53	1
5	Theni	236	129	2
6	Virudhunagar	208	139	1

from Ariyalur district followed by Tiruchirapalli (53%), Perambalur (95%), and Thanjavur (42%), respectively.

11.4.9 Affected, Recovered, and Deaths in Zone 5

The districtwise classification of data affected, recovered, and deaths was presented in the table and the same has been shown in Table 11.9.

According to the Health Department Data, Madurai district had reported 849 cases affected in total, of which 389 has been recovered and discharged. However, the number of recovered cases significantly rises in other district also such as Dindigul (213), Virudhunagar (139), Ramanathapuram (131), and Theni (129). However, the number of deaths is seen at eight in Madurai district, followed by four in Dindigul districts, two each in Ramanathapuram and Theni districts; finally one death is registered each in Sivaganga and Virudhunagar, respectively.

11.4.10 Affected, Recovered, and Deaths in Zone 6

The data availability of affected, recovered, and deaths among zone six districts is also measured and the same has been shown in Table 11.10.

TABLE 11.10
Affected, Recovered, and Deaths in Zone 6

		Zone 6		
S.No.	District	Affected	Recovered	Deaths
1	Kanyakumari	178	99	1
2	Tenkasi	261	108	0
3	Thoothukudi	636	391	4
4	Tirunelveli	644	428	4

TABLE 11.11
Affected, Recovered, and Deaths in Zone 7

		Zone 7		
S.No.	District	Affected	Recovered	Deaths
1	Chengalpattu	3872	1968	53
2	Kancheepuram	1215	600	12
3	Thiruvallur	2645	1427	42

The above table reveals that Tirunelveli district records high number (644) of affected cases followed by Thoothukudi (636), Tenkasi (261), and Kanyakumari (178) whereas in the recovered status 428 (67%) recovered in Tirunelveli district followed by Thoothukudi 391 (62%), Tenkasi 108 (41%), and Kanyakumari 99 (56%) districts. In Thoothukudi and Tirunelveli districts only four deaths were cross-notified.

11.4.11 Affected, Recovered, and Deaths in Zone 7

The districtwise classification of data for affected, recovered, and deaths in zone seven was presented in the table and the same has been shown in Table 11.11.

In the above table, Chengalpattu district topped zone 7 with 3872 cases affected followed by Thiruvallur (2645) and Kancheepuram (1215) districts. The districts also reported 53 (1.36%) deaths in Chengalpattu followed by Thiruvallur 42 (1.58%) and Kancheepuram 12 (9.8%) districts.

11.4.12 Affected, Recovered, and Deaths in Zone 8

The highest affected zone in Tamil Nadu is Chennai and same is shown in Table 11.12.

The above table reveals that Chennai reported 42,752 positive cases, bringing the city's huge total and 23,756 cases were recorded as recovered from viruses; deaths confirmed by the Health Department took the Chennai's death toll due to the virus to 623.

TABLE 11.12
Affected, Recovered, and Deaths in Zone 8

		Zone 8		
S.No.	**District**	**Affected**	**Recovered**	**Deaths**
1	Chennai	42752	23756	623

11.5 AGE AND GENDER DISTRIBUTION

The age and genderwise distribution of corona virus affected cases is presented in three categories of age group such as 0–12, 13–60, and 60 above, and the same has been shown in Table 11.13.

It is observed that from the above age and gender distributions table, 1711 male cases come under the age group between 0 and 12 whereas 34,966 male cases fall under 13–60 age group; however, 5001 cases were found in the 60+ age group. When compared with male, nearly 60% female (21,138) fall between the 13 and 60 age group and 93% (1606) of the female come under 0–13 age group. 61% (3026) of the female fall under 60+ age group. It was fascinating that 20 third gender also got affected by corona virus in the age group between 13 and 60 and the same is shown in Figure 11.2.

TABLE 11.13
Age and Gender Distribution

		Age		
S.No	**Gender**	**0–12**	**13–60**	**60 and above**
1	Male	1711	34966	5001
2	Female	1606	21138	3026
3	Third Gender	0	20	0
	Total	3317	56124	8027

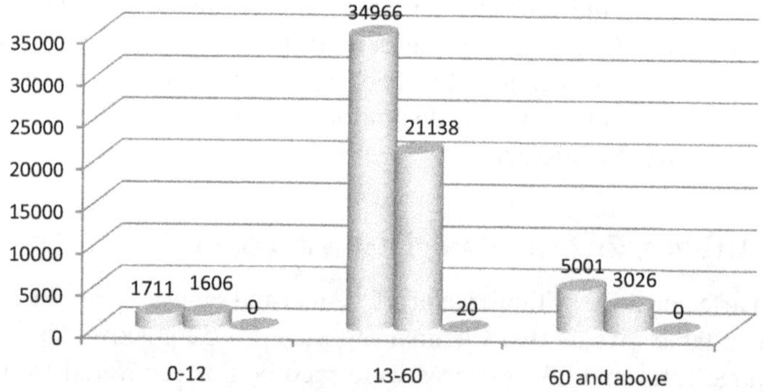

FIGURE 11.2 Age and gender distribution.

11.6 SYMPTOMS OF CORONA VIRUS

National Institutes of Health (NIH) proposes that few gatherings of individuals have the most noteworthy danger of creating complexities due to COVID-19.

These gatherings include as follows:

1. Little youngsters
2. Individuals matured 65 years or more seasoned
3. Ladies who are pregnant

11.6.1 General Preventive Measures from Corona Viruses

People over 65 years old, people with comorbidities, and pregnant ladies are encouraged to remain at home, aside from fundamental and well-being purposes. The conventional preventive measures incorporate basic general well-being estimates that are to be followed to lessen the danger of disease with COVID-19 [13]. These measures should be seen by all consistently. These include as follows:

1. Individuals must keep up a base separation of 6 feet out in the open spots similar to plausible.
2. Use of face covers/veils to be obligatory. Practice visit handwashing with cleanser (for in any event 40–60 seconds) in any event, when hands are not noticeably grimy. Utilization of liquor-based hand sanitizers (for at any rate 20 seconds) can be made plausible at any place.
3. Respiratory behaviors to be carefully followed. This includes severe act of covering one's mouth and nose while hacking/sniffling with a tissue/hanky/flexed elbow and arranging off utilized tissues appropriately.
4. Self-observing of well-being by all and detailing any ailment at the soonest to the quick administrative official.
5. Spitting will be carefully precluded.
6. Installation and utilization of Aarogya Setu App by workers.

11.6.2 Specific Preventive Measures for Workplaces

1. Entrance to have obligatory hand cleanliness (sanitizer distributor) and warm screening arrangements.
2. Only asymptomatic staff/guests will be permitted.
3. Any official and staff living in control zone ought to illuminate the equivalent to administrative official and not go to the workplace till regulation zone is denotified. Such staff ought to be allowed to telecommute and it will not be considered as leave period.
4. Drivers should keep up social distancing and will follow the required do's and don'ts identified with COVID-19. It will be guaranteed by the specialist co-ops/officials/staff that drivers dwelling in control zones will not be permitted to drive vehicles.

5. There should be arrangement for cleansing of the inside of the vehicle utilizing 1% sodium hypochlorite arrangement/shower. A legitimate purification of controlling, entryway handles, keys, and so on ought to be taken up.
6. Advise all workers who are at higher hazard, for example, more seasoned representatives, pregnant workers, and representatives who have fundamental ailments, to play it safe. They ought to ideally not be presented to any cutting edge work requiring direct contact with general society. Offices and the board should encourage telecommute at any place achievable.
7. All officials and staff/guests to be permitted partly just if utilizing face spread/covers. The face spread/cover must be worn consistently inside the office premises.
8. Routine issue of guests/brief passes ought to be suspended and guests with appropriate consent of the official, who they need to meet, ought to be permitted in the wake of being appropriately screened.
9. Meetings, to the extent achievable, ought to be done through video conferencing.
10. Posters/standees/AV media on preventive measures about COVID-19 to be shown noticeably.
11. Staggering of available time, lunch hours/quick rests to be done, similar to plausible.
12. Proper groups: the executives in the parking areas and outside the premises properly following social separating standards are guaranteed.
13. Valet stopping, if accessible, will be operational with working staff wearing face covers/covers and gloves as fitting. An appropriate cleansing of controlling, entry-way handles, keys, and so on, of vehicles ought to be taken up.
14. Any shops, slows down, cafeteria, and so on, outside and inside the workplace premises will follow social separating standards consistently.
15. Specific markings might be made with adequate separation to deal with the line and guarantee social removing in the premises.
16. Preferably independent section and exit for officials, staff, and guests will be composed. Appropriate cleaning and continuous disinfection of the work environment, especially of the every now and again contacted surfaces must be guaranteed.
17. Ensure standard flexibly of hand sanitizers, cleanser, and running water in the washrooms.
18. Required precautionary measures while taking care of provisions, inventories, and products in the workplace will be guaranteed.
19. Seating game plan to be made so that satisfactory social separating is kept up.
20. Number of individuals in the lifts will be limited, appropriately keeping up social separating standards.
21. For cooling/ventilation, the rules of The Central Public Works Department (CPWD) will be followed which accentuates that the temperature setting of all cooling gadgets ought to be in the scope of 24–30°C, relative dampness ought to be in the scope of 40%–70%, admission of natural air ought to be

however much as could reasonably be expected and cross-ventilation ought to be sufficient.

22. Large social occasions keep on staying disallowed.
23. Effective and visit sanitation inside the premises will be kept up with specific spotlight on restrooms, drinking, and handwashing stations/zones.
24. Cleaning and normal sterilization (utilizing 1% sodium hypochlorite) of much of the time contacted surfaces (door handles, lift catches, hand rails, seats, washroom apparatuses, and so on) will be done in office premises and in like manner territories.
25. Proper removal of face covers/veils/gloves left over by guests and additionally workers will be guaranteed.

11.6.3 Preventive Measures in Cafeteria/Container/Feasting Lobbies

1. Adequate group and line the board to be guaranteed to guarantee social separating standards.
2. Staff/servers to wear veil and hand gloves and take other required prudent steps.
3. The guest plan to guarantee a separation of at any rate 1 meter between supporters to the extent plausible.
4. In the kitchen, the staff to follow social removing standards.

11.7 CONCLUSION

Government of Tamil Nadu has strengthened the surveillance and control measures against the disease, as per the national guidelines. The government has taken several steps to prevent from corona viruses; general public is advised to adhere to health advisories and travel advisories issued by the state Government. Public should follow the cough etiquette by covering the face using handkerchief/towel while sneezing/coughing. Avoid close contact with anyone showing symptoms of respiratory illness such as coughing and sneezing and do frequent handwashing with soap and water. The State Government has taken numerous steps to prevent the public; the public has to follow the government rules and regulation, so that definitely public will overcome the corona virus completely.

REFERENCES

1. Unhale S. S., Ansar Q. B., Sanap S., Thakhre S., Wadatkar S. (2020). A review on corona virus (Covid-19). *World Journal of Pharmaceutical and Life Sciences*, 6(4), 109–115.
2. Gralinski L., Menachery V. (2020) Return of the corona virus, 2019- nCoV. *Viruses*, 12(2), 135.
3. Yin Y., Wunderink R. G. MERS, SARS and other (2018) Corona viruses as causes of pneumonia. *Respirology*, 23(2), 130–137.
4. Luk H. K., Li X., Fung J., Lau S. K., Woo P. C.(2019) Molecular epidemiology, evolution and phylogeny of SARS corona virus. *Infection, Genetics and Evolution*, 71, 21–30.

5. Zaki A. M., van Boheemen S., Bestebroer T. M., Osterhaus A. D., Fouchier R. A. (2012) Isolation of a novel corona virus from a man with pneumonia in Saudi Arabia. *The New England Journal of Medicine*, 367, 1814–1820.
6. Zhao L., Jha B. K., Wu A., Elliott R., Ziebuhr J., Gorbalenya A. E., Silverman R. H., Weiss S. R. (2012) Antagonism of the interferon-induced OAS-RNase L pathway by murine corona virus ns2 protein is required for virus replication and liver pathology. *Cell Host & Microbe*, 11(6), 607–616.
7. Barcena M., Oostergetel G. T., Bartelink W., Faas F. G., Verkleij A., Rottier P. J., Koster A. J., Bos B. J. (2009) Cryo-electron tomography of mouse hepatitis virus: Insights into the structure of the corona virion. *Proceedings of the National Academy of Sciences of the United States of America*, 106(2), 582–587.
8. Woo P. C., Huang Y., Lau S. K., Yuen K. Y. (2010). Corona virus genomics and bioinformatics analysis. *Viruses*, 2, 1804–1820.
9. Peiris, J. S. M., Lai S. T., Poon L. et al. (2003) Corona virus as a possible cause of severe acute respiratory syndrome. *The Lancet*, 361(9366), 1319–1325.
10. Worldometer. https://www.worldometers.info/coronavirus/country/india/ (Accessed on 20 June 2020).
11. Covid19 statewise status. https://www.mygov.in/corona-data/covid19-statewise-status/ (Accessed on 20 June 2020).
12. Tamil Nadu COVID-19 Unlock 5.0: State divided into 8 zones; check what's allowed. https://scroll.in/announcements/963606/tamil-nadu-covid-19-unlock-5-0-state-divided-into-8-zones-check-whats-allowed (Accessed on 20 June 2020).
13. Health & family welfare department government of Tamil Nadu. https://stopcorona.tn.gov.in (Accessed on 20 June 2020).

12 The Impact of COVID-19 on Consumer

Digital Marketing of the New Normal in Indonesia

Seprianti Eka Putri
Business University of Bengkulu

CONTENTS

12.1 INTRODUCTION

Economic activity in various business sectors in the country has experienced a decline in sales due to the COVID-19 pandemic. Meanwhile, diseases such as pneumonia that are highly contagious have created opportunities for some of these businesses, such as health products and services, while other businesses including airlines, hotels, retail, food, and beverage industries have taken a decline in profits. Moreover, the COVID-19 outbreak has caused an inevitable surge in the use of digital technology due to the norms of social distance and national locking. People and organizations around the world must adapt to new ways of working and living which is an important issue. People and organizations around the world must be able to adapt to new ways of doing work and addressing important life problems and regarding the problems of COVID-19 against the business sector and the number of cases of COVID-19 in Indonesia.

12.2 THE IMPACT OF COVID-19 ON BUSINESS SECTORS

Today, the number of cases of COVID-19 in Indonesia is recorded to continue to grow, and there were 579 people who confirmed positive COVID-19 (Kompas.com). The workers are asked to do activities at home subsequently the tools needed to support Work From Home (WFH) are an application that allows someone to have meetings online (Mervosh et al., 2020). The interesting thing in the process of running WFH is the increase in virtual meetings by utilizing a lot of media, such as Zoom, Google Meeting, etc. This phenomenon makes the use of Zoom and other applications loved by workers. Almost all sectors, including the government, also use this media to support performance during the pandemic. Probably, employees will not want to return to the physical office after the pandemic has passed. Moreover, the marketer or manager can estimate that the impact of COVID-19 on the office may occur longer with a flexible working hours system, and also the virtual is expected to further change the image of traditional companies (Xu & Sundar, 2016). Businesses appear to be experimenting with decentralized decisions and/or new software to create a new digital work culture to be effectively productive if they work in a physical office and to make this transition and several other consequences that are also needed to get business continuity attention are often overlooked in discussion related to COVID-19's influence on consumers.

12.3 IMPACT OF BUSINESS ON CONSUMER

McKinsey & Company has surveyed in early April 2020 with 700 respondents in Indonesia. The survey results show 49% of respondents believe the Indonesian economy will recover in 2–3 months and grow at the same pace or even faster than before COVID-19. This consumer confidence contrasts sharply with what people feel in a more developed economy. Furthermore, surveys of consumer sentiment around the world have found only 11% of South Korean respondents, 24% of Germans, 37% of Americans, and only 4% of Japanese people expect their economy to recover within 3 months approximately. The optimism of Indonesia is extraordinary and they have assumed financial difficulties will recover well soon, even though the situation is getting worse. The majority of respondents said that they expected the crisis to hurt their ability to meet their needs and nearly two-thirds said they agreed or strongly agreed that their work felt less safe with the highest number of scores among Asian countries surveyed.

In India, South Korea, and China, only 35% said they felt more of their insecurities and 83% said they were very careful about spending their money. Nearly two-thirds of Indonesians also expect work restrictions and social interaction to drain large savings in consumer spending intentions in the category of choices, such as eating out of the house (70), staying in a hotel (83), clothing (61), and mat feet (65). On the other hand, consumer goods have increased, including food (+45), household equipment (+36), and home entertainment (+21). Further, the survey conducted by the Mobile Marketing Association (MMA) has shown that 79% of local business uses digital media as a means to reach consumers during the health crisis. The MMA survey has found that nearly 70% of customers have used new digital services during the pandemic, such as fitness applications, digital entertainment, work-from-home software, and online education services. Similar trends have also been found in several countries, such as Singapore, India, and Vietnam, obtained 95% of nearly 400 businesses have stated that digital platforms are priority over television, radio, billboards, and print. In the various forces of technological evolution, digital transformation, growing consumer demand, and changing demographics (Grewal et al., 2017), the dominant structure seems to be diminishing rapidly. The involvement of producers, consumers, and third parties in the retail function has increased, resulting in the traditional value given by retail institutions. Meanwhile, institutional (physical) retail must struggle in an intermediary position and this position will fade, especially as producers continue to reach end-consumers through integrated distribution. For example, the Adidas sports apparel manufacturer has planned to monitor 60% of the global retail brand in 2020, a huge improvement in the concept of the stores (Kell, 2016).

Apple as a well-known manufacturer of IT and cellular phones has reached the highest-selling retail business in the world (Eadicicco, 2016). Internet of Things (IoT) technology that allows producers to continue to build communication with consumers through the life cycle of their products. Retail producers have encouraged them to get involved such as brand management, comprehensive customer information about all brands, assessing consumer choice, and opportunities to obtain higher profits (Osegowitsch & Madhok, 2003; Teece, 2010). Since IoT technology functions as an interactive product platform, brands can engage with individual consumers dynamically through high value-added products (Ramaswamy & Ozcan, 2018). Moreover, digital transformation in the retail value chain has an impact on consumers. In particular, authority over the main point of interaction with consumers when preparing and realizing purchasing decisions tends to shift in many cases. This can be seen in physical retail stores and their digital equivalents (e.g., online stores, smartphone applications, IoT devices), which can serve as a key information and customer transaction points, such as Amazon not only as an online retailer and the largest retail platform and Amazon is famous as the widest product search engines (McGee, 2017). As important elements in digital marketing, value creation is the core of all activities carried out by companies to make a profit. When understanding the value creation of an online business that is done will be able to help well. Value creation can be created through products, marketing creativity, and distribution channels. Value creation in customers can occur at every stage of the consumer decision-making process (Sweeney & Soutar, 2001; Woodruff, 1997). The sources of value creation can be concurrent in customer interaction when it occurs at the prepurchase stage (introduction of needs, an effort to find information or alternative evaluation), the purchasing level (selection, order, payment), and also the level of postpurchase (consume, use, involvement, and service) (Lemon & Verhoef, 2016; Puccinelli et al., 2009).

12.4 E-COMMERCE

E-commerce companies have recognized that an attractive shopping experience for customers is very important in today's digital world. A big change has occurred in online shopping attitudes because customers use a lot of devices supported by the internet (Wagner et al., 2013). Today, corporations have been led to participate in environmental competition related to various formats of online channels, including e-commerce websites, and social media. Consumers can use smartphones for various things such as viewing the company website by using their computers to view e-commerce sites at other times. This connected customer experience includes every contact point (social media, website, application) that the customer decides to synergize with the company. This modern social web has advanced opportunities to access various viewpoints and customer thinking for firms to encourage novelty (Oinas-Kukkonen & Oinas-Kukkonen, 2013).

Channel interaction between customers and companies is a valuable point as a stimulating shopping experience that has been universally connected. As an example, Amazon customers who can send a product inquiry to other customers who have that product can comment on those questions. Likewise, consumers can communicate with vendors at the same time. Rogers views the diffusion of innovation as a social phenomenon, including a new product being spread to the dominant because of the characteristics of enthusiastic consumers, who called early adopters by promoting the brand new to their social environment and dispersion can be generated through network effects (Rogers et al., 1983). Initial research on diffusion theory found that consumers could be assorted into two groups: early and late adopters with certain characteristics such as renewed eagerness.

Then, a survey administered in 1998 when eBay imported online shopping revealed that 46% of early users often used e-commerce while only 8% of end-users had a web shopping experience. However, recent surveys on online shopping have shown a boost that can occur only because of the influence of connections while the COVID-19 outbreak can also be a generation for e-commerce growth. Mintel's research shows consumers in the Asia-Pacific have been active online through various things. Based on survey results in 13 Asia Pacific countries out of 12,500 consumers who have any digital devices and most are already active in many fields of digital involvement, and they appear to be open to brand involvement, especially when they have free time.

Furthermore, Mintel research in the 3 months to December 2019 on each device has shown that those who read news websites/blogs 93%; shop online 93%; access social networks 93%; use instant messages 89%; watch video streaming 87%; managing finances, site/application price comparison used 83%; photos/videos together 83%; listen to streaming 83%; and play games 82%. Likewise, three-quarters of respondents in the Asia-Pacific have opinions agree that they want to discover more broadly about everything than before (e.g., brands, social problems, etc.).

This shows that there is a need for reliable news and information sources, and brands can also play a role in offering not only human interaction but also authoritative expertise. Moreover, when consumers begin opportunities for cooperation in overcoming any crisis, they also expect more things to work within the community to find social solutions for the virus outbreak. The COVID-19 which has disorganized various business sectors in the country has resulted in a positive effect in the e-commerce field because this virus has discovered new habits in online purchasing habits for customers.

The results of a survey by the management consulting firm Redseer have shown that Indonesia's e-commerce progress path will continue to exist positively with an increase to the next year which is expected to reach a figure of 23 billion in 2019 to $35 billion this year. A report from economy SEA 2019 Google, Temasek, and Bain & Company even predicted that Indonesia's internet economy was well on track to cross the $130 billion mark by 2025. In 2019, the economic reports of the SEA,

Google, Bain & Company even forecast that conditions will have been on a good track of $ 130 billion in 2025 in the development of the internet economy in Indonesia.

It is important for further analysis of the growth of e-commerce in Indonesia because the number of consumers who shop online has shown an increase. In Indonesia, that new users for online shopping were launched in the first quarter of 2020. Further, involving the concept of multichannel retailing is represented as an integration of a cross-channel business model that retailers use to enhance the impression of the customer (Avery et al., 2013; Verhoef et al., 2015).

Multichannels are identified as an unintegrated way to approach customers but omnichannel requires coherent integration (Staflund & Kersmark, 2015). Cellular technology alteration, i.e., smartphones and other types of electronic devices, has enabled retailers to reach a wider range of customers through online and offline channels, in case their market share will be broad as well (Brynjolfsson et al., 2013).

Thus, the unlimited universal buying mode in this omnichannel has become a game-changer for the retail industry by changing the independent track in the digital world technology (Bell et al., 2014). A product and service offered through interactions with customers observed through this contact point can improve business performance through the creation of value-added and personal shopping impressions (Herhausen et al., 2015; Lewis et al., 2014). Omnichannel in the retailing business allows customers to deliver precise information such as product availability and price so that they can make a good purchasing decision (Cao, 2014).

12.5 AN INNOVATIVE ONLINE SHOPPING EXPERIENCE THROUGH MULTICHANNELS

This growing online shopping experience arises because consumer depends on social relations, opinion leaders, online recommendation engines, and technical capabilities that support purchasing decisions, repurchases, subsequent use of digital devices to make it easy in these purchasing activities (Cheung et al., 2015). Industry research has shown that mobile devices are used by consumers to shop either outside or vice versa. Based on industry reports that technology adoption occurs on tablets and smartphones (Siwicki, 2014). It is found that 80% of tablet buyers and 67% of smartphone buyers have used shopping devices at dwelling (Nielsen, 2014).

Portable devices have become a daily necessity in the lives of connected people and this represents a good opportunity for the service industry through the use of various platforms for communication and maintaining online relationships with an increasing number of connected customers. These important technological advances are needed in communication networks, in which the user and a computer system interact, in particular the use of input devices and software with designs that are headed to the Web, online publishing, and e-commerce. Customers who have access to extensive information storage from any location. Thus, there are different device restrictions such as smartphones, tablets, PCs, or other types of devices that cause the presentation of information to be adjusted again.

Innovations that point in the direction of online shopping trends are likely to lead better during the coronavirus crisis and the new normal era. While the COVID-19 pandemic undoubtedly put e-commerce strategies into sharper focus, which has led to approaches through consumer behavior that have continued to shift with more digital activity in the situation. In this case, consumer skepticism about making online purchases is a challenge in digital sales (Stouthuysen et al., 2018). Trust is determining the factor for predicting consumers is intended to buy or repurchase in e-commerce (Bilgihan, 2016).

Consumer characteristics tend to be credence, faithfulness, and brand advocacy, such as expressing a sense of emotion or friendliness (Aggarwal & McGill, 2012). Brand personalization on social media and reaction involvement can be a manager's attention because it can cause emotional closeness for them (Xu & Sundar, 2016).

Previous research said that if you want to reduce doubts when shopping online can be seen from the reviews or testimonials from those who have made online purchases (Kim, 2020). Afterward, digital technology gives rise to a new pattern for market behaviors, relationship or experiences (Lamberton & Stephen, 2016) and reshape customer relationships, internal processes, and value propositions (Westerman et al., 2011) or the economic value-creation process as a whole (Reinartz & Imschloß, 2017).

In this case digital technology has drastically changed the balance of power between companies and customers who get the power of information and choice. The role of digital transformation for many businesses has become a major driver in the business world and digital business leaders apply these rules to engage, grow, and compete. Subsequently, Indonesia is listed as an exited user of digital technology in the world for the average number of people who have spent 4 hours a day to access the internet on mobile devices twice at the average number of US (McKinsey & Company).

Further, businesses need to be observant in observing consumer behavior in the new normal era. After 3 months of staying at home with all its dynamics, where we have entered a new normal business landscape completely different from before the COVID-19 pandemic situation. Hereinafter, COVID-19, this pandemic has caused regions throughout Indonesia to have implemented Pembatasan Sosial Berskala Besar actions that cover public areas, schools, and the closure of places of worship, except for important needs. COVID-19 has changed the way we live our daily lives including consumer behavior.

Moreover, major changes have occurred in the transformation to face the new normal through the consumer shift crisis of COVID-19, which consists of the following:

Emphatic Society: COVID 19 which has caused lives, has created an empathetic new society with a high level of solidarity. Go Virtual: this pandemic causes consumers to avoid physical contact with humans who have shifted to virtual/digital for their daily activities. Stay at Home Lifestyle: Lifestyle living at home with working, living, and playing activities because there is social distancing. Back to Bottom of the Pyramid: this refers to the Maslow pyramid that consumers shift their needs from the top of the pyramid which is self-actualization/self-esteem to the basic pyramid as well as the need for food, drink, health, and safety. Thus, a business transformation is a holistic approach that adopts digital change and initiatives at several levels in various industries.

The digital transformation includes a large number of transaction interaction processes from technological evolution, change, internal and external factors, industry, stakeholders. When reading suggestions about digital transformation or reading reports and predictions, different perspectives such as challenges, goals, and common traits in organizations are seen throughout the world. Other differences include the industry, region, and organization. Furthermore, things that make sense in one temporary area do not have to make sense in another, even if we only study the regulatory environment.

This integrated digital transformation is needed to develop core capabilities in business (i-scoops.eu.digital transformation) consisted of (1) business activities including marketing, operations, human resources, administration, customer service, etc.; (2) business processes represent one or more operations, activities, and sets that are connected to achieve certain business objectives, business process management, business process optimization, and business automation appearing in the picture (with new technologies such as robotic process automation); optimal business processes are important in digital transformation strategies and most industries and cases are a mix of the goals facing customers and current internal goals; (3) business models: businesses that function from market entry approaches and value propositions make money and effectively change their core business, take advantage of new sources and revenue approaches, sometimes even leaving the traditional core business after a while; the business ecosystem is a network between partners and stakeholders, as well as contextual factors related to business through priorities and regulations and economic evolution. The new ecosystem that has been built within the company will certainly relate to the organizational structure that transforms digital information with the data intelligence that can be followed up into capital innovation; (4) business asset management: focus on traditional assets, but which are less tangible assets in which source information of customers (improving customer experience is the main purpose of digital and information transformation projects that is the source of business life, technological evolution, and any human relationship). Customer information is treated as tangible assets from a holistic perspective; (5) organizational culture: there must be clear customer-centric, agile, and hyperconscious goals that are achieved by gaining core competencies in all fields like leadership, skill, and so on that enable organizational development. The business process, collaboration, and IT side of digital transformation are related to organizational culture. A faster application of the market changes is needed and that is the essence of DevOps: development and operation. Furthermore, efforts to develop IT can be done by working together in business /processes/activities, changes (operational information and technology, culture), and others; (6) ecosystem models and partnerships, including the emergence of co-operative, collaborative, co-create, and, last but not lost, a completely new business ecosystem approach, which leads to new business models and sources of income and ecosystems will be one of the keys to the success of the digital transformation.

This country has attracted the attention of several business conglomerates and investors due to various factors including (1) increased internet and cellular penetration at 80% in which Indonesians prefer to use smartphones to access the internet; (2) the High e-commerce adoption rate at 90% of 152 million internet users in Indonesia have bought online before; (3) increased e-commerce: the Indonesian e-commerce market reaches a gross market value of USD 21.0 billion in 2019. Further, Indonesia is a target for omnichannel marketers and as a home for the largest viewers in the world. In January 2020, it was recorded that online penetration in the country had reached 60% and 170 million have been registered as social media users. For most Indonesians, social media is an easy way to contact family in remote locations in the islands, enabling them to connect with friends and also keep up with the daily news. The popular social networks in Indonesia at the adoption stage include YouTube at 88%, Facebook at 81%, WhatsApp at 83%, Instagram at 80%, and Line at 59% (Surveysensum, 2020). However, social media viewers from Indonesia are not only large but also they are also very active, including the average Indonesian who spends 2–3 hours on social media every day, Then it can be said that Indonesian is a heavy mobile internet user. Mintel's research reports that dominant consumers in the Asia-Pacific have actively on various occasions. 12,500 consumers who own any digital devices have been surveyed by Mintel in 13 Asia-Pacific countries and the majority of consumers who are active in digital engagement and these consumers are already open to brand involvement, especially when they have free time. McKinsey & Company notes that organizations must devise an effective marketing approach by encouraging scale revenue. An effective organization builds the SHAPE model with five core elements consisting of as follows:

1. Start-up mindset: This is indicated by organizational agility.
   Some companies that have adapted global health have successfully relocated their marketing budgets within days and launched e-commerce on a weekly basis.
2. Humans at its core:
   As the global workforce begins to shift to remote employment, companies also need to rethink their business operations model and find out how their people work best at home. This can be done by allowing employees with new skills and the right tools to work remotely.
3. Digital acceleration, technology, and analytics:
   Forward-thinking organizations understand that the transition to digital is a necessity at this time. Successful companies have improved and expanded their digital channels and they have also successfully used sophisticated analytics to combine new data sources for their consumer insights.
4. Purpose-driven customer handbooks:
   This COVID-19 pandemic has caused businesses to reassess consumers' purchasing decisions. Companies need to review their buyer data and new insights about customer experiences in the post-COVID-19 world.
5. Ecosystems that drive abilities:
   Disruptions in the supply chain and physical stores have forced organizations not only to adapt and survive but also to assess new opportunities that can be done through collaboration among organizational stakeholders.

12.6 CONCLUSION AND RECOMMENDATION

The papers in this issue provide a valuable contribution to understanding digital marketing that can help marketers and managers deal with new normal situations in Indonesia.

1. Development and innovation are very fast in every aspect, making business today launching the era of VUCA (Volatility, Uncertainty, Complexity, and Ambiguity) where the business world today is forcing companies to be creative and able to adapt to be competitive and successful in the business. In Indonesia businesses must be able to adapt to changes in consumer behavior and other changes, including in terms of marketing. Consumer needs through the use of digital marketing as a marketing strategy through online media are needed so those companies can reach consumers widely and effectively even amid a pandemic and new normal.
2. Digital marketing has become a tool and strategy in the industrial era with the lifestyles of people throughout the world who have experienced drastic changes in the era. The computer and internet industries have become part of all aspects of everyday human life, ranging from communicating, working, studying, shopping, entertainment, and others so that changing consumer behavior has led to changes in the marketing world as well. This digital revolution can be an opportunity and a necessity because customers have easy access to information by seeking greater value. The role of the company is required to increase customer satisfaction with technological capabilities to create a digital marketing strategy through innovation that will be aligned with their needs.
3. The implementation of digital marketing needs to pay attention to the digital transformation of business and organizational activities, processes, competencies by utilizing the changes and opportunities for the mix of digital technology and the impact of their acceleration on society in a strategic and prioritized manner with this new normal situation. Marketers or managers may want to utilize the latest technology to help consumers easier decisions when shopping online concerning whether they are offline or not by considering their omnichannel strategy.

REFERENCES

Aggarwal, P., & McGill, A. L. (2012). When brands seem human, do humans act like brands? Automatic behavioral priming effects of brand anthropomorphism. *Journal of Consumer Research 39*(2), 307–323.

Avery, J., Steenburgh, T. J., Deighton, J., & Caravella, M. J. M. I. R. (2013). Adding bricks to clicks: On the role of physical stores in a world of online shopping. *Marketing Intelligence Review 5*(2), 28–33.

Bell, D. R., Gallino, S., & Moreno, A. J. M. S. M. R. (2014). How to win in an omnichannel world. *MIT Sloan Management Review 56*(1), 45.

Bilgihan, A. J. C. i. H. B. (2016). Gen Y customer loyalty in online shopping: An integrated model of trust, user experience, and branding. *Computers in Human Behavior 61*, 103–113.

Brynjolfsson, E., Hu, Y. J., & Rahman, M. S. (2013). Competing in the age of omnichannel retailing. *MIT Sloan Management Review 54*(4), 23–29.

Cao, L. (2014). Business Model Transformation in Moving to a Cross-Channel Retail Strategy: A Case Study. International Journal of Electronic Commerce 18(4), 69–96.

Cheung, C. M., Liu, I. L., Lee, M. K. J. o. t. A. f. I. S., & Technology. (2015). How online social interactions influence customer information contribution behavior in online social shopping communities: A social learning theory perspective. *Journal of the Association for Information Science and Technology 66*(12), 2511–2521.

Eadicicco, L. J. T. (2016). Apple stores make an insane amount of money. *Time*. May 18.

Grewal, D., Roggeveen, A. L., Sisodia, R., and Nordfält, J. (2017), Enhancing customer engagement through consciousness. *Journal of Retailing 93*(1), 55–64.

Herhausen, D., Binder, J., Schoegel, M., & Herrmann, A. (2015). Integrating bricks with clicks: retailer-level and channel-level outcomes of online–offline channel integration. Journal of Retailing 91(2), 309–325.

Kell, J. (2016). How sporting giants Nike and Adidas are pushing the future of retail. *Fortune*, December 14. http://fortune.com/2016/12/14/nike-adidas-retail-future, Accessed date: 2 February 2018.

Kim, R. Y. J. E. C. R. (2020). When does online review matter to consumers? The effect of product quality information cues. *Electronic Commerce Research*. doi:10.1007/s10660-020-09398-0

Kompas.com. (2020). Update tambah data sebaran 10.3843 kasus COVID-19 di 34 Provinsi Indonesia. https://nasional.kompas.com/read/2020/03/23/15460721/update-tambah-65-pasien-kini-ada-579-kasus-covid-19-di-indonesia

Lamberton, C., & Stephen, A. T. J. o. M. (2016). A thematic exploration of digital, social media, and mobile marketing: Research evolution from 2000 to 2015 and an agenda for future inquiry. *Journal of Marketing 80*(6), 146–172.

Lewis, J., Whysall, P., & Foster, C. (2014). Drivers and Technology-Related Obstacles in Moving to Multichannel Retailing. International Journal of Electronic Commerce, *18*(4), 43–68.

Lemon, K. N., & Verhoef, P. C. (2016). Understanding customer experience throughout the customer journey. Journal of marketing 80(6), 69–96.

McGee, C. (2017). Amazon is becoming a 'more important search engine than Google,' says NYU professor. CNBC, April 4 https://www.cnbc.com/2017/04/04/nyuscott-galloway-amazon-becoming-bigger-search-engine-than-google.html

Mervosh, S., Lu, D., & Swales, V. (2020). See which states and cities have told residents to stay at home. *New York Times*, 3. https://www.nytimes.com/interactive/2020/us/coronavirus-stay-at-home-order.html.

Mobile Marketing Association. www.mmaglobal.com

Nielsen, F. (2014). The US digital consumer report. New York, NY: The Nielsen Company.

Osegowitsch, T., & Madhok, A. (2003). Vertical integration is dead, or is it? *Business Horizons*, *46*(2), 25–34.

Oinas-Kukkonen, H. and Oinas-Kukkonen, H. (2013). Humanizing the *web*: Change and *social innovation*, Basingstoke: Palgrave Macmillan.

Puccinelli, N. M., Goodstein, R. C., Grewal, D., Price, R., Raghubir, P., & Stewart, D. (2009). Customer experience management in retailing: Understanding the buying process. Journal of Retailing 85(1), 15–30.

Ramaswamy, V., & Ozcan, K. (2018). Offerings as digitalized interactive platforms: A conceptual framework and implications. *Journal of Marketing*, *82*(4), 19–31.

Reinartz, W., & Imschloß, M. J. M. I. R. (2017). From point of sale to point of need: How digital technology is transforming retailing. *Marketing Intelligence Review 9*(1), 42–47.

Rogers, E. M., Burdge, R. J., & Korsching, P. F. (1983). *Diffusion of innovations* (3rd editions) New York: The Free Press A Division of Mc Millan Publishing Co. Inc.

Siwicki, B. J. I. R. (2014). E-commerce and m-commerce: The next five years. www.Internetretailer.com/commentary//04/28/e-commerce-and-m-commerce-next five-years.

Stouthuysen, K., Teunis, I., Reusen, E., Slabbinck, H. J. E. C. R., & Applications. (2018). Initial trust and intentions to buy the effect of vendor-specific guarantees, customer reviews, and the role of online shopping experience. Electronic Commerce Research and Applications 27, 23–38.

Sweeney, J. C., & Soutar, G. (2001). Consumer perceived value: The development of multiple item scale. Journal of Retailing *77*(2), 203–220.

Surveysensum. https://www.surveysensum.com/blog/indonesia-customer-experience-trends-2020/

SEA Report. (2019). Google & Temasek Bain & Company. https://www.bain.com/insights/... economy-sea-2019/

Staflund, L., & Kersmark, M. (2015). Omni-channel retailing: Blurring the lines between online and offline. *Independent thesis Advanced level (degree of Master)*, 1–69.

Teece, D. J. (2010). Forward integration and innovation: Transaction costs and beyond. *Journal of Retailing 86*(3), 277–283.

Verhoef, P. C., Kannan, P. K., & Inman, J. J. J. o. r. (2015). From multi-channel retailing to omnichannel retailing: introduction to the special issue on multi-channel retailing. *Journal of Retailing 91*(2), 174–181.

Wagner, G., Schramm-Klein, H., & Steinmann, S. (2013). "Effects of cross-channel synergies and complementarity in a multichannel e-commerce system – An investigation of the interrelation of e-commerce, m-commerce and IETV-commerce", The International Review of Retail, Distribution and Consumer Research *23*(5), 571–581.

Westerman, G., Calméjane, C., Bonnet, D., Ferraris, P., McAfee, A. J. M. C. f. D. B., & Consulting, C. (2011). Digital Transformation: A roadmap for billion-dollar organizations. *MIT Center for Digital Business and Capgemini Consulting 1*, 1–68.

Woodruff, R. B. (1997). Customer value: The next source of competitive advantage. Journal of the Academy of Marketing Science *25*(2), 139–153.

Xu, Q., & Sundar, S. S. J. C. i. H. B. (2016). Interactivity and memory: Information processing of interactive versus non-interactive content. *Computers in Human Behavior 63*, 620–629.

[illegible]

[illegible]

[illegible]

[illegible]

[illegible]

[illegible]

[illegible]

[illegible]

[illegible]

13 Epidemiological Analysis of an Outbreak of Coronavirus (COVID-19) Disease in the World

Gaurav Kumar
Vidya College of Engineering

Krishan Arora
Lovely Professional University

CONTENTS

13.1 INTRODUCTION

In the month of December 2019, this epidemic comes into picture at wet market (Wuhan city) in China. The first recognized cases have found some symptoms like problem in respiratory system, sneezing, dry coughing and fever. But there was no any proof that this virus spread due to this wet market in Wuhan city [1]. Some researcher also said that this virus can also come due to taking the soup of bats but it is not proven yet. But after sometimes this virus spreads slowly in more than 34 outlying areas in China and 25 other countries, resulting in 75,199 definite cases and 2009 deaths [2]. Then, World Health Organization (WHO) has given a statement that this virus has a tendency to spread from human to human and also called this disease as Coronavirus (COVID-19) [3]. After increasing the number of patients, China started some action plan against to fight this disease like lockdown in whole country, sanitized all necessary places and complete quarantine of every citizen. Today till 25 June 2020, the world corona meter showing that the total number of cases in China is 83,449 from which 4634 have lost their life and 78,443 have recovered from this dangerous virus.

Due to mobility of the people from one country to another this virus had been spread almost all over the world. And total number of cases due to this virus is 9,543,163 from which 485,294 have lost their life and 5,187,645 have recovered from this virus till 25 June 2020. Nowadays, this virus has become a huge threat to all over the world. Because there is no proper drugs and vaccine prepared yet. This virus also causes a huge impact on the world economy which is continuously decreasing. The pathogenic features, symptoms, protection, treatments and statics have been described in this present research.

13.2 PATHOGENIC FEATURES OF CORONAVIRUS

Coronavirus has a shell, the elements are round or oval, frequently pleomorphic and 50–200 nm in diameter as shown in Figure 13.1. S protein lies on the virus surface, creating a rod-designed arrangement [4]. As one of the virus' most essential antigenic proteins, the S protein gene is the primary goal applied for typing. Xu et al. also stated that SARS-CoV-2 S-protein ropes a robust connotation with human angiotensin-converting enzyme 2 (ACE2) particles, which signifies that S-protein-ACE2 binding pathway stances a major public health hazard to human spread [5]. Awareness of coronavirus physical and chemical characteristics mainly comes from SARS-CoV and MERS-CoV studies. The coronaviruses are heat-sensitive and can be destroyed for 30 min at 56°C. In addition, the virus can be efficiently deactivated by ether, 75% ethanol, chlorine disinfectant, peracetic acid and chloroform, but not by chlorhexidine [6].

13.3 SYMPTOMS OF COVID-19

According to WHO, there are some following symptoms of COVID-19.

- Problems in respiratory system like breathing problem in the human body.
- Chest pain
- Dry cough
- Sneezing
- High fever
- Fatigue
- Diarrhea
- Lymphoma
- Headache and pain in the body
- Hemoptysis [7]

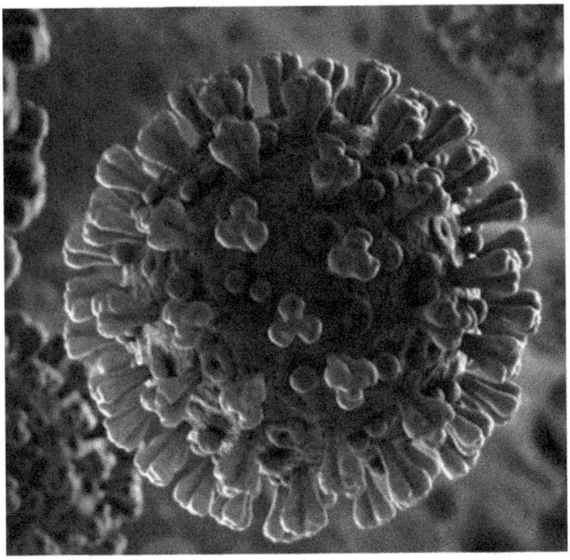

FIGURE 13.1 Image of coronavirus.

It takes 5–6 days on average for someone being diagnosed with the virus to show symptoms, but it can take up to 14 days [8,9]. There is no vaccine for the prevention from this virus. And more than 70 year citizens have large threats from this virus than below 70 years of age. People can save their life from this virus by a good immune system which will be made by some healthy food, fruits and drinks.

13.4 PROTECTION FROM THE SPREAD COVID-19

There are some safety measures and precautions to prevent COVID-19 from spreading and follow the advice given by WHO:

- Keep at least 1 meter (3 feet) social distance from each other during talking, sneezing and coughing.
- During coughing and sneezing do not use the direct hand always, try to cover the nose and mouth with the elbow or a tissue paper and, after using this, throw it into the dustbin.
- For destroying the virus from the hands always use alcohol-based sanitizers for cleaning the hands frequently.
- Never try to touch the face, hand and eyes because through hands they can easily catch the virus which will be very dangerous for the body.
- Stop going to overcrowded places because it is more difficult to maintain 1 m (3 feet) of social distance.
- Try to increase the immune system of the body by taking hygienic food, taking hot water, taking more vitamin C and also with meditation.

- In case of feeling slight symptoms of COVID-19 then, immediately keep himself isolated and stay at home only.
- In case of feeling symptoms like fever, chest pain, cough and difficulty in breathing then immediately call local helpline in advance and try to follow the instructions given by the local health authority. This will be very helpful to protect you to fight against the infections.
- Keep yourself up-to-date regarding all the necessary information against the COVID-19. As in India an Aarogya Setu app is used for giving the information about the infected person through their mobile phones.
- Try to drink more hot water and keep doing exercise and meditation which will increase the immune system of the body.

13.4.1 Advice on the Safe Use of Alcohol-Based Hand Sanitizers

- Try to wash your hand with soap or water. In case of absence of this then only use alcohol-based sanitizers.
- Avoid going near fire after using alcohol-based sanitizer because it contains flammable property.
- Only use a small amount of hand sanitizer. Never use a large amount of this.
- Keeping alcohol-based hand sanitizers out of reach of children.
- Escape stirring your eyes, nose and mouth after using alcohol-based hand rub because it may cause irritation.
- Do not try to drink this orally because it can be toxic.

13.5 TREATMENT OF COVID-19

13.5.1 Treatment Area Decision as per Disease Severity

With the effective protection and isolation, the confirmed and suspected cases must be treated in hospitals. Separation of confirmed and suspects cases of corona is very essential during treatment to reduce the exposure. The patient must be admitted to ICU in case of serious conditions.

13.5.2 General Treatment

A. Proper rest, support auxiliary care, confirm adequate energy, pay consideration to control hydrolysis and electrolyte, sustain internal environmental constancy, monitoring of signs which may be fatal and measurement of oxygen level in human body regularly, etc.

B. Track urinary and blood flows with all the necessary tests of the body like blood test and chest imaging and closely watch all the health signs of a patient.

C. Successful oxygen treatments are provided which includes oxygen provided by a nasal catheter or mask, depending on changes in oxygen saturation. High-flow oxygen treatment is carried out through nose, noninvasive or aggressive mechanical ventilation, etc., if necessary.

D. Antiviral therapy: No operative vaccine and antiviral drugs are available currently. Inhalation of IFN-alpha aerosols (five million U per time for adults, two times per day) and/or oral Lopinavir/Ritonavir (two tablets per time, two times per day) is treated.
E. Antibiotic treatment: Evade blind and unnecessary usage of antibiotics, particularly the usage of wide-spectrum antibiotics in combinations. Fortify microbiological surveillance. In secondary bacterial infections antibiotics should be used on time. One more medicine, hydroxychloroquine sulfate which is an antimalaria medicine has also been used for the treatment of COVID-19 patients in USA. But medically it is not proven that this medicine will give the best result in the handling of COVID-19 victim.

13.5.3 Treatment of Severe and Critical Cases

A. **Principle of treatment**: Symptom-based dealing, active anticipation and treatment of difficulties, handling of basic diseases, prevention of secondary contaminations and timely application of organ functional support.
B. **Respiratory support:** Application of ventilation without inserting tube in the trachea for two hours, if the illness does not improve or the infected person is not responding to ventilation without tube, and suffering from the increase in airway secretions, severe cough or hemodynamic instability; the patient should be timely admitted into mechanical ventilation with the help of tube interface. The mechanical ventilation with an endotracheal tube should follow the "lung-protective ventilation technology" with low tide volume to minimize ventilator-related lung damage. Ventilation in unstable locations, recruitment techniques or extracorporeal membrane oxygenation may be used as appropriate.
C. **Circulation support:** Encourage microcirculation based on complete resuscitation of the blood, use vasoactive medicines and use hemodynamic monitoring wherever appropriate.
D. **Others:** Depending on the degree of dyspnea and duration of the chest scan, using glucocorticoids effectively for a brief period of time (3–5 days) with a prescribed dosage no more than what is equivalent to methylprednisolone of 1–2 mg/kg·day [10].

13.6 STATISTICS

The data are available on the corona meter till 25 June 2020. Table 13.1 is showing the total number of confirmed cases, recovered cases and deaths across the world.

But the effect in India is less in comparison to the other countries like America, China, Italy, etc., due to lockdown (see above Table).

13.7 CONCLUSION

After Chinese New Year celebrations, a new type of epidemic virus called "COVID-19" has become a new challenge for the people of the country and around the world.

TABLE 13.1
Overall World Corona Meter Report

Total Confirmed	**Total Recovered**	**Total Deaths**	
9,543,028	5,187,624	485,294	
Location	**Confirmed**	**Recovered**	**Deaths**
USA	2,462,708	1,040,608	124,282
Brazil	1,192,474	649,908	53,874
Russia	613,994	375,164	8,605
India	*474,272*	*271,934*	*14,914*
United Kingdom	306,862	N/A	43,081
Spain	294,166	N/A	28,327
Peru	264,689	104,107	8,586
Chile	254,416	215,093	4,731
Italy	239,410	186,111	34,644
Iran	212,501	172,096	9,996

Because after people move from one place to another, this virus spreads around the world because it can be transferred from human to human. In order to defeat this COVID-19, no effective vaccine has been prepared for treatment. The only effective treatment is to maintain self-isolation, stay at home, regularly use alcohol disinfectants to clean your hands, always wear masks which are prescribed by government and also follow the instruction made by the local health authorities. If the people feel uncomfortable or unwell, then try to isolate them. And in case of feeling any symptoms such as respiratory problems, pain in chest, sneezing and dry coughing call your local health authorities immediately and do not take antibiotics and medicines on your own. More health education and awareness-raising programs will go a long way toward fighting this disease, known as COVID-19. Many epidemiological threats, such as Severe Acute Respiratory Syndrome (SARS) and Middle East Respiratory Syndrome (MERS), have been defeated in the past by effective vaccines and drugs. Therefore, in order to defeat the epidemic virus, known as coronavirus (COVID-19), it is necessary to obtain equally effective vaccines and beneficial drugs as soon as possible. In future, the whole world must prepare itself to fight this kind of diseases with more safety precautions.

CORRESPONDING AUTHORS PROFILE

Krishan Arora is presently working as Assistant Professor and Head in School of Electronics and Electrical Engineering in Lovely Professional University, Phagwara (Punjab). He did his PhD degree in Electrical Engineering from Maharishi Markandeshwar (Deemed to be University), Mullana-Ambala. He has published more than 30 papers in various International/National Journals and Conferences. His area of research is load frequency control and power systems.

REFERENCES

1. Wuhan Municipal Health Commission. Report on the current situation of Pneumonia in Wuhan (2019-12-31). Available online: http://wjw.wuhan.gov.cn/front/web/showDetail/2019123108989 (accessed on 19 February 2020).
2. World Health Organization. (2020). Coronavirus disease 2019 (COVID-19): Situation report, 70. Available online: https://www.who.int/emergencies/diseases/novel-coronavirus-2019/situation-reports
3. Alexander, E. G. (2020) Severe acute respiratory syndrome-related coronavirus—The species and its viruses, a statement of the Coronavirus Study Group. *bioRxiv*. [Google Scholar] [CrossRef]
4. Knoops, K., Kikkert, M., Worm, S. H., Zevenhoven-Dobbe, J. C., van der Meer, Y., Koster, A. J., Mommaas, A. M., & Snijder, E. J. (2008) SARS-coronavirus replication is supported by a reticulovesicular network of modified endoplasmic reticulum. *PLoS Biology*, 6, e226. [Google Scholar] [CrossRef] [PubMed]
5. Xu, X. T., Chen, P., Wang, J. F., Feng, J. N., Zhou, H., Li, X., Zhong, W., & Hao, P. (2020) Evolution of the novel coronavirus from the ongoing Wuhan outbreak and modeling of its spike protein for risk of human transmission. *Science China Life Sciences*, 63(3), 457–460. [Google Scholar] [CrossRef] [PubMed].
6. National health commission of the people's republic of China and national administration of traditional Chinese medicine. Diagnosis & treatment scheme for novel coronavirus pneumonia (trial) 6th edition. Available online: http://www.nhc.gov.cn/xcs/zhengcwj/202002/8334a8326dd94d329df351d7da8aefc2.shtml (accessed on 19 February 2020).
7. Carlos, W. G., Dela Cruz, C. S., Cao, B., Pasnick, S., & Jamil, S. (2020). Novel wuhan (2019-nCoV) coronavirus. *American Journal of Respiratory and Critical Care Medicine*, 201(4), P7–P8.
8. Li, Q., Guan, X., Wu, P., Wang, X., Zhou, L., Tong, Yang. & Xing, X. (2020). Early transmission dynamics in Wuhan, China, of novel coronavirus–infected pneumonia. *New England Journal of Medicine*, 382, 1199–1207.
9. Wang, W., Tang, J., & Wei, F. (2020). Updated understanding of the outbreak of 2019 novel coronavirus (2019 CoV) in Wuhan, China. *Journal of Medical Virology*, 92(4), 441–447.
10. Deng, S. Q., & Peng, H. J. (2020). Characteristics of and public health responses to the coronavirus disease 2019 outbreak in China. *Journal of Clinical Medicine*, 9(2), 575.

14 Prediction of COVID-19 in India Using Machine Learning Tools

A Case Study

Pooja and Karan Veer
DR B R Ambedkar National Institute of Technology

CONTENTS

14.1 INTRODUCTION

2019-nCoV is a new coronavirus which affects the respiratory system of the infected persons/patients. It enters from bat to human being. Globally there are 13,616,593 confirmed cases in which 585,727 people had died due to this pandemic. Also in India confirmed cases are increasing with time, which is 10,40,457 confirmed cases and 26,285 deaths that are recorded till July 18, 2020. It depicts how cases increased in India from 1 to 10,40,457. It also states that the growth rate of disease with time is rising. At first, symptoms of pneumonia and common cold will be seen, and with time other severe symptoms make the condition uncontrollable [1]. World Health Organization declared COVID-19 as a pandemic on March 11, 2020. Coronaviruses have different categories of their self. Health experts named as 2019-nCoV also as it belongs to the class of SARS and MERS viruses which have a large family. Various countries are working on a vaccine for a long time, but a reliable vaccine is not obtained until now to stop it. Government labs and institutes in India like Indian Council of Medical Research (ICMR), All India Institute of Medical Sciences (AIIMS), National Institute of Virology (NIV), National Center for Disease Control (NCDC) are working together with the private labs for the vaccine of COVID-19. People are getting infected can be of low, medium and robust immunity; it affects equally. But some reports say that people with low immunity cannot recover as fast as of strong immunity [2]. People with weak immunity such as the elderly, pregnant women and patients with chronic diseases are prone to severe acute symptoms after contracting novel coronavirus. It can be spread by self-infection if hands are contaminated, by a handshake, by droplets, cough and sneezing if the healthy one is in the vicinity of one meter of the infected one. As China declared the spreading of any virus infection, they found that it enters in human beings from bats. How it spreads from primary host to human being and then human-to-human transmission is depicted in Figure 14.1. Centre for Control and Disease defined this pneumonia as novel coronavirus pneumonia.

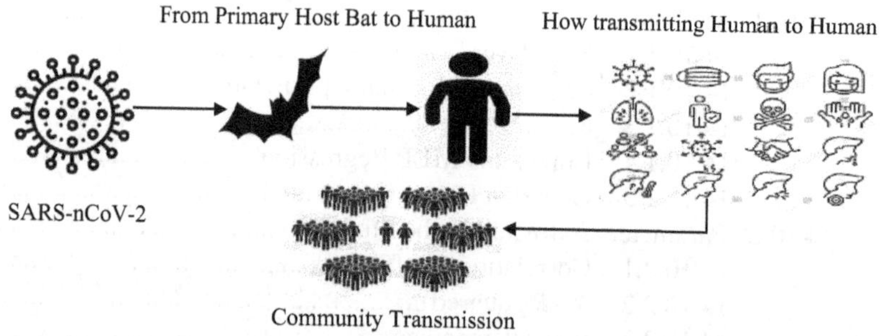

FIGURE 14.1 Transmission of COVID-19 from human to human.

International Committee on Taxonomy named it as SARS-CoV-2 (severe acute respiratory syndrome coronavirus-2). With time, India enhanced the production of personal protective equipment (PPE) with the COVID-19 dedicated hospitals. The government takes various technical steps like "Aarogya Setu app" which is a significant action by the Indian government. To help the administration the forecasting has been done using WEKA software to predict the COVID-19 cases. Real data of few days are compared with the predicted data. After that same data are predicted for 10 days ahead, it gives information about the approximate number of COVID-19 cases. If the public is aware of precautions and measures, it can be easy for any government to handle the situation in any natural calamities. The effects of lockdown in India are enlightened in the study. As there are four stages of COVID-19, India is safe from community transmission nowadays.

This chapter contains the stages of coronaviruses and their categories in Sections 14.2 and 14.3, respectively. Further sections carry the symptoms, causes and effects of COVID-19 in India. Sections 14.7 and 14.8 contain the detection and diagnosis of COVID-19 with the remedies of government at an early stage which are described in Section 14.9. The significant part of this study is a case study, as shown in Section 14.10. The knowledge about preventive measure published by ICMR, AIIMs and Ministry of Health and Family Welfare is described in Section 14.11. At last, the conclusion and references are shown.

14.2 VARIOUS STAGES OF COVID-19

It is categorized mainly into four stages as shown below in Figure 14.2. The first stage, which is considered as a safe stage, means no one is corona positive. The next step is the isolation stage; it occurs when cases are imported, and they need isolation. When three cases were reported in India from China, and it was a simple thing to tackle them. The third stage is the cluster stage, which is more challenging to handle than the second one because cases will increase in it. The fourth and last stage is community transmission when thousands and millions of people get the infection, which is the severe phase for the disease; in this stage the number of positive cases increases in a logarithmic way [3]. At this stage, it becomes challenging to provide diagnosis and healthcare facilities for patients and medical staff [4].

14.3 CATEGORIES OF CORONAVIRUSES

14.3.1 SARS-Cluster Coronaviruses

The first-time coronavirus was detected by two groups of researchers separately in China, and they found that it has approximately the same genome identity (92%) as

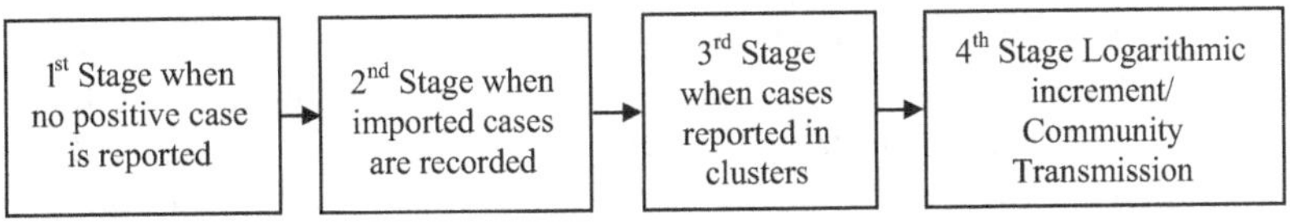

FIGURE 14.2 Different stages of COVID-19.

human SARS-CoV. These groups concluded that the virus could penetrate the human body through ACE-2 (human Angiotensin-Converting Enzyme 2), and these viruses can become the cause of various pandemics like SARS. SARS was also detected in China in 2012.

14.3.2 MERS-Related Coronaviruses

After an enormous outbreak which happened due to MERS, the related CoV was observed also in bats. In 2006, in China, two coronaviruses (Tylonycteris HKU4 and Pipistrellus HKU5) were recognized by scientists in bats and were categorized in 2c group of CoV. The two newly identified viruses were relating to MERS-CoV, and they found the cause of pandemic 2012. Most of the MERS-cluster coronaviruses are produced by bats and are detected in China. They have a relation with the Vespertilionidae family.

14.3.3 HKU2 (SADS)-Associated Coronaviruses

This virus belongs to Rhinolophus bats family which are found in Hong Kong, Tibet and Yunnan. This type of virus can grow in animals and can affect humans and also give birth to SARS-CoV. SARS has many characteristics, similar to a virus named swine acute diarrhea syndrome coronavirus (SADS-CoV), which was the cause of a vast epidemic spread from the pig in China. Though the receptor needs by the virus to penetrate the human body is not discovered [5].

14.4 SYMPTOMS

For the public awareness Indian government has published the advisory in the early phase of a pandemic and modifying it with new research and guidelines provided by the WHO. Symptoms in a suffering person can be mild and severe. In the mid of March, it was reported that elder, kids and people suffering from lungs or heart disease have more chances of getting affected. But a study [6] concluded that younger people are also infected equally as elderly or children. Most common symptoms are cough, fever/chills and tiredness. Mild symptoms of COVID-19 are fatigue, sore throat, headache, nausea, body aches, rashes on the skin, discoloration of fingers and toes. Severe symptoms of virus are difficulty in breathing, loss of voluntary movement, loss of speech and chest pain/pressure [7]. In random testing, few cases found positive without any symptoms also.

14.5 CAUSES

In the late of 2019, a group of people with similar symptoms was observed by doctors in Wuhan, and they informed that some virus from bats is infecting to humans. A few days later, doctors understand the way of infection from one human to others. The way can be handshaking, sneezing droplets and the vicinity of an infected person. The cause of this pandemic is the nCoV from bats, which is originated from Wuhan, a city of China. In this way, the virus was imported from China to other parts of the world. In India, the first case reported was a student of Wuhan University, Wuhan, returned to the home on January 30, 2020.

14.6 EFFECTS OF PANDEMIC

This virus is spread in 212 countries throughout the world. It is affecting humans in several ways as given below.

14.6.1 Human Life

In the world, 13,616,593 confirmed cases, and 585,727 deaths are reported due to pandemic. Because of the nonavailability of vaccine, social distancing is the only option to remain safe from the disease. To decrease the infection rate long-time lockdown is applied in various countries [8–10] and due to this, people are staying at home which increased health issues like obesity, depression, loneliness and anxiety. In a few countries, divorce and domestic violence cases also increased than regular days [11].

14.6.2 Nature

Opposite to human being, nature is regaining itself during this pandemic. Pollution throughout the world is decreasing with the biggest fall of the decade. The ozone layer is healing itself day-by-day. Some database published by NASA on how airborne particles level in India is decreased in 2020, which reduces human visibility typically [12]. Due to lockdown, pollution is decreasing as well as airborne particles which also enhance visibility. Global temperature is reduced by approximately 2°C [13].

14.6.3 Education

Due to lockdown school, colleges and universities are closed. Online classes and examination are going on, but because of less physical activity, children are suffering from anger, anxiety and mental health issues. Few boards and universities where the infection rate is high comparatively they are upgrading the students on behalf of previous results [14].

14.6.4 Economy

Due to lockdown, inflation rate throughout the world is at a higher level in recent years. To remain safe from the financial emergency and inflation in India in the coming month/year the cabinet of India makes the ordinance to deduct 30% remuneration of members of parliament for 1 year which will be paid into consolidated fund of COVID-19 in India. About PM (Prime Minister) CARES account is generated for the public to collect donation from the people of India. Governors, President and Vice President also taking it as a severe condition for the country and deciding to give 30% amount from their earned income, accepting it as social responsibility [15]. MPLAD (Members of Parliament Local Area Development Schemes) funds suspended temporarily for 2 years for the treatment of people suffering from COVID-19. On May 12 2020, the Indian government announced a unique package of 20 lakhs (approximately equal to 10% India GDP) to make the country independent against the global competition in the supply chain. This campaign is named "Aatam Nirbhar Bharat."

14.6.5 Healthcare Facility

All over the world, countries either developed or developing suffering with facilities required for the treatment of COVID-19 are ventilators, mask, dedicated hospitals and PPE kits. Somehow, developed countries have completed their demand and developing countries are doing by enhancing the production in their own country or by importing from other countries. In India, 586 hospitals which are dedicated to COVID-19 are established, and PPE kits production is increased at a large scale [16].

14.7 DETECTION OF COVID-19

Antibody test was used previously for the detection of COVID-19. Because of unreliable results, reverse transcription-polymerase chain reaction came into effect. In the early stage of this disease, there was a deficiency of testing kits and other healthcare facilities in India, even the same was happening with developed countries like America and Italy. To fulfil the deficiency government ordered millions of kits from abroad and increased the manufacturing in-country even [17,18].

14.8 DIAGNOSIS OF COVID-19

Till now, there is no effective vaccine came into the market. So the primary diagnosis used nowadays is social distancing. Throughout the world, countries are trying to stop the infection before community transmission. Health experts say that after getting infected, quarantine is the significant solution under the supervision of doctors only.

14.9 INITIAL REMEDIES BY INDIA

From the day when the first case was reported in India, the government became active about the disease. The Indian government made and executed the strategies by learning from more affected countries. AIIMS New Delhi published a report that in India 80% of people suffering by COVID-19 are getting recovered with isolation only [19]. Less than 20% needs to be hospitalized. A tiny ratio may need an Intensive Care Unit. People with low immunity such as the elderly, pregnant women and patients with chronic diseases (High Blood Pressure, Heart Disease, Lung Disease or Diabetes) are prone to symptoms after contacting with the suffering person. Fever, Sore throat, Cough and Shortness of Breath are common symptoms which are similar to the signs of any viral infection like the common cold, influenza, etc. AIIMS published the guidelines regarding the precautions to avoid COVID-19 are as regular handwash, maintain 2-meter distance from anyone who is coughing or sneezing, etc. AIIMS's report also said that COVID-19 could not transmit via cooked food, pets and dead body [20,21].

14.10 CASE STUDY

Despite various control measures, the disease is surging very fast and spreading in more than 212 countries which infected more than 13,616,593 around the world [22]. In India, the first case of COVID-19 was reported on January 30 2020, and 10,40,457 cases have been reported till July 18, 2020 [23]. The coming days contain whether India will remain away from community transmission; presently, which is in the acceleration stage [24,25]. For the help of administration, several models are used to predict the number of confirmed cases, death rate and recovery rate. Here, the seven different prediction models Multilayer Perceptron (MLP), and Sequential Minimal Optimization (SMO), M5, Hot Winters, Linear Regression, Gaussian Progression and MLP regression are performed for finding the best model fit, and it was concluded that Hot Winters outperformed the other one [26–28]. The predictive data of the respective region are compared with the data considered to be adequate in practice. From the analysis, it can be said that the methodology of Hot Winters can be adopted for possible pandemic prediction such as COVID-19.

14.10.1 Methods

The data used in forecasting have been taken from the official dashboard made by the Indian government which includes the number of confirmed cases and deaths in India (from March 24, 2020 to June 27, 2020) as shown in Table 14.1 [23]. The primary sources for the database are (https://www.mohfw.gov.in) and (https://www.covid19india.org/). Then the data are forecasted by applying two models. The results of this study are predicted using software Weka 3.8.4; a machine learning tool. The data set is trained and tested in ratio 70:30.

14.10.1.1 Multilayer Perceptron

In general, it is a grid of layers with connections between individual neurons in each layer and then with subsequent layers, so its name is given as MLP. During data processing, connection weights will change so that perceptron supervised learning can occur. The node weight can be adjusted to minimize the error with the help of the below equation

$$-\frac{d\varepsilon(n)}{dv_j(n)} = \Phi'(v_j(n))\sum_k -\frac{d\varepsilon(n)}{dv_k(n)} w_{kj}(n)$$

14.10.1.2 Sequential Minimal Optimization

SMO is d acquired with the idea of decomposition method with its extreme and optimizes at least a subset of two points at individual iteration. For finding the optimum solution of the problem, the following function should be maximized.

$$\max W(\alpha) = \sum_{i=1}^{l} \alpha_i - \frac{1}{2}\sum_{i=1}^{l}\sum_{j=1}^{l} \alpha_i \alpha_j y_i y_j K(x_i, x_j)$$

TABLE 14.1
Daywise COVID-19 Data of Confirmed, Death and Recovered Cases in India [23]

Date	Confirmed IND	Death IND	Recovery	Date	Confirmed IND	Death IND	Recovery	Date	Confirmed IND	Death IND	Recovery
3/24	400	6	3	4/24	24277	777	5496	5/25	144776	4167	60706
3/25	486	7	3	4/25	26112	821	5938	5/26	150683	4340	64291
3/26	559	12	3	4/26	27719	877	6523	5/27	157929	4528	67725
3/27	712	15	5	4/27	29287	935	7103	5/28	165183	4704	70896
3/28	848	20	5	4/28	31189	1004	7739	5/29	173321	4973	82631
3/29	968	23	99	4/29	32894	1075	8429	5/30	181685	5178	86934
3/30	1155	37	141	4/30	34695	1150	9059	5/31	190474	5400	91862
3/31	1464	43	160	5/1	37091	1227	10021	6/1	198197	5601	95744
4/1	1888	49	169	5/2	39655	1319	10852	6/2	207012	5823	100275
4/2	2374	65	191	5/3	42607	1459	11763	6/3	216701	6082	104064
4/3	2934	79	230	5/4	46263	1562	12845	6/4	226548	6356	108454
4/4	3513	92	286	5/5	49234	1690	14140	6/5	236020	6642	113224
4/5	4122	114	329	5/6	52836	1781	15301	6/6	246428	6939	118657
4/6	4606	130	394	5/7	56180	1885	16776	6/7	257310	7200	123848
4/7	5179	157	469	5/8	59519	1982	17887	6/8	265846	7471	129019
4/8	5744	177	565	5/9	62694	2097	19301	6/9	275827	7743	134653
4/9	6557	223	635	5/10	67005	2209	20970	6/10	286983	8101	140928
4/10	7428	245	786	5/11	70597	2290	22549	6/11	298118	8495	146972
4/11	8282	286	972	5/12	74159	2410	24454	6/12	309424	8883	154235
4/12	9040	328	1086	5/13	77885	2547	26417	6/13	321463	9207	162327
4/13	10283	355	1198	5/14	81876	2644	28011	6/14	332868	9603	169685
4/14	11314	392	1365	5/15	85684	2748	30245	6/15	342900	11607	180324

(Continued)

TABLE 14.1 (*Continued*)
Daywise COVID-19 Data of Confirmed, Death and Recovered Cases in India [23]

Date	Confirmed IND	Death IND	Recovery	Date	Confirmed IND	Death IND	Recovery	Date	Confirmed IND	Death IND	Recovery
4/15	12200	419	1509	5/16	90476	2866	34257	6/16	353990	11921	187553
4/16	13261	445	1767	5/17	95525	3018	36277	6/17	367263	12262	194439
4/17	14183	483	2040	5/18	100153	3149	39277	6/18	381092	12605	205180
4/18	15554	518	2466	5/19	106307	3295	42309	6/19	395832	12969	214209
4/19	17134	556	2854	5/20	112027	3429	45422	6/20	411750	13277	228183
4/20	18373	589	3273	5/21	118050	3577	48553	6/21	426901	13703	237253
4/21	19910	642	3976	5/22	124586	3719	51833	6/22	440461	14015	248137
4/22	21202	678	4370	5/23	131251	3861	54409	6/23	456117	14483	258599
4/23	22869	718	5012	5/24	138362	4017	57694	6/24	472985	14907	271688

$$\sum_{n=1}^{l} \alpha_i y_i = 0;\ 0 \le \alpha_i \le C;\ \text{where } i = 1,2\ldots l.$$

where α_1 and α_2 are authorized to change.

14.10.1.3 M5

M5 model tree is a classifier which uses linear regression functions and segregates the whole problem into various subproblems depending upon the input domain. This technique deals with continuous type problem, not with discrete ones, and gives the information of each linear model which are made to estimate the nonlinear relationship of data. Error is measured using standard deviation at each node which is the criteria to divide the problem into subproblems. This error can be reduced by testing individual variables at that particular node. This standard deviation reduction is computed as

$$SDR = sd(k) - \sum \frac{|k_i|}{|k|} sd(k_i)$$

where sd is the standard deviation, k is the number of nodes and k_i denotes the nodes that have the ith outcome.

14.10.1.4 Linear and MLP Regression

Linear regression is a statistical model which is used to model the simulated data. No iteration is required for it as it is a connectionist model. Gauss formalized the equation of linear regression by applying least square approximation having the minimum variance among all unbiased approximation. It can be defined as

$$Y_p = X_p^T B$$

Here $X_p^T B_p$ is the mean and $\sigma\sqrt{(X_p^T (X^T X)^{-1} X_p)}$ is the variance of normally distributed mean. Generally, MLP contains three layers, input, hidden and output of neurons. Each layer of the model has multiple processing units which are connected to succeeding layers. It is the generally used in a connectionist model for biomedical signals where biasing is given on hidden layer. MLP can approximate any function and can attain the required accuracy using one hidden layer and squashing transfer function if this function is nondecreasing and bounded.

14.10.1.5 Gaussian Regression

It is a probabilistic model for finding the weights w for the features Φ. Gaussian distribution is calculated as the minimizer of a loss function

$$L(w) = w^T \sum^{-1} w + \sum_{n=1}^{N} (y_n - \Phi(x_n) w)^2$$

where $L(w)$ is loss function and $\sum$- is the least square estimator for w.

14.10.2 Parameters Extracted in the Study

The study is analyzed using some basic parameters such as mean values of real and predicted data, correlation coefficients, R^2 values and error percentage [29,30].

14.10.2.1 Correlation Coefficients

It is a statistical strength measurement variable which measures the relationship between two variables. It gives relative movements between variables. $Cov(x, y) = p_{xy}(\sigma_x\sigma_y)$, where p_{xy} is the personal product-moment correlation coefficient and σ_x, σ_y are the standard deviation of x and y.

14.10.2.2 R^2 (*R*-squared)

In the regression model, R^2 represents the ratio of variance for a dependent variable which is explained by the independent variable. $R^2 = 1 -$ (Unexplained Variation)/(Total Variation).

14.10.2.3 Error Percentage

It measures the difference between the actual number and estimated number with the reference of the actual number and expressed in the form of a percentage.

14.10.3 Results and Discussion

Coronavirus, which affects the respiratory system severely, is spreading throughout the world. The severity of this disease is increasing in India. The proposed work is a pioneer study which focuses on forecasting of spreading of COVID-19 in India. These results are concluded by using MLP, SMO, M5, Hot Winters, Linear Regression, Gaussian Progression and MLP regression models. The best-fit model is stable in all variables in the coming weeks and assured that India might halter SARS-CoV-2 pandemic. These results related to strict rules, strategies and quarantine by the Indian government.

Table 14.2 describes seven prediction models in India. The author calculated correlation coefficient, error percentage and R^2 for each method. The prediction of a cumulative number of confirmed cases in India with various predictive models is shown in Table 14.2. Comparison of real data and predicted data with different models had been done, with individual means, error percentage and R^2 values. During analysis for each model, the confidence interval (CI) is taken 95%. By comparing the hot winters and MLP method with others gives low error percentage with more R^2 (*R*-squared) value. But the error percentage is minimum (i.e., 1.86) for Hot Winters which is performing the best-fit model for the prediction. In the same way value of R^2 (0.8563) which is maximum for the Hot Winters also proves it best-fit predictive model. MLP is also giving the approximate results. If correlation coefficient is focused, which is maximum (i.e., 0.9229) for Hot Winters model in Table 14.2 also assuring it as the best predictive model.

In Figure 14.3, the graph of various techniques is depicted with the horizontal and vertical axes represented by date and a cumulative number of cases, respectively. Here the real-time data set is processed and analyzed with a predictive number of

TABLE 14.2
Comparison of Prediction Models with Calculated and Real Mean of Confirmed Cases (95% CI)

Date	6/25	6/26	6/27	Mean	Error Percentage (%)	R^2	Correlation Coefficient
Real Confirmed Cases	**482962**	**509445**	**529579**	**133260**			
Prediction of confirmed cases by MLP	482962	496629	510350	136873	2.71	0.8401	0.9163
SMO	489869	507005	524488	137387	3.09	0.8364	0.9149
M5	462547	474153	485760	135903	1.98	0.8462	0.9189
Hot winters	334678	340551	354834	130589	1.86	0.8563	0.9229
Linear regression	485147	499725	518136	137113	2.89	0.8384	0.9157
Gaussian progression	497322	520588	546294	138165	3.68	0.8300	0.9128
MLP regression	465761	473559	480014	135758	1.87	0.8471	0.9190

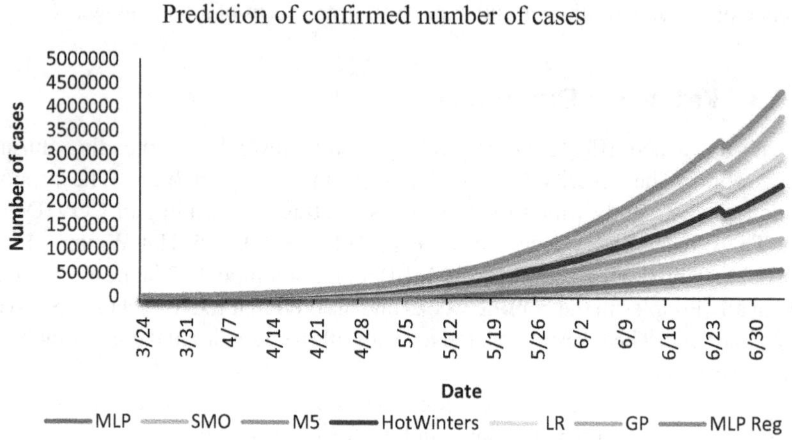

FIGURE 14.3 Prediction of cumulative number of confirmed cases in India.

confirmed cases in the near future. Various color lines show the respective technique graph. On behalf of the graph line of Hot Winters, i.e., black is showing stability in a cumulative number of confirmed cases of COVID-19 in coming weeks.

The best-fit model is helpful to several dynamic inputs and can process the data set of different length. Tables 14.3 and 14.4 are comparing the real and predicted data of death and recovered cases, respectively. As shown in Table 14.3 and Figure 14.4, Hot Winters performed well for death cases in India in terms of error percentage and R^2 values with the CI of 95%. In the same table, MLP regression model gives error rate and R^2 values 1.8 and 0.8272, respectively, which are also approximate results of the best-fit model. So for death prediction, Hot Winters method is the best fit, but MLP regression can also be used.

TABLE 14.3
Comparison of Used Prediction Models with Calculated and Real Mean of Death Cases (95% CI)

Date	6/25	6/26	6/27	Mean	Error Percentage (%)	R^2	Correlation Coefficient
Real Confirmed Cases	**15308**	**15689**	**16103**	**4079**			
Prediction of confirmed cases by MLP	14650	14783	14919	4164	2.0	0.8260	0.9069
SMO	15455	16033	16648	4226	3.6	0.8129	0.9019
M5	14483	14821	15158	4170	2.2	0.8249	0.9068
Hot winters	11692	11928	12724	4062	0.4	0.8361	0.9117
Linear regression	15664	16320	16940	4237	3.8	0.8102	0.9007
Gaussian progression	15410	16013	16698	4228	3.6	0.8124	0.9019
MLP regression	14591	14633	14683	4156	1.8	0.8272	0.9074

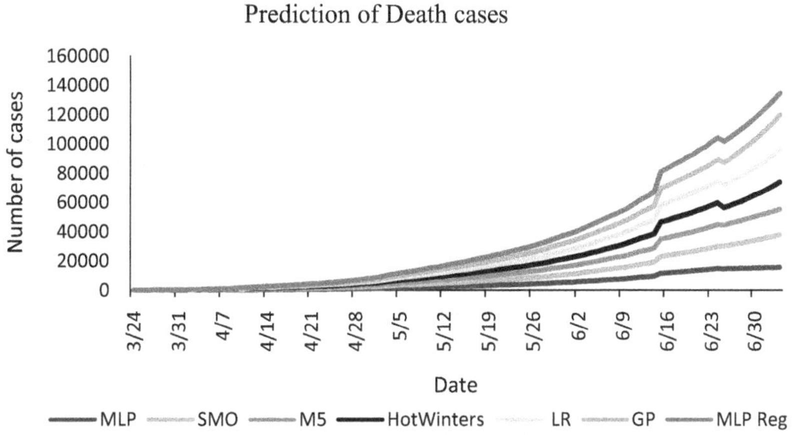

FIGURE 14.4 Prediction of cumulative number of death cases in India.

Recovery rate is an essential parameter for this pandemic. As no vaccine is available in the market, then there are two options to make the people safe from pandemic; the first is to stop the infection and second is if someone is already infected then enhance the recovery rate.

To evaluate the results of death and recovery rate dashboard of the Indian government is followed. Then the real values are compared with prediction values. As shown in Table 14.4, Hot Winters model again gives more approximate results than others in terms of all three variables that are error percentage, R^2, and correlation coefficient attained 0.10, 0.7876 and 0.8875, respectively. Prediction of recovery cases in India is shown in Figure 14.5. For recovery rate also Hot Winters is showing the stability in the near future. So for a recent study, Hot Winters can be concluded as the best-fit model for predicting any future pandemics like COVID-19.

TABLE 14.4
Recovery Rate Comparison of used Prediction Models of Predicted and Real Mean of (95% CI)

Date	6/25	6/26	6/27	Mean	Error Percentage (%)	R^2	Correlation Coefficient
Real Confirmed Cases	**285671**	**295917**	**310146**	**64819**			
Prediction of confirmed cases by MLP	285782	295827	304587	64761	0.08	0.7672	0.8759
SMO	283091	295573	308493	64771	0.07	0.7669	0.8757
M5	273377	282594	291811	64361	0.70	0.7731	0.8792
Hot winters	172567	175521	185117	64750	0.10	0.7876	0.8875
Linear regression	280108	291985	304786	64664	0.23	0.7686	0.8767
Gaussian progression	285422	300100	315620	64917	0.15	0.7645	0.8743
MLP regression	270504	274575	277439	64700	0.18	0.7767	0.8813

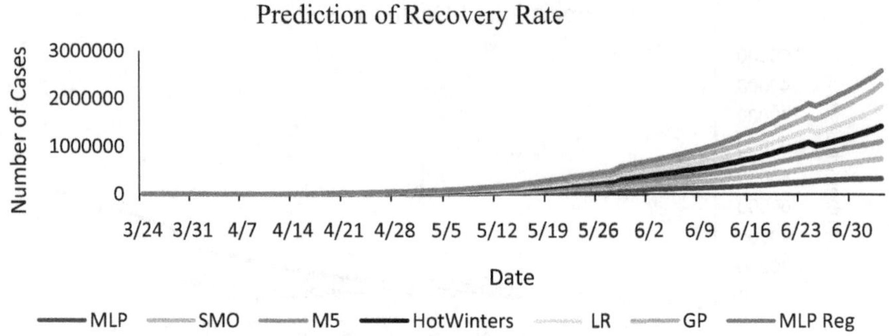

FIGURE 14.5 Prediction of cumulative number of recovered cases in India.

This study shows that although India achieved some results to stop the infection health experts say there is need to make tough decisions to prevent the spreading of COVID-19 so that increasing death rate can be decreased shortly. Based on this prediction, public health employees should focus more toward to avoid the rapid infection of the virus, which is mandatory to control community transmission in India. Although the Hot Winters exhibits some limitations, even though the prediction using this model may be helpful in future strategies regarding health decisions taken by experts. This work suggests that Hot Winters is easy to use for prediction of COVID-19 spread in the coming months.

14.11 ACTION AND PREVENTIVE MEASURES BY INDIAN GOVERNMENT

Indian government applied the full country lockdown of 21 days (March 25–April 14 2020) as the experts said that this is the minimum time required to break the chain of COVID-19. But again this period was increased in three phases up to May 31 2020 [3].

During this long period of lockdown, Indian people did not face domestic violence, anxiety and depression, as a couple of cases were reported in China and other countries. India is the second most populated country in the world, so to maintain hospital facilities is itself defiance. When the first case of COVID-19 was imported in India by a student who was studying in Wuhan University on January 30, 2020, the Indian government started work against it about all medical equipment like mask, ventilators and COVID-19-dedicated hospitals. A few days before, the first COVID-19 specialized hospital had been started in Odisha, India. A report says that on June 25, 2020, India has 586 COVID-19-dedicated hospitals and is able to treat 10 lakhs patients simultaneously [8]. The Indian government bought one million testing kits from other countries and also enhanced the manufacturing of PPE in India. Due to lockdown, inflation rate throughout the world is at a higher level of recent years [9]. On May 12, 2020, the Indian government announced a unique package of 20 lakhs (approximately equal to 10% India GDP) to make the country independent against the global competition in the supply chain. 13,36,82,275 samples are tested until now in India. 735 confirmed cases per million people are reported in India, which is 10,896 in America, 10,507 in Peru, 9534 in Brazil and 5,028 in Russia. The number of deaths in India is 17.2 per million people which is 660 in Britain, 607 in Spain, 406 in America, 336 in Brazil and 269 in Mexico.

14.12 CONCLUSION

This case study provides a model to predict the situation of the pandemic in the coming days. In literature, different prediction algorithms are available. Still, a more accurate algorithm was required, which can help more in the prediction of deaths due to this virus and for any other pandemics like COVID-19 in the future. Various preventive measures taken by the government in India from the first day are shown and how improvement in each action was reported. This study focused on the effects of the virus on education, economy, nature and healthcare in India and how India is fighting with economic and social losses during the lockdown. The significant efforts by the Indian government for COVID-19-dedicated hospitals are described here. The measures taken by the government to return the economy and to refrain from financial issues are an essential part of the study. Some suggestions by health experts might become useful for India to stop community transmission. The ventilators, mask and PPE kits production and import from other countries have been increased at a large scale. Because of all these remedies, in India, death rate is comparatively lower than other countries. Physiological and psychological impacts in India are found lesser than the rest of the nations.

REFERENCES

1. Guide on Anti-COVID-19 by Beijing magazine_compressed.pdf. https://s.docwork-space.com/d/ADktrB6H8Jg1w ZOI7bedFA
2. The Jewish Voice. http://thejewishvoice.com/wp-content /uploads/2020/01/FOR-E-MAIL-BLAST-5-Coronavirus-Cases-article-1.jpg
3. Tomar A, Gupta N. (2020). Prediction for the Spread of COVID-19 in India and the Effectiveness of Preventive Measures. *Science of Total Environment*, 728, 1–6.

4. COVID-19, Get the latest information from the MoHFW about COVID-19. https://www.youtube.com/watch?v=n_EFJI6zvB8 (Accessed March 18, 2020).
5. Fan Y, Zhao K, Shi Z-L Zhou P. (2019). Bat Coronaviruses in China. *MPDI Viruses*, 11, 1–14. doi: 10.3390/v11030210.
6. Sharma P, Veer K. Action and Problems of Covid-19 Outbreak in India. *Infection Control & Hospital Epidemiology*. doi: 10.1017/ice.2020.186
7. Coronavirus disease – symptoms https://www.google.com/search?q=symptoms+of+corona&oq=symptms+&aqs=chrome.1.69i57j0l7.5224j0j7&sourceid=chrome&ie=UTF-8 (Accessed July 13, 2020).
8. COVID-19 Pandemic Lockdown in Italy. https://en.wikipedia.org/wiki/COVID-19_pandemic_lockdown_in_Italy (Accessed June 24, 2020).
9. The Guardian International Edition. https://www.theguardian.com/world/2020/may/11/russia-to-end-national-covid-19lock down -as-cases-hit-record-high (Accessed June 24, 2020).
10. Coronavirus Lockdown: All You Need to Know About New Measures. https://www.bbc.com/news/explainers-52530518. (Accessed June 25, 2020).
11. Belfin R V, Piotr B, Radhakrishan B L, Rejula V. COVID-19 Peak Estimation and Effect of Nationwide Lockdown in India. *MedRxiv*. doi: 10.1101/2020.05.09.20095919.
12. Airborne Particle Levels Plummet in Northern India, https://earthobservatory.nasa.gov/. https://eoimages.gsfc.nasa.gov/images/imagerecords/146000/146596/india_tmo_2016-2020_lrg.png
13 Land Surface Temperature Anomaly. https://earthobservatory.nasa.gov/global-maps/MOD_LSTAD_M
14. Brooks SK, Webster RK, Smith LE, Woodland L, Wessely S, Greenberg N, Rubin GJ. (2020) The Psychological Impact of Quarantine and How to Reduce It: Rapid Review of the Evidence. *Lancet*, 395, 912–920.
15. PRS Legislative Research, the PRS Blog. https://www.prsindia.org/theprsblog/does-changing-mp-salaries-and-mplad-entitlements-raise-resources-fight-covid-19 (Accessed June 22, 2020).
16. Hindustan Times Coronavirus Outbreaks. https://www.hindustantimes.com/india-news/india-now-has-infrastructure-to-treat-10-lakh-covid-19-patients-report/story-IWPjAlQL4xrZx215Qw6GLJ.html (Accessed June 23, 2020).
17. The Hindu COVID-19: What are Different Types of Tests? https://www.thehindu.com/sci-tech/science/covid-19-what-are-the-different-types-of-tests/article31507970.ece.
18. Arino J, Portet S. (2020). A Simple Model for COVID-19. *Infectious Disease Modelling*. doi: 10.1016/j.idm.2020.04.002.
19. The Economic Times. https://economictimes.indiatimes.com/industry/healthcare/biotech/virus-diagnostic-tests- sensitive -gene-sequencing-possible-only-at-governmentlabs/articleshow/744676 70.cms?from=mdr
20. All India Institute of Medical Sciences, New Delhi. https://www.aiims.edu/images/pdf/notice/COVID_19_HIC_SUPPLEMENT_VERSION_1.2__26 _3_20.pdf
21. The Economic Times. https://economictimes.indiatimes.com/news/politics-and-nation/john-hopkins-varsity- disassociates from-study-on-indias-possible-covid-19cases/articleshow /74872 298.cms
22. WHO Coronavirus Disease (COVID-19) Dashboard. https://covid19.who.int/.
23. COVID-19 India Dashboard. https://www.covid19india.org/.
24. Tran TT, Pham LT, and Ngo QX. Forecasting Epidemic Spread of SARS-CoV-2 Using ARIMA Model (Case study: Iran). *Global Journal of Environmental Science and Management* doi: 10.22034/GJESM.2019.06.SI.01.
25. Alazab M, Awajan A, Mesleh A, Abraham A, Jatana V, Alhyari S. (2020). COVID-19 Prediction and Detection Using Deep Learning. *International Journal of Computer Information Systems and Industrial Management Applications*, 12, 168–181.

26. Chen J, Fu MC, Zhang W, Zheng J. (2020). Predictive Modeling for Epidemic Outbreaks: A New Approach and COVID-19 Case Study. doi: 10.1142/S0217595920500281.
27. Khan ZS, Bussel FV, Hussain F. (2020). A Predictive Model for Covid-19 Spread Applied to Six US States.
28. Kumar S, Kumar K, Bajpai MK. (2020). Kalman Filter Based Short Term Prediction Model for COVID-19 Spread. doi: 10.1101/2020.05.30.20117416 medRxiv preprint.
29. Sree PK, Usha Devi Nedunuri SSSN. (2020). A Novel Cellular Automata Classifier for COVID-19 Prediction. *Journal of Health Sciences*. doi: 10.17532/jhsci.2020.907
30. Morozova O, Li ZR, Crawford FW. (2020). A Model for COVID-19 Transmission in Connecticut. doi: 10.1101/2020.06.12.20126391.

15 Preventive Measures for Corona Virus Considering Different Perspectives in Indian Conditions

Saumyadip Hazra, Abhimanyu Kumar, and Souvik Ganguli
Thapar Institute of Engineering and Technology

Sahil Virk
Mentor Graphics Pvt. Ltd.

CONTENTS

15.1 INTRODUCTION

Prevention is better than cure, so goes the saying. With corona virus engulfing the entire world, we need to look for the remedies to keep ourselves fit and healthy. People throughout the world are often puzzled with the changing guidelines, newly added symptoms and the treatment protocols of the World Health

Organization (WHO). There are different strategies that can be thought to combat this pandemic.

Corona virus damages the lungs of human beings, hence, regular exercise, yoga, running and even regular walking (for the elderly people) may help in smooth functioning of lungs and develop sufficient immunity to combat COVID-19. A person getting cold and flu twice or thrice a year is a good sign toward immunity elevation. Consumption of food supplements like vitamin B, C and especially A is highly important. Other supplements like iron and zinc work equally well for immunity development. It is worth mentioning that the people residing in suburban and rural areas are likely to have better immunity power as compared to the people living in the metro cities.

Even the use of turmeric, cumin seeds, coriander and fenugreek which are typical food ingredients in Indian diet is likely to inculcate better immunization [1]. Almost the entire India, except the eastern and north eastern part, relies on vegetable diet rather than animal food. Plant protein is very effective as an immunity supplement. Attempts have been made to improve the concentration of amino acids [2]. Moreover, the pulse food in plentiful amounts can not only add good calorific value to our diet but also develop excellent disease-withstanding capability. Consumption of some fluid-based diets like coconut water, watermelon, uncooked tomato, cucumber, fruit juices, hot lemon water, hot water and milk in proper amount is also helpful.

Less reliance on medicines, ventilators and life support systems has also been emphasized because such systems are dependent on the infrastructure available in a country and the way it is delivered to the people [3–5]. Sanitization of oneself, appropriate use of masks in crowded places, maintaining social distancing, avoiding social functions and ceremonies, use of disinfectants in homes and work area, etc., are some of the preventive measures to fight against corona virus. These are included in detail within this chapter.

The rest of the chapter constitutes the following parts. Section 15.2 considers the sanitization, cleanliness and socializing issues under COVID-19 environment. The importance of vitamins and other food supplements is highlighted in Section 15.3 which can raise the immunity level of the individuals to a great extent. Finally, in Section 15.4, the Indian cuisines are compared with the Spanish, Italian, French, American and Chinese dishes to draw certain conclusions. The entire chapter is summed up in Section 15.5.

15.2 SANITIZATION, CLEANLINESS AND SOCIALIZING UNDER COVID-19 AMBIENCE

People around the world are troubled as the corona virus is spreading its tentacles to almost each and every country. Rather than getting distressed with the disease, people should take certain basic precautions to avoid the danger of attack from the virus. In this section, we will discuss mostly the different issues related to sanitization, cleanliness and socializing during the COVID-19 situation.

Usually, any virus remains in the inactive state while it lies on a surface and only gets active when it enters a host (human body) through mouth or eyes. So, one should ensure two things: cleaning the surface you regularly touch, and not to touch mouth or facial parts directly with contaminated hands. These two steps, if followed on a regular basis, the chances of infection due to corona virus can be avoided to a great extent.

Thus, cleanliness must be maintained both at home and at the work place. Regular cleaning of floors with disinfectants is a must these days. Also, the children must not be allowed to crawl, sit and play on the floors. Replacing and washing bedsheets and bedcovers and even pillow covers should also be done regularly.

Metallic doorknobs, toilet taps and electronic switches should be disinfected after use. Cleaning of tables, chairs, laptops, desktops, printers and last but not least mobile phones that we frequently use must be cleaned from time to time. Other regular objects at home and work place can be cleaned with the disinfectant liquids [6]. An aqueous solution of potassium hypochloride can help us clean the different surfaces likely to get contaminated by the virus.

Use of separate hand towels and also normal towels and other toiletry items for individuals can minimize the risk. If possible, use of separate toilets is also recommended on availability. Humid climate is very much conducive for the infection, so additional precautions must be taken during rainy seasons. Avoid sitting in the office and sleeping under air-conditioners (AC) is a good practice. At least, the continuous or prolonged usage of ACs can be avoided to be make individual less susceptible to the infection.

The cleaning of vegetables, brought from the market, with slightly warm water before refrigeration or cooking is necessary. Even people can avoid cleaning it by letting them lie in one corner of the house just after bringing then from the market. If possible, the people can also keep their vegetable bags in a balcony or some space immediately outside their house.

Most people these days are using elevators in their home apartments or in their offices; therefore use of bare hands to operate the lift can be a risky proposition. If no hand gloves are there, some people prefer to use their elbow to operate the elevator buttons. It is worth-mentioning here that one should avoid overcrowded lifts.

Time to time cleaning of hands with soap or hand sanitizers is mandatory [7,8]. Washing hands with soap for 20–30 seconds can easily get rid of most of the germs that have gathered on the hands due to contact with surfaces like tables, chairs, etc. If cleaning with soaps is not always possible or convenient, it is preferable to use pocket-sized sanitizer.

Maintaining social distancing is a compulsory norm being followed everywhere throughout the world. In India, there is a huge population and great population density in big cities, thus it is difficult for individuals to maintain distance between each other while walking down the streets, travelling in a market place, availing public transport or at the work place. The distance of six feet between individuals is to be maintained. If this is not possible, then it is preferable to take certain precautions while in a crowded place. People should get habitual of wearing masks and hand gloves on a regular basis. Head cover can also give an additional benefit. It is also advised to wash our clothes when we come from outside. In addition, taking a shower is also a good safety measure.

Unless not required, it is advisable to avoid public toilets, public places like shopping malls, movie theatres, paying a visit to friend/relatives' residences, etc. It is also preferred to avoid public transport like buses, trains, flights, etc., unless it is a necessity. Public ceremonies must be avoided as far as possible. In Indian conditions, almost 60%–70% people share their rooms and toilets in their houses. It is also preferable to maintain health and hygiene properly to keep ourselves fit and fine.

15.3 VITAMINS AND FOOD SUPPLEMENTS TO BOOST IMMUNITY

Immunity system is present in every organism, from bacteria to primates. The problems related to immunity system are faced by all forms of life [9]. Cells, tissues and organs form a complex network in an immune system which collectively works to protect the body from harmful foreign organisms and defends the body against diseases [10]. Substantial decrease in the quantity of vitamins may increase the susceptibility to infection and diseases.

Deficiency of vitamin D may increase autoimmunity. Autoimmunity is the condition when the body of the organism starts immunizing against its own healthy cells and tissues. Vitamin D promotes calcium homeostasis and bone health. It also enhances absorption of calcium in small intestine. When antibiotics were not invented, vitamin D unknowingly treated several infections. The patients were told to sit in sunlight. Significant reduction in the amount of vitamin D has been observed with ageing process, which means there is insufficient vitamin D supply when a person gets older [11]. Also, the older individual is more prone to corona virus and hence there are more chances that the people with less proportion of vitamin D are more susceptible to have infection from the new corona virus. In Indonesia, 780 people who died of COVID-19 had vitamin D below the normal levels [12].

Corona virus has a major impact on physical activity. Due to the virus outbreak, everyone has been locked inside their homes. The WHO has put several recommendations for different age groups regarding physical workout, for instance, it has been recommended to have moderate to vigorous physical activity of about 60 minutes/day for 6–17 years old age group, and for adults and elderly, it has been recommended to have physical activity of 75 minutes/week to 150 min/week including muscle and bone strengthening of 2–3 days/week. Adapting physical training program at home may decrease negative physiological and psychological impacts of sedentary behavior.

According to the reports of UNESCO (The United Nations Educational Scientific and Cultural Organization), there are around 861.7 million students who are not able to attend their schools due to the corona virus pandemic. Therefore, physical activity can prove one of the efficient ways to keep this young generation active at home. As staying home can lead to building up of stress, bodyweight as well as anxiety, therefore, the optimum way to fight these barriers is to undergo a home-based physical regime which boosts our immune system and keeps us active and motivated [13].

Iron in the human body is also another essential nutrient. It exists in two oxidation states. It must be carefully controlled as its redox potential can also contribute to toxicity [14]. In erythrocytes, major proportion of the iron is usually found. The amount of iron in the human body should be balanced as either iron deficiency or excess can both lead to harmful effects on the functions of cells, tissues and organs. Iron deficiency can be caused due to the improper intake of balanced diet and such people are more susceptible to infections. Excess of iron in some people can be observed in comparison to their calorie requirements [15]. Iron helps in fighting immunity by boosting the production of white blood cells or lymphocytes which plays an imperative role in fighting diseases and infections by protecting the body from harmful invaders like bacteria and viruses. Therefore, iron deficiency may lead to suboptimal production of the red blood cells and can also cause anemia [16]. Consuming iron rich foods can also boost the immunity.

The importance of zinc and its essentiality was known by the humans 40 years ago only in the Middle East. Zinc deficiency affects cognitive development, retards growth and results in severe immune dysfunctions. Zinc has anti-inflammatory actions. It is also an antioxidant. Zinc is responsible in decreasing oxidative stress and inflammatory cytokines. A reprogramming of the immune system occurs in chronic zinc deficiency [17]. We have no storage system in our body to store zinc; therefore, we have to consume it every day. Zinc could suppress inflammation and can inhibit the binding of cold virus. It ensures proper functioning of taste buds and smell senses. It is also responsible for the normal development during different growth stages like childhood and adolescence and also during pregnancy. Zinc also plays an important part in wound healing. It is also responsible for the working of around 300 enzymes in the human body [18].

Many food spices like turmeric, cumin seeds, etc., enhance our immunity. Turmeric boosts the human body's immunity, and the core ingredient in it which is responsible for making the immunity stronger is curcumin which is a phytoderivative and has great healing properties. Curcumin is present in around 3%–5% of turmeric. Antibacterial, anti-inflammatory and antiseptic properties are constituted in turmeric. It helps in fighting cough and cold by cleaning the respiratory tract. Curcumin present in turmeric relieves congestion and pain, inhibits inflammation and also improves breathing. Turmeric helps in fighting virus and helps in reducing their replication [19].

Cumin seeds also known as "jeera" helps to bolster the immune system due to the presence of vitamin C. It is a good source of iron and fiber. It has antibacterial properties that help in keeping infections and diseases at bay and thus making the immune system stronger [20].

Another Indian ingredient which is known to amplify immunity is coriander, because it is rich in anti-oxidants. They prevent cellular damage which is caused by the free radicals. According to the studies, it contains compounds like terpinene, quercetin, and tocopherols which constitute neuroprotective, immune-boosting and anti-cancer effects [21].

Cinnamon spice is a native spice to the Southeast Asia and has also been referred to as the Chinese medicine in old books. It has been used in the treatment of arthritis, nausea, cough, indigestion, abdominal cramps and in many other medical treatments. It has antimicrobial, anti-inflammatory, antioxidant, immunomodulatory and antitumor properties. It also prevents the spoilage of food to some extent. The presence of cinnamaldehyde in cinnamon prevents the growth of fungi including yeasts [22].

It would sound weird but dark chocolates also help to ameliorate immunity as the cocoa present in them and the phenolic compounds found in cocoa can provide a defense mechanism against foreign particles. Dark chocolate is more effective in suppressing a cough than codeine due to the presence of theobromine which is an alkaloid found in cocoa. It can also reduce the risk of cardiovascular disease and also boosts up mood by relaxing blood vessels, improving circulation and blood flow. Cocoa also has good amounts of manganese, iron, copper, magnesium and riboflavin [23].

Nuts like almonds, pistachios and walnuts have high quantities of vitamin E, niacin and riboflavin which are required by the immune system to fight with the bacteria [24]. Nuts are a good source of fat and proteins as well.

Beetroot is a great source of vitamin C which further enhances the immune system. It protects cells from harmful free radicals. It also helps in wound healing and reducing cholesterol levels, thus controlling weight. The consumption of beetroot juice shows an increased flow of blood in the frontal lobes, which is associated with cognitive thinking and behavior [25].

The status of health and hygiene is also one of the important elements in human development of a region. WHO says health should not only be measured by the absence of diseases, rather it should be considered as a complete state of physical, social and mental well-being [26]. WHO recommended two kinds of hand sanitizers to be used during the corona pandemic: the sanitizer must either contain 80% ethanol, 1.45% glycerol and 0.125% hydrogen peroxide or 75% 2-propanol, 1.45% glycerol and 0.125% hydrogen-peroxide. The presence of ethanol and 2-propanol has been proven to reduce the virus levels in 30 seconds to significant amounts [27].

Corona virus has been caused due to a group of a large family of viruses which can result into less severe cold and can further lead into more chronic diseases like Severe Acute Respiratory Syndrome. Wearing masks is thus important as coughing and sneezing can spread microorganisms in the air from an infected person to a healthy person thus spreading this communicable disease. For instance, coughing can produce up to 3000 droplets from the mouth. The size of the corona virus is around 0.07–0.09 μm; therefore, proper masks have to be used in order to restrict the entry of these particles from entering our respiratory tract [28].

Disinfecting our environment can also protect us from the virus. It is a process to reduce the number of harmful microorganisms. According to WHO instructions, high-touch surfaces like door knobs, window handles, kitchen and food preparation areas, bathroom surfaces, work surfaces, touchscreen personal devices, etc., should be disinfected on a regular basis. Disinfectants such as sodium hypochlorite and alcohol at 70%–90% can be used for surface disinfection [29].

Vitamin C neutralizes free radicals which are unstable compounds and hence prevent us from the cellular damage caused by these compounds. One should consume 90 mg of vitamin C in a day. It acts as a barrier for the harmful compounds which try to enter the human body. It fosters the activity of phagocytes which swallow harmful bacteria and develops lymphocytes which are known for increasing antibodies in the body [30].

Folic acid also known as folate is a water-soluble vitamin and also called vitamin B-9. The red blood cells become larger than the normal size if there is a deficiency of folic acid in the human body. This disease is called macrocytic anemia. Folate boosts immunity and also useful in treating depression. An adult should consume 400 μg of folic acid [31].

Selenium is an anti-oxidant and an essential micronutrient. It is a vital component of selenocysteine. Selenium helps in redox control of metabolic functions in cells and in thyroid hormone metabolism [32]. Deficiency of selenium can lead to high risk of thyroid, autoimmune disease, frequent colds and flu and many such illnesses. It acts as a great detoxifier because it is essential for the production of glutathione peroxidase in the liver which protects against the free radical damage. Selenium protects the immune system from the attack of foreign particles [33].

15.4 FOOD HABITS OF DIFFERENT COUNTRIES AND ITS COMPARISON WITH INDIAN CONTEXT: A CASE STUDY

In this section, the popular cuisines of different countries along with their food values are deliberated. Mainly some of the worst corona-affected countries and their food habits are discussed. The ingredients of those cuisines are also provided. The nutritional values of these popular dishes are mentioned. Finally, the Indian food habits are discussed so that a comparative study can be developed. Thus, six subsections are created for the readers to get acquainted with each of the cuisines separately.

15.4.1 Spanish Cuisines

Spanish cuisine has historical roots and has played an important role in the development of food habits in many parts of Europe. The traditional Spanish foods have seen many changes which it underwent due to different conquests. The present food consumed has been derived from them and slight modifications have also been added. The meal routine of Spanish people includes a breakfast in the morning and then a quick snack before lunch. Lunch is an essential meal and included man courses. After that they consume dinner and it is the last meal of the day and may include appetizers before dinner. The famous Spanish cuisines along with their ingredients and food values have been presented in the following table.

Table 15.1 shows the famous cuisines of Spain. Out of all the dishes presented above, only two of them are completely vegetarian, out of them Gazpacho is a famous soup. All the dishes have the nutrients which are required by human body and covers mostly all the vitamins. The foods are rich source of fats and cholesterol and may prove harmful to the body. Although the dishes are full of the types of metals, elements and other nutritional values required by humans but they are present in small quantities hence they fail to impart much immunity and resistance power to the human body.

15.4.2 French Cuisines

The French cuisine was developed over the years and was influenced from its neighboring countries along with its own food habits. The French food has been acclaimed by many travelers who visited there from all over the world. The French food has still a lot of popularity among the people in different countries and some of the dishes have become an important part of many food lovers. One of such dishes is the French fries which has become a dish being served with almost all type of snacks throughout

TABLE 15.1
Famous Spanish Cuisines with Their Ingredients and Nutritional Values [34–36]

Cuisine Name	Ingredients	Nutritional values
Paella	Olive oil, Paprika, Chicken, Garlic, Cloves, Red pepper, Rice, Saffron, Chicken stock, Lemons, Onion, Chorizo sausage, Shrimp	Protein, Dietary fiber, Fat, Cholesterol, Vitamins (A, B6, C), Calcium, Iron, Folate, Magnesium, Potassium, Sodium, Thiamine
Patatas Bravas	Potatoes, Olive oil, Onion, Clove, Garlic, Chilli, Tomatoes, Mayonnaise	Protein, Carbohydrate, Dietary fiber, Fat, Calcium, Iron, Sodium, Magnesium
Gazpacho	Tomatoes, Cucumber, Red bell pepper, Onion, Jalapeno, Cloves, Garlic, Cayenne pepper, Black pepper, Lemon, Vinegar	Calories, Fat, Carbohydrate, Proteins, Vitamins (A, C), Iron, Calcium
Tortilla Espanola	Olive oil, Potatoes, Pepper, Onion, Eggs, Parsley	Calories, Fat, Cholesterol, Potassium, Carbohydrate, Dietary fibers, Sugar, Proteins, Vitamins (A, C, D, E, B-12), Copper, Folate, Iron, Niacin, Phosphorous, Selenium
Chorizo	Pork, Aleppo pepper, Chilli, Garlic, cloves, Black pepper, Coriander, Vinegar	Calories, Fat, Cholesterol, Dietary fibers, Vitamins (A, C, D, E, B-6), Proteins, Calcium, Iron, Magnesium, Niacin, Pantothenic acid, Thiamine, Zinc
Croquettes	Olive oil, Flour, Chicken, Eggs, Breadcrumbs, Butter, Milk	Carbohydrates, Calories, Fat, Sodium, Proteins, Sugars, Vitamins (A, C), Iron, Calcium
Empanada	Flour, Baking powder, Eggs, Chicken stock, Olive oil, Beef, Garlic Vinegar, Cumin, Chilli, Vinegar, Cloves garlic, Bell pepper, Onion, Mayonnaise	Calories, Fat, Cholesterol, Sodium, Potassium, Proteins, Carbohydrates, Vitamins (A, B-6, B-12, E), Copper, Magnesium, Manganese, Iron, Riboflavin, Phosphorous, Zinc
Gambas al Ajillo	Cloves garlic, Shrimp, Paprika, Olive oil, Sherry, Parsley,	Calories, Fat, Cholesterol, Sodium, Potassium, Dietary fibers, Carbohydrates, Vitamins (A, C, D, E, B complex), Calcium, Copper, folate, Iron, Phosphorous, Thiamine, Zinc

the world. French gastronomy was also added to the UNESCO's list of "intangible cultural heritage" in the year 2010.

Table 15.2 showed the famous cuisines eaten by the French people along with their ingredients and nutritional values. The tables showed that mostly the food consumed by the French is nonvegetarian food, and hence, the protein supplement is perfectly fulfilled by the food. Other than that, some of the ingredients such as garlic, cloves, black pepper, thyme, basil and onions help in boosting their resistance power but not too much extent. The French food habits are also rich with cholesterol and some of the dishes contain alcoholic beverages.

TABLE 15.2
Famous French Cuisines with Their Ingredients and Nutritional Values [34–37]

Cuisine Name	Ingredients	Nutritional Values
Soupe a l'oignon	Butter, onion, salt, black pepper, sherry, bay leaf, chicken stock, fish, vinegar, clove garlic, cheese	Calories, fat, cholesterol, sugars, sodium, potassium, vitamins (A, B-12, B-6, C), iron, copper, magnesium, folate, niacin, selenium
Coq au vin	Olive oil, bacon, chicken, onion, carrot, black pepper, garlic, red wine, thyme, garlic, butter, flour	Calories, fat, cholesterol, carbohydrates, sodium, vitamins (A, B-6, B-12, C, D, E), iron, calcium, folate, manganese, riboflavin
Cassoulet	Meat/sausage, thyme, garlic, peppercorns, clove, onions, tarbais, pancetta, bay leaf, black pepper, pork, carrots, tomatoes	Calories, fats, cholesterol, sodium, potassium, proteins, vitamins (A, B-complex, C, D), iron, folate, magnesium, zinc, phosphorous
Boeuf bourguignon	Olive oil, bacon, beef, carrot, onion, cloves garlic, flour, red wine, beef stock, tomatoes, thyme, parsley, mushrooms, butter	Calories, fat, cholesterol, sodium, potassium, carbohydrate, proteins, vitamins (A, B-complex, C, D, E), copper, magnesium, manganese, iron, calcium, thiamine, zinc
Flamiche	Butter, leek, eggs, salt, black pepper, milk	Carbohydrates, fat, protein, vitamins (A, C), cholesterol
Confit de canard	Duck meat/ pork, clove garlic, salt, peppercorns, bay leaves, thymes, butter	Calories, fat, carbohydrates, proteins, vitamins (A, C), iron, copper
Ratatouille	Eggplants, tomatoes, squashes, zucchinis, olive oil, onion, clove garlic, red bell pepper, yellow bell pepper, basil, salt, black pepper, parsley	Calories, sodium, potassium, proteins, vitamins (A, C, E), copper, calcium, iron, magnesium, manganese
Croquet monsieur	Butter, flour, milk, kosher salt, black pepper, nutmeg, gruyere, cheese, ham, mustard, bread	Calories, fat, cholesterol, sodium, potassium, carbohydrates, vitamins (A, b-12, D, E), calcium, iron, phosphorous, pantothenic acid, riboflavin

15.4.3 Italian Cuisines

Italian cuisine is Mediterranean type of cuisine and has been inspired by the development of Italian peninsula over the years. Many food products such as maize, potatoes, tomatoes, capsicums, etc., were introduced into their foods when they colonized over America. Italian cuisine has diversified culture over the country and is one of the most copied and acclaimed cuisine cultures in the world. Italian foods are known for their simplicity in preparation and largely depend on the ingredients rather than the preparation strategies.

Table 15.3 shows the famous Italian cuisines with their ingredients and nutritional values. The Italian cuisines mostly do not contain any animal food and are rich with

TABLE 15.3
Some Popular Italian Cuisines with Their Ingredients and Nutritional Values [36–38]

Cuisine Name	Ingredients	Nutritional Values
Caprese salad with pesto sauce	Tomatoes, holy basil, mozzarella cheese, basil, pine nuts, parmesan cheese, olive oil, lemon	Calories, fat, cholesterol, sugars, sodium, potassium, vitamins (A, B-12, B-6, C, E), iron, calcium, magnesium, folate, niacin, selenium
Bruschetta	Bread, cloves of garlic, tomatoes, basil, olive oil, mushroom, red chilli, butter, salt, black pepper	Calories, sodium, potassium, vitamins (A, B-6, C, E), iron, copper, magnesium, folate, niacin, thiamin
Focaccia	Gram flour, olive oil, sugar, yeast, olives, parmesan cheese, onion, tomatoes, black olives, basil, thyme, black pepper	Calories, sodium, potassium, carbohydrates, vitamins (B-6, E), iron, folate, manganese, niacin, thiamine, phosphorous
Pasta carbonara	Spaghetti, bacon, eggs, parmesan cheese, olive oil, black pepper	Calories, fat, cholesterol, proteins, sodium, potassium, vitamins (A, B-12, B-6, D, C, E), iron, calcium, magnesium, folate, pantothenic acid, manganese
Mushroom risotto	Olive oil, butter, onion, rice, white wine, chicken stock, parmesan cheese, Italian parsley, salt, black pepper	Calories, fat, cholesterol, sugars, sodium, potassium, vitamins (A, B-12, B-6, C, E), iron, calcium, magnesium, folate, riboflavin, phosphorous
Pasta con pomodoro E basilica	Tomatoes, extra virgin olive oil, garlic cloves, basil, chilli flakes, parmesan cheese, salt	Calories, fat, potassium, proteins, vitamins (A, C, E), iron, calcium, phosphorous, pantothenic acid, selenium
Lasagna	Bacon, onions, carrots, celery sticks, olive oil, lamb, tomatoes, sea salt, pepper, basil, pasta sheets, fresh cream, parmesan cheese	Calories, fat, cholesterol, carbohydrates, sodium, proteins, potassium, vitamins (A, B-12, B-6, C, E), iron, calcium, copper, magnesium, manganese, folate, niacin, selenium, zinc
Pistachio panna cotta	Cream, sugar, gelatin, milk, pistachio	Calories, fat, cholesterol, sodium, potassium, vitamins (A, B-12, D, E), calcium, phosphorous, riboflavin

vegetarian components. The cuisines contain mostly the unhealthy ingredients and hence the nutritional values contained in them are very less. The nutritional components are present in a less amount which makes the Italians prone to any diseases. The cuisines are rich with cheese and vegetables which acts as an important aspect for the growth of the body but do not much contribute toward the immunity boost of the body.

15.4.4 Chinese Cuisines

Chinese cuisine has played an important role in the history of China and has been mostly inspired from the traditional food. China being one of the most powerful countries in the history has inspired the food cultures of many countries. Chinese food has been traditionally eaten in a bowl using chopsticks and now is being followed worldwide. Chinese cuisine is known for their cooking techniques and seasoning and is judged by color, smell and taste. These features also describe the nutritional values and meaning of the cuisine prepared.

Table 15.4 presented the famous cuisines of China along with ingredients and their nutritional information. It can be clearly seen from the table that the Chinese mostly consume animal food and is the largest consumer of pork in the world. In China besides, many types of other animals have also been consumed and Chinese are famously known for this. All the Chinese foods consist of some of the essential ingredients which have been proved important for the growth of immunity in

TABLE 15.4
Few Famous Chinese Cuisines with Their Ingredients and Nutritional Values [36–39]

Cuisine Name	Ingredients	Nutritional Values
Almond and chicken momos	Chicken, garlic, carrots, spring onions, ginger, soya sauce, oyster sauce, pepper, eggs, almonds, oil	Calories, fat, cholesterol, sodium, potassium, dietary fibers, proteins, vitamins (A, B6, B12, D), calcium, copper, iron, zinc
Peri peri chicken satay	Chicken, salt, pepper, yogurt, chilli, garlic, coriander, oil, peri peri sauce, potatoes	Fat, cholesterol, sodium, carbohydrates, proteins, iron, calcium
Hakka noodles	Noodles, oil, salt, garlic, ginger, beans, cabbage, carrot, spring onion, bell pepper, dark soy sauce, chilli, tomatoes, beef/chicken/pork (optional)	Calories, fat, potassium, sodium, carbohydrate, vitamins (A, B-6, C, D), calcium, copper, folate, iron, thiamine, proteins
Sweet and sour pork	Soy sauce, rice wine, pepper, garlic, eggs, flour, oil, bell pepper, onion, pork	Fat, cholesterol, sodium, protein, carbohydrates, vitamins (A, B-12, C, D, E), folate, iron, copper, calcium, zinc
Char siu	Pork, soy sauce, honey, ketchup, sugar, rice wine, oil, pepper, hoisin sauce, food colors, Chinese traditional spices	Proteins, carbohydrates, fats, vitamins (A, B-6, C), niacin, folate, calcium, iron, sodium, magnesium, thiamine
Beef and broccoli	Beef, soy sauce, sherry, oil, onion, garlic, ginger, broccoli, beans, chicken stock, sugar, pepper, corn starch	Calories, carbohydrates, proteins, fat, cholesterol, sodium, fibers, vitamins (A, C, D, E), calcium, iron
Sweet and crispy beef	Beef, potato starch, onion, garlic, ginger, soy, sauce, honey, oil, vinegar, sesame oil, edible wine, corn starch, food color	Calories, carbohydrates, fat, sodium, proteins, potassium, dietary fibers, vitamins (A, B-complex, C, D, E), iron, copper, manganese, magnesium, calcium, phosphorous
Ginger soy steamed fish	Fish, spring onions, cilantro leaves, ginger, clove garlic, canola oil, soy sauce, oyster sauce, sesame oil, sugar, corn starch, pepper	Calories, fat, carbohydrates, sodium, cholesterol, proteins, vitamins (A, B-complex, C, D, E), calcium, iron, manganese, magnesium, phosphorous

humans. The foods are also rich with much required nutrients required by human body. The disadvantage faced by Chinese food is extensive consumption of animals, and these animals may carry different microorganisms which can also prove out to be the pathogens of some diseases, and particularly during the pandemic situation, people must avoid the consumption of animal-based cuisines.

15.4.5 American Cuisines

American cuisines are the blends of the various groups who resided in America and belonged to various parts of the world. The complete food culture of the Americans is the fusion of the various colonists such as the European cuisines along with the native Latin American food habits. This has also been the reason of introducing many new vegetables and fruits in the food culture as they were cultivated for the first time by the colonist Europeans.

Table 15.5 shows the main cuisine of America and the ingredients required for them along with their nutritional values. The American cuisine mostly consists of animal food and is junk in nature. Due to these reasons, the cuisine is not much healthy for the humans particularly during the pandemic situation although they consist of the required nutrients. Besides, the animal food has also proved less effective and people should minimize the intake of these products.

15.4.6 Indian Cuisines

Indian cuisine developed over the years and depends on the type of soil it is grown, the ethnic group consuming the food and based on the regional diversity. Indian cuisine has been known for its rich heritage and culture and has changed very less. Indian cuisine is known for the spices and additives added to enhance the taste of the food and is considered as very healthy food consisting of all the required nutrients in adequate amount for the growth of body. Currently, the food habits have been changed a bit due to the cultivation of food habits of other countries.

Table 15.6 shows the famous Indian cuisines along with their ingredients and nutritional information. It can be noted very clearly that the Indian cuisine is rich in nutrients essentially required by the body for growth and other purposes. All the Indian cuisine consists of food items such as ginger, garlic, turmeric, cloves, cardamom, cinnamon, etc. These items play an important role in boosting the immunity of the body and also have many other benefits. During global pandemic situation, where the sole purpose of every human has become to keep the body healthy, Indian cuisine is the best choice present before them. Some of the cuisines consist of animal food which may not prove much healthier during pandemic situation and a person must refrain from consuming them. Other than that, the cuisine consists essentially of vegetarian foods which contains authentic Indian spices and proves very helpful in these pandemic situations.

15.5 CONCLUSIONS

This chapter provides a comprehensive set of information to prevent and fight against the pandemic that brought the world to almost a standstill. Several aspects starting

TABLE 15.5
Some Popular American Cuisines with Their Ingredients and Nutritional Information [36–38,40]

Cuisine Name	Ingredients	Nutritional values
Creamy potato salad with bacon	Potatoes, black pepper, kosher salt, bacon, mayonnaise, sour cream, wine vinegar, celery sticks, parsley, tarragon	Calories, fat, cholesterol, sodium, potassium, dietary fibers, proteins, vitamins (A, B12, C, D, E), calcium, iron, zinc, magnesium, niacin, pantothenic acid, phosphorous, riboflavin, selenium
Baked beans	Soldier beans, molasses cider vinegar, mustard, brown sugar, black pepper, onion, bacon	Calories, cholesterol, sodium, potassium, proteins, vitamins (A, B-6, B12, E), calcium, iron, manganese, niacin, pantothenic acid, phosphorous, riboflavin
Meat loaf	Beef, onion, onion soup, Italian cheese, parsley, eggs, saltines, barbeque sauce	Calories, fat, cholesterol, sodium, potassium, carbohydrates, proteins, vitamins (A, B-6, B12, C, D, E), calcium, copper, iron, zinc, magnesium, manganese, niacin
Not-so-sloppy joes	Olive oil, onion, garlic, bell pepper, beef, tomato paste, chilli powder, cumin, kosher salt, black pepper, cinnamon, hamburger buns, cheddar cheese, sour cream	Calcium, calories, carbohydrates, cholesterol, fibers, fat, iron, protein, sodium
Hamburger	Potatoes, olive oil, kosher salt, pepper, beef, hamburger buns, mayonnaise, butter, lettuce, tomatoes, olives	Calories, fat, cholesterol, sodium, proteins, carbohydrates, sugar, fibers, iron, calcium, vitamins (A, B-6, C), folate
Fried chicken	Chicken, buttermilk, garlic, gin, breadcrumbs, parmesan cheese, salt, lemon zest, thyme leaves, olive oil	Calcium, calories, carbohydrates, cholesterol, fat, fiber, iron, protein, sodium, potassium, vitamins (A, B-6, C), folate
Hotdogs	Beef, hotdog bun, yellow mustard, green pickle, onion, tomato, white sauce, ketchup, spicy chilli sauce, celery stick, lettuce	Protein, carbohydrate, fat, cholesterol, vitamins (A, B-6, C), folate, calcium, iron, potassium, sodium, thiamine, fibers

TABLE 15.6
Some Popular Indian Cuisines with Their Ingredients and Nutritional Values [37,41,42]

Cuisine Name	Ingredients	Nutritional Values
Malai kofta	Potatoes, cottage cheese, flour, coriander leaves, onions, garlic, ginger, tomatoes, cream, raisins, cashew nuts, turmeric, fenugreek, black pepper	Calories, fat, cholesterol, sodium, potassium, dietary fibers, proteins, vitamins (A, B-6, B12, C, D, E), calcium, copper, iron, zinc, magnesium, manganese, niacin, pantothenic acid, phosphorous, riboflavin, selenium, thiamine

(Continued)

TABLE 15.6 (*Continued*)
Some Popular Indian Cuisines with Their Ingredients and Nutritional Values [37,41,42]

Cuisine Name	Ingredients	Nutritional Values
Palak Paneer	Spinach, cottage cheese, oil, cumin seeds, bay leaf, ginger, garlic, onion, tomato, garam masala, red pepper, coriander, cardamom, cloves, nutmeg, cream	Calories, fat, cholesterol, sodium, dietary fibers, proteins, vitamins (A, C, D, E), calcium, iron, zinc, magnesium, manganese, niacin, pantothenic acid, phosphorous, riboflavin, selenium, thiamine
Rajma masala	Kidney beans, onion, tomatoes, garlic, ginger, chillies, coriander powder, coriander leaves, red chilli powder, garam masala, cumin seeds, fenugreek, turmeric, cloves, cream, cardamom	Calories, fat, cholesterol, sodium, potassium, dietary fibers, proteins, vitamins (A, B6, B12, C, D, E), calcium, copper, iron, zinc, magnesium, manganese, niacin, pantothenic acid, phosphorous, riboflavin, selenium, thiamine
Chole masala	Chickpeas, Indian gooseberries powder, onion, tomatoes, garlic, cloves, cardamom, ginger, turmeric, chilli, garam masala, dried mango power, oil, cinnamons, Indian bay leaf, carom seeds, cumin seeds, fenugreek seeds, coriander powder, coriander leaves, lemon	Calories, fat, cholesterol, sodium, potassium, dietary fibers, proteins, vitamins (A, B6, B12, C, D, E), calcium, copper, iron, zinc, magnesium, manganese, niacin, pantothenic acid, phosphorous, riboflavin, selenium, thiamine
Matar paneer	Cottage cheese, green peas, cumin seeds, carom seeds, cloves, cardamom, turmeric, garlic, ginger, garam masala, oil, coriander leaves, coriander powder, chilli, tomato, onion, cashews, peppercorns, cinnamon	Calories, fat, cholesterol, sodium, potassium, dietary fibers, proteins, vitamins (A, B6, B12, C, D, E), calcium, copper, iron, zinc, magnesium, manganese, niacin, pantothenic acid, phosphorous, riboflavin, selenium, thiamine
Butter chicken	Chicken, chillies, garlic, ginger, salt, lemon, juice, yogurt, garam masala, mustard oil, fenugreek leaves, oil, butter, cloves, cinnamon, cardamom, nutmeg, tomatoes, onion, carom seeds, coriander powder, cream	Calories, fat, cholesterol, sodium, potassium, dietary fibers, proteins, vitamins (A, B6, B12, C, E), calcium, copper, iron, zinc, magnesium, manganese, niacin, pantothenic acid, phosphorous, riboflavin, selenium, thiamine
Mutton curry	Mutton yogurt, ginger, garlic, garam masala, mustard oil, bay leaf, cardamom, cinnamon, cloves, onion, tomatoes, potatoes, coriander powder, coriander leaves, turmeric powder, red chilli powder, cumin seeds, cumin powder, lemon juice	Calories, fat, cholesterol, sodium, potassium, dietary fibers, proteins, vitamins (A, B6, B12, C, D, E), calcium, copper, iron, zinc, magnesium, manganese, niacin, pantothenic acid, phosphorous, riboflavin, selenium, thiamine
Kadha (herbal tea)	Clove, ginger, black pepper, holy basil, honey, cinnamon, Indian ginseng, heart leaved moonseed	Calories, sodium, potassium, vitamins (A, B6, B12, C, D, E), zinc, magnesium, manganese, niacin, pantothenic acid, riboflavin, thiamine

from sanitization, maintenance of cleanliness and social distancing, utilities of vitamins and other food supplements, and last but not the least, the Indian food ingredients in comparison to some other popular food cuisines of the world have been covered in this chapter. Maintaining a healthy regime through daily exercise, intake of good foods with adequate nutritional value, keeping our homes clean, basic sanitization steps of individuals, taking immunity boosting food supplements and vitamins can prevent us to a considerable extent. It is also possible to stay indoors throughout our lives safely. So, everyone needs to reach their work place, schools and colleges following daily routine within a month or so making use of public transport system. Further, people also need to travel to places to attend meetings and conferences, public functions in distant places, as well as their home towns. Naturally they have to take adequate precautions to avert any chances of danger. Further, cold and flues are never to subside from our lives permanently. So, individuals should learn to live with it and fight it out suitably. Medicines and vaccines are expected to come in due time but the disease cannot be evaded out totally.

REFERENCES

1. Coriander, Cumin and Fenugreek: Add These Three Spices to Your Diet for Healthy Living. *NDTV Food.* Retrieved from https://food.ndtv.com/food-drinks/coriander-cumin-and-fenugreek-add-these-three-spices-to-your-diet-for-healthy-living-2004664
2. Tabe, L. (1998). Engineering Plant Protein Composition for Improved Nutrition. *Trendws in Plant Science*, 3(7), 282–286. doi: 10.1016/s1360–1385(98)01267-9
3. Patrone, D., & Resnik, D. (2011). Pandemic Ventilator Rationing and Appeals Processes. *Health Care Analysis*, 19(2), 165–179. doi: 10.1007/s10728-010-0148-6
4. Scheunemann, L. P., & White, D. B. (2011). The Ethics and Reality of Rationing in Medicine. *Chest*, 140(6), 1625–1632. doi: 10.1378/chest.11–0622
5. Ozawa, S., Evans, D. R., Bessias, S., Haynie, D. G., Yemeke, T. T., Laing, S. K., & Herrington, J. E. (2018). Prevalence and Estimated Economic Burden of Substandard and Falsified Medicines in Low- and Middle-Income Countries. *JAMA Network Open*, 1(4), e181662. doi: 10.1001/jamanetworkopen.2018.1662
6. World Health Organization Writing Group. (2006). Nonpharmaceutical Interventions for Pandemic Influenza, National and Community Measures. *Emerging Infectious Diseases*, 12(1), 88–94. doi: 10.3201/eid1201.051371
7. Hermann, C. US9123233B2 – Systems for Monitoring Hand Sanitization. Granted on 1 September 2015.
8. Sultan, S., Lim, J. K., Altayar, O., Davitkov, P., Feuerstein, J. D., Siddique, S. M., Falck-Ytter, Y., El-Serag, H. B. (2020). AGA Institute Rapid Recommendations for Gastrointestinal Procedures during the COVID-19 Pandemic, *Gastroenterology.* doi: 10.1053/ j.gastro.2020.03.072.
9. Nicholson, L. B. (2016). The Immune System. *Essays in Biochemistry*, 60(3), 275–301. doi: 10.1042/ebc20160017
10. Immune System – Latest Research and News | Nature. Retrieved from https://www.nature.com/subjects/immune-system
11. Biesalski, H. K. (2020). Vitamin D Deficiency and Co-morbidities in COVID-19 Patients – A Fatal Relationship? *NFS Journal*, 20, 10–21. doi: 10.1016/j.nfs.2020.06.001
12. Vitamin D Affects Covid-19 Mortality Pharmaceutical Technology. Retrieved from https://www.pharmaceutical-technology.com/comment/vitamin-d-covid-19/

13. Hammami, A., Harrabi, B., Mohr, M., & Krustrup, P. (2020). Physical Activity and Coronavirus Disease 2019 (COVID-19): Specific Recommendations for Home-Based Physical Training. *Managing Sport and Leisure*, 1–6. doi: 10.1080/23750472.2020.1757494
14. Cassat, J. E., & Skaar, E. P. (2013). Iron in Infection and Immunity. *Cell Host & Microbe*, 13(5), 509–519. doi: 10.1016/j.chom.2013.04.010
15. Cherayil, B. J. (2010). Iron and Immunity: Immunological Consequences of Iron Deficiency and Overload. *Archivum Immunologiae et Therapiae Experimentalis*, 58(6), 407–415. doi: 10.1007/s00005–010–0095–9
16. Iron for Immunity: How Food Rich in Iron Can Help You Combat Diseases | Inquirer Lifestyle. Retrieved from https://lifestyle.inquirer.net/361760/iron-for-immunity-how-food-rich-in-iron-can-help-you-combat-diseases/
17. Prasad, A. S. (2008). Zinc in Human Health: Effect of Zinc on Immune Cells. *Molecular Medicine*, 14(5–6), 353–357. doi: 10.2119/2008–00033.prasad
18. te Velthuis, A. J. W., van den Worm, S. H. E., Sims, A. C., Baric, R. S., Snijder, E. J., & van Hemert, M. J. (2010). Zn^{2+} Inhibits Coronavirus and Arterivirus RNA Polymerase Activity in Vitro and Zinc Ionophores Block the Replication of These Viruses in Cell Culture. *Plos Pathogens*, 6(11), e1001176. doi: 10.1371/journal.ppat.1001176
19. Use Power of Turmeric and Boost Your Immunity with Curcumin in the Time of Viral Infections. *The Financial Express.* Retrieved from https://www.financialexpress.com/lifestyle/health/use-power-of-turmeric-and-boost-your-immunity-with-curcumin-in-the-time-of-viral-infections/1899124/
20. 7 Benefits of Cumin (Zeera) You Must Know. *NDTV Food*. Retrieved from https://food.ndtv.com/food-drinks/7-benefits-of-cumin-zeera-you-must-know–1826677
21. 8 Surprising Health Benefits of Coriander. Retrieved from https://www.healthline.com/nutrition/coriander-benefits
22. Hamidpour, R., Hamidpour, M., Hamidpour, S., & Shahlari, M. (2015). Cinnamon from the Selection of Traditional Applications to Its Novel Effects on the Inhibition of Angiogenesis in Cancer Cells and Prevention of Alzheimer's Disease, and a Series of Functions Such as Antioxidant, Anticholesterol, Antidiabetes, Antibacterial, Antifungal, Nematicidal, Acaracidal, and Repellent Activities. *Journal of Traditional and Complementary Medicine*, 5(2), 66–70. doi: 10.1016/j.jtcme.2014.11.008
23. Wellbeing – Chocolate: A Delicious Way to Boost Immunity – The Healthy Chef. Wellbeing – Chocolate: A Delicious Way to Boost Immunity – The Healthy Chef. Retrieved from https://thehealthychef.com/blogs/wellbeing/5-reasons-why-chocolate-is-good-for-you
24. 5 Powerful Foods to Boost Your Immune System This Cold and Flu Season. *Daily News*. Retrieved from https://www.dailynews.com/2015/10/20/5-powerful-foods-to-boost-your-immune-system-this-cold-and-flu-season/
25. Beetroot Juice: 12 Health Benefits. Retrieved from https://www.healthline.com/health/food-nutrition/beetroot-juice-benefits
26. Sharma, A. & Bhuyan, B. (2018). A Study on Health and Hygiene Practices among the Tea Garden Community of Dibrugarh District, Assam. *Indian Journal of Applied Research*, 8, 489–490.
27. Studies: Hand Sanitizers Kill COVID-19 Virus, E-consults Appropriate | CIDRAP. Retrieved from https://www.cidrap.umn.edu/news-perspective/2020/04/studies-hand-sanitizers-kill-covid-19-virus-e-consults-appropriate
28. Why We Should All Be Wearing Face Masks. *BBC Future*. Retrieved from https://www.bbc.com/future/article/20200504-coronavirus-what-is-the-best-kind-of-face-mask
29. Q&A: Considerations for the Cleaning and Disinfection of Environmental Surfaces in the Context of COVID-19 in Non-health Care Settings. Retrieved from https://www.who.int/news-room/q-a-detail/q-a-considerations-for-the-cleaning-and-disinfection-of-environmental-surfaces-in-the-context-of-covid-19-in-non-health-care-settings

30. Can Vitamin C Protect You from COVID-19? Retrieved from https://www.healthline.com/nutrition/vitamin-c-coronavirus
31. Content—Health Encyclopedia—University of Rochester Medical Center. Retrieved from https://www.urmc.rochester.edu/encyclopedia/content.aspx
32. Arthur, J. R., McKenzie, R. C., & Beckett, G. J. (2003). Selenium in the Immune System. *The Journal of Nutrition*, 133(5), 1457S–1459S. doi: 10.1093/jn/133.5.1457s
33. Strengthen Your Immune System with Selenium! | Liver Doctor. Retrieved from https://www.liverdoctor.com/strengthen-immune-system-selenium/
34. Expatica | The Largest Online Resource for Expat Living. Retrieved from https://www.expatica.com/
35. Allrecipes | Food, Friends, and Recipe Inspiration. Retrieved from https://www.allrecipes.com
36. Retrieved from https://recipes.sparkpeople.com
37. Food News, Health News, Indian Recipes, Healthy Recipes, Vegetarian Recipes, Indian Food recipes. *NDTV Food*. Retrieved from https://food.ndtv.com
38. Retrieved from https://www.thespruceeats.com
39. Travel The World Through Your Taste Buds! *The Daring Gourmet*. Retrieved from https://www.daringgourmet.com
40. Real Simple: Home Decor Ideas, Recipes, DIY & Beauty Tips. Retrieved from https://www.realsimple.com
41. Indian Curries Recipes. Retrieved from https://www.vegrecipiesofindia.com
42. Tried & Tested Step by Step Indian and International Recipes. *WhiskAffair*. Retrieved from https://www.whiskaffair.com

Measures [illegible] in [illegible] conditions

26. [illegible] Vitamin C [illegible] [illegible]
27. [illegible] Health Encyclopedia – University of Rochester Medical Center [illegible]
28. [illegible] Selenium in the [illegible] [illegible]
[illegible]
[illegible] Retrieved from [illegible]
[illegible] Retrieved from [illegible]
[illegible] Retrieved from [illegible]
[illegible] Retrieved from [illegible]
41. Indian [illegible]
42. [illegible] Step by Step Indian and International Recipes [illegible]

16 IoT-Based Automatic Corona Virus Detection and Monitoring System

Dipak P. Patil and Amit Mishra
SIEM Sandip Foundation

Tushar H. Jaware
R C Patel Institute of Technology

CONTENTS

16.1 INTRODUCTION

16.1.1 OVERVIEW

Nowadays global lockdown is there due to pandemic and epidemic caused by COVID-19. To detect COVID-19 Polymerase Chain Reaction (PCR) tests are carried out but are expensive and cannot be afforded by most of the countries [1]. So to have a solution for this issue, there is a need to develop a cost-effective solution. For accurate detection of corona virus it is very much necessary to design and develop an efficient framework based on Internet of Things (IoT) technique [6].

COVID-19 is a disease caused by corona virus. People those who get infected with corona experience fever, cough, etc. Especially seniors of underlying medical problems such as blood pressure, lung disease and low immune system have chances of developing severe illness.

The best way of stopping and reducing its spread is by protecting ourselves and by washing hands or using alcohol-based sanitizer; hands need to be rubbed frequently. Scientists and doctors are trying to find vaccines or treatments for this disease but till date nothing has been found.

16.1.2 Preventions

To reduce the spread of corona virus, we should do the following things:

a. Rub your hand with alcohol-based sanitizer
b. Cover mouth and face with mask
c. Stay home and only come out if it is very essential or if you are feeling unwell then stay home
d. Maintain social distance

16.1.3 Symptoms

The common symptoms of COVID-19 are as follows [10]:
Clean Toast
Fatigue

Less Frequent Symptoms
Diarrhea
Checkers
Lose taste

Serious Symptoms
Difficulty in Breathing
Chest Pain
Its takes around one week to find the symptoms of COVID-19 [8].

16.1.4 Present Situation

There were 9,236,128 confirmed COVID-19 cases worldwide [7]. Figures below depict the present scenario of COVID-19 around the world (Figure 16.1).

The above graphical interface is showing the current scenario of COVID-19 around the globe. This helps people from all over the world to get updates of COVID-19

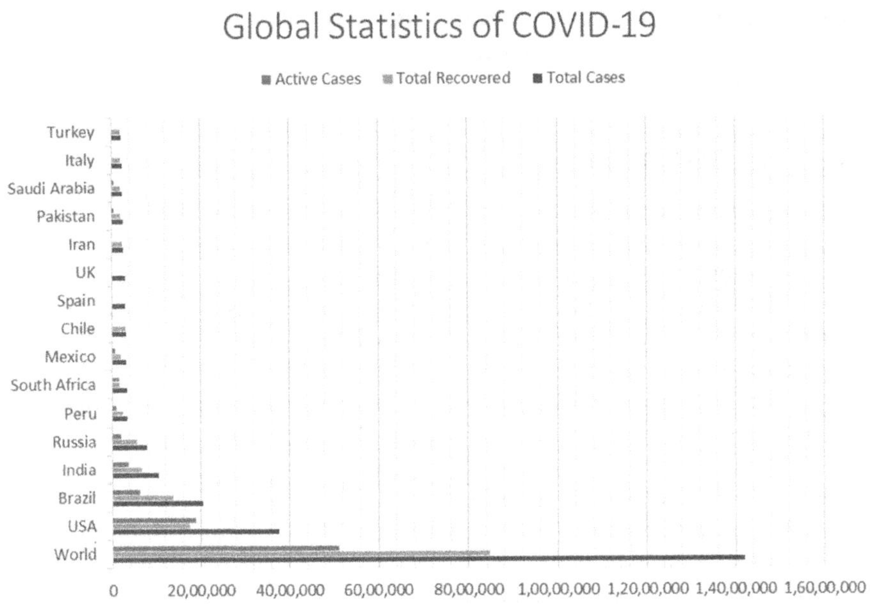

FIGURE 16.1 Global statistics of COVID-19.

cases; like number of cases, spreading frequency and region of infected patients. This provides useful information to the society about COVID-19.

Figure 16.2 gives regionwise data of patients suffering from novel corona virus disease in various regions identified by World Health Organization [7]. Six regions

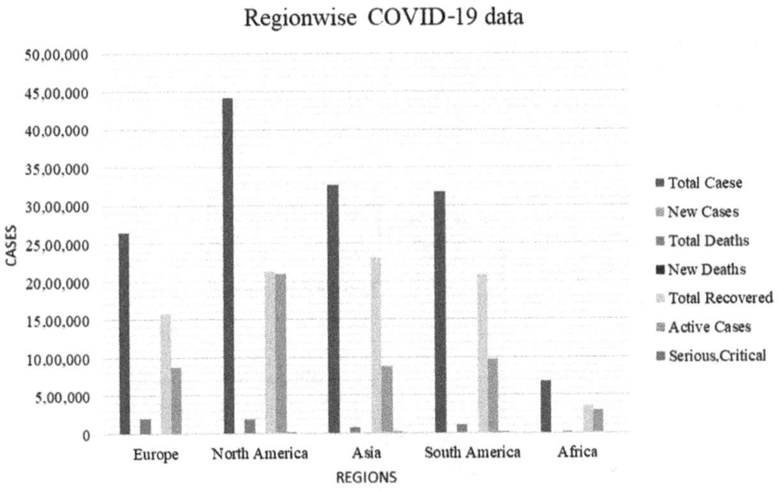

FIGURE 16.2 Region wise COVID-19 data.

such as America, Europe, Asia, South-east Asia, Western pacific and Eastern Mediterranean are considered and live updates of these regions are provided through this graphical dashboard [11].

16.2 RELATED WORKS

Thermal imaging technique consisting of thermal camera mounted on smart helmet is proposed by Abdulrazaq et al. [2]. The system uses IoT to store database at the cloud end. The system is costly as it involves thermal camera and smart helmet. Md. Siddikur Rahman et al. [3] proposed IoT-based health monitoring system; the system may use IoT but it is in the early stage of designing. The concept is in womb stage.

Ravi Pratap Singh et al. [4] highlights the whole implementation of the well-known theory of IoT by providing the peek guide for tackling the pandemic called COVID-19. They discussed various IoT applications for healthcare. Ravi Pratap Singh et al. [5] discussed the possibility of tackling the current COVID-19 pandemic through the introduction of the Internet of Medical Things (IoMT) method, thus providing orthopedic patients care. Swati and Chandana enlighten [10] review and analysis which includes numerous novel applications of Cognitive IoMT to address the global health crisis, COVID-19. From the study of extensive literature review, its immediate need is to develop an IoT-based cost-effective and powerful solution to combat the COVID-19 crisis.

16.3 SYSTEM DESIGN

16.3.1 Hardware Implementation

We have implemented a prototype of this system on Node MCU. Block diagram of this has been shown below (Figure 16.3):

This system will detect the temperature and humidity of environment. For this digital temperature and humidity (DHT) sensor has been used. Here we will see about various parts of this system in detail (Figure 16.4).

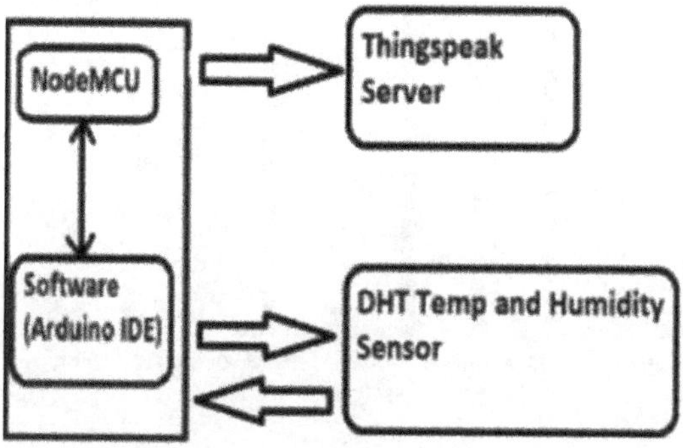

FIGURE 16.3 Block diagram of "Temperature and Humidity" detection from DHT sensor—prototype.

FIGURE 16.4 Node MCU.

16.3.1.1 Node MCU

Following are the specifications of Node MCU ESP8266:

Microcontroller: RISC Xtensa LX106 32-bit Tensilica CPU.
Voltage to operate: 3.3 V.
Voltage to input: 7–12 V.
Digital pins I/O (DIO): 16.
Input analog pins (ADC): 1.

16.3.1.2 DHT 11 Sensor

The above system consists of Node MCU which works with Arduino software (Arduino IDE; Figures 16.5 and 16.6).

The following are the specifications of DHT 11 sensor:

PCB dimensions: 22.0 mm × 20.5 mm × 1.6 mm
Voltage service: 3.3 or 5 V DC
Range of measurement: 20%–95% RH; 0°C–50°C
Resolution: 8-bit (humidity), 8 bit (temperature)

FIGURE 16.5 DHT sensor.

FIGURE 16.6 Interfacing of DHT sensor with Node MCU.

16.3.1.3 ThingSpeak Server

ThingSpeak is an IoT analytics platform which allows visualizing, segregating, visualizing and evaluating live cloud data sources. We can send out data to ThingSpeak device and also send alerts with instant visualization of live data (Figures 16.7 and 16.8).

Insertion into ThingSpeak: ThingSpeak is a free IoT database. Using a RESTful API, your computer or application will communicate with ThingSpeak, and either keep your data private or make it available. Also, use ThingSpeak to analyze the data and act on it.

FIGURE 16.7 ThingSpeak server webpage.

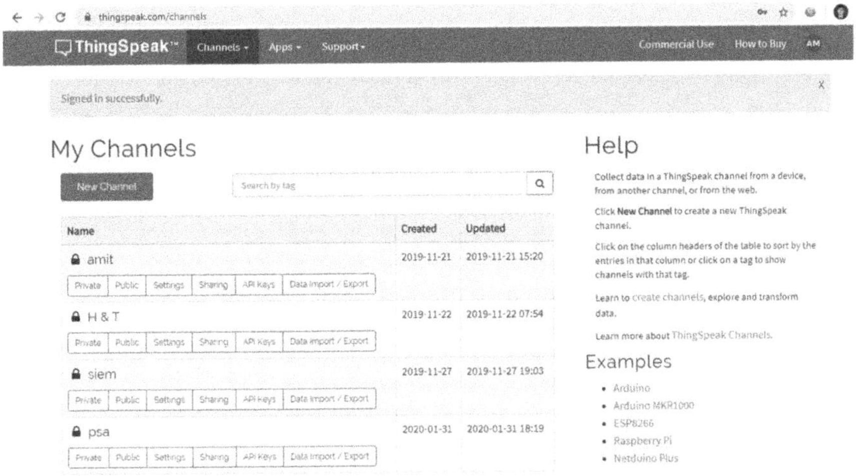

FIGURE 16.8 Channel creation on ThingSpeak server.

Channels on ThingSpeak. You start by creating a ThingSpeak channel once you have signed up and signed into ThingSpeak. There is a channel where you send the data to store. Each channel contains eight fields for any data type, three location fields and one status field.

Steps involved in channel creation.

Build a Channel:

1. Sign in to ThingSpeakTM or create a new account using your MathWorks® account Account to MathWorks:
2. Tap on > MyChannels.
3. Click on New Channel, on the Channels page.
4. Check Fields 1–3 boxes next door. Enter the setting values for these channels:
5. Click at the bottom of the options, Save Tab. You will see those tabs now (Figure 16.9).

You will need proper read and write permissions when reading or writing data to your server using the ThingSpeak™ API (application programming interface) or MATLAB® code. Because of API 16-digit key, we are reading and writing from a private computer to a server.

16.3.1.3.1 API Keys

- Click on > My Channels.
- Choose the Update channel.
- Pick API Tab Keys (Figure 16.10)

After sign-in using MathWorks account and creating channel we can have this window to set parameters of channel.

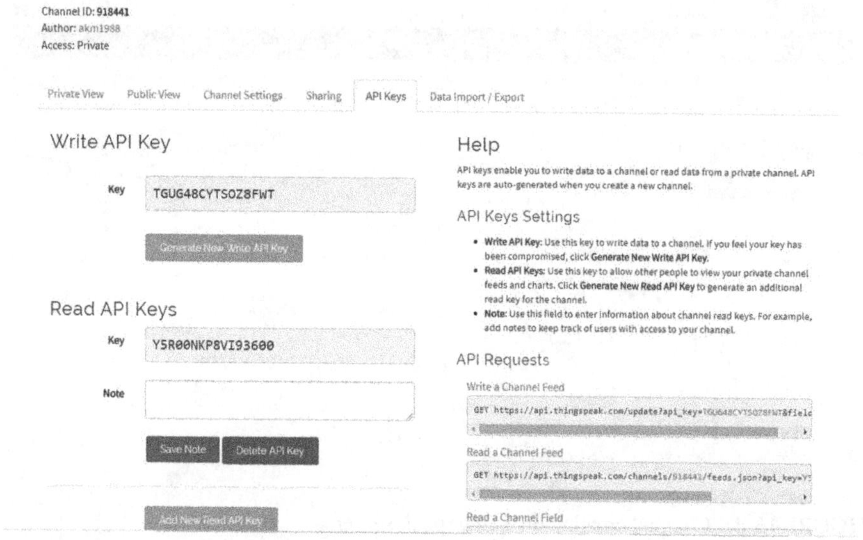

FIGURE 16.9 API key of channel on ThingSpeak.

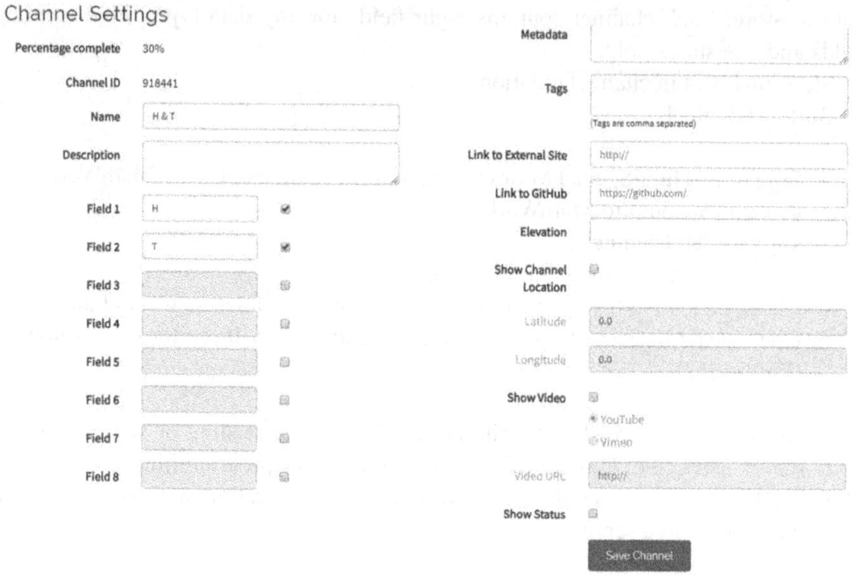

FIGURE 16.10 Channel settings on ThingSpeak.

16.3.1.4 Arduino IDE

Arduino IDE is a software in which we write program and upload programs to Arduino boards or Node MCU (Figure 16.11).

Arduino Software (IDE) allows code writing and submission to server. This is operating on Windows, Linux, Mac OS X.

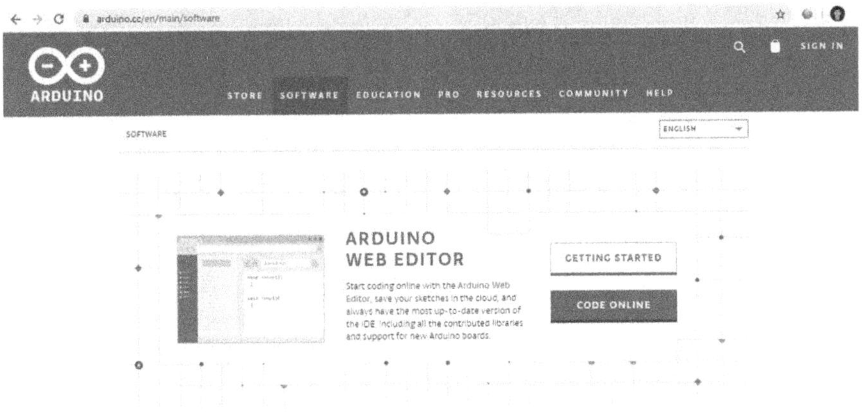

FIGURE 16.11 Arduino IDE.

16.3.1.4.1 Reading on ThingSpeak

The above reading is observed by DHT11 sensor and recording on ThingSpeak server (Table 16.1).

Figure 16.12 gives the graphical representation of humidity and temperature taken by DHT11 sensor and uploaded on ThingSpeak server. The above prototype design is very much effective and readings taken are almost 100% accurate. Now we will see the system that will detect corona-effective patient.

Note: Operation of prototype is available on YouTube.

https://www.youtube.com/watch?v=UH-QgFmGe6I.

TABLE 16.1
Reading on ThingSpeak Server

Created_at	Entry_id	Humidity	Temp in C
2019-11-22 08:13:14 IST	1	nan	nan
2019-11-22 08:13:31 IST	2	56	24
2019-11-22 08:13:48 IST	3	55	24
2019-11-22 08:14:06 IST	4	56	24
2019-11-22 08:14:24 IST	5	56	24
2019-11-22 08:14:41 IST	6	55	24
2019-11-22 08:15:01 IST	7	55	24
2019-11-22 08:15:18 IST	8	56	24
2019-11-22 08:15:38 IST	9	56	25
2019-11-22 08:15:56 IST	10	56	25
2019-11-22 08:16:13 IST	11	56	25
2019-11-22 08:16:31 IST	12	56	25
2019-11-22 08:16:49 IST	13	55	25
2019-11-22 08:17:06 IST	14	55	25
2019-11-22 08:17:23 IST	15	55	25

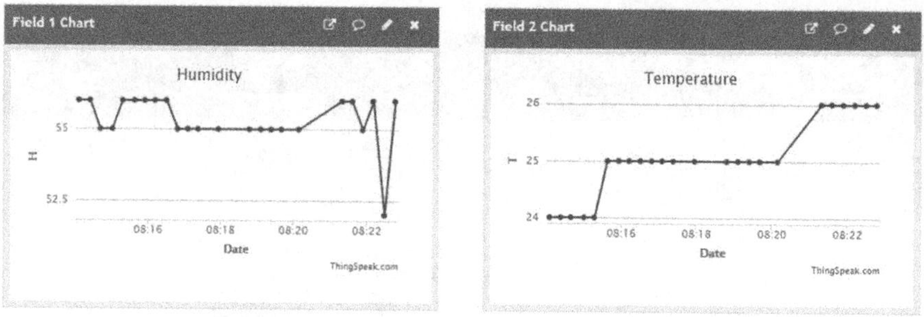

FIGURE 16.12 Graphical representation of humidity and temperature reading on ThingSpeak server.

16.4 PROPOSED SYSTEM FOR DETECTION OF CORONA PATIENTS

16.4.1 Introduction

Nowadays global lockdown is there due to pandemic and epidemic caused by COVID-19. To detect COVID-19 PCR tests are carried out but are expensive and cannot be afforded by most of the countries. So to have a solution for this issue, there is a need to develop cost-effective solution. For accurate detection of corona virus, it is very much necessary to design and develop an efficient framework based on IoT technique.

COVID-19 was first found in China and has affected public health and global economy, causing global lockdown. COVID-19 is highly infectious and spreads rapidly worldwide making its early diagnosis of great importance. The novel corona virus causes damage inside lungs of COVID-19 patients. Damages are mainly located in the lung periphery, subpleural area and both lower lobes. To find the exact damage location and to help doctors save time, lives screening system is necessary.

Nowadays the whole world is under scariness of the virus named as COVID-19 (corona virus). Its effect is so lethal that it has put most of the countries into lockdown. Corona virus spreads by the discharge of nose saliva from an infected person sneeze.

Till now there are no clear COVID-19 vaccines or therapies. There are several experiments going on to evaluate vaccine for COVID-19 [8]. We have also worked in that direction and come up with the system named as "IoT-based Automatic Corona Virus Detecting and Monitoring System," which can detect corona-affected person and also generate its database through which monitoring will be easy.

IoT can be defined in many ways like:

- anything connected through Internet
- connectivity of physical world with the computer system

Examples of IoT are Wearable Technologies, Health Care, Smart Appliances, etc.

There are different hardwares available like Rasberry Pi, Arduino, etc., for the implementation of IoTs. Our system has been implemented on Node MCU because Arduino has several limitations like Separate ESP Module which is required for

Wi-Fi Connectivity; due to this connectivity issue occurs. Node MCU is a hardware platform which has in-built ESP module (8266 module). We must be aware of the specifications of Node MCU platform used in our system. These are as follows:

10 input-output pins, input voltage—3.3 V, DC Current—250 mA, in-built wi-fi, RAM of 32 KB, DRAM of 80 KB, flash—200 KB, 16 MHz clock frequency.

16.3.2 Arduino IDE

An open-source, Arduino IDE is a platform which supports Node MCU. File required for interfacing Node MCU with Arduino IDE is hidden. Hence path to unhide is as follows: C drive – Users – App Data – local – Arduino 15. In India all devices have 9600 baud rate; transmitter and receiver should have the same baud rate. So in programming Serial begin (9600) has to be mentioned.

16.3.3 Hardware Implementation

The system shown in this section will detect the corona-affected person (Figure 16.13).

Components required to implement in our system are as follows:

Breadboard: 01
Node MCU: 01
Connecting wires: 20
Potentiometer: 01
MOSFET: 01
Resistors -1k: 1 piece
10K: 2 pieces
UJT: 01
IR thermometer: 01[9]

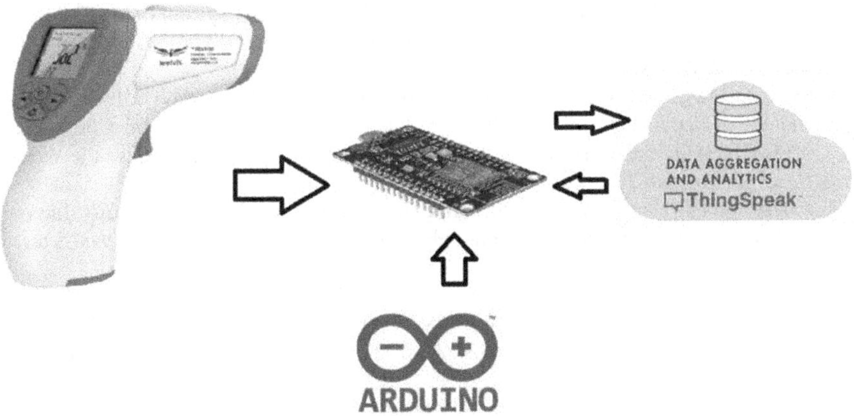

FIGURE 16.13 Block diagram of the proposed system of "Corona Patient Detection and Monitoring" [12].

It will work as follows: Our system will detect the temperature of a person with the help of infrared (IR) thermometer which is interfaced with Node MCU. Data will be uploaded on ThingSpeak server (cloud) through which it can be monitored. We connect IR sensor, silicon-controlled rectifier (SCR)-triggered circuit and IR thermometer together so that IR thermometer will take readings automatically when someone crosses it.

System is quite simple and very effective, especially at places like Industries, Offices and Colleges. The prototype of this system is ready with us and the complete system costs around Rs. 5000 (Rupees Five Thousand Only).

16.5 SUMMARY

Our system named as "IoT-Based Automatic Corona Virus Detecting and Monitoring System" which can detect the corona-affected person and also generate its database through which monitoring will be easy. Our system will detect the temperature of a person with the help of IR thermometer which is interfaced with Node MCU. Data will be uploaded on ThingSpeak server (cloud) through which it can be monitored. We connect IR sensor, SCR triggered circuit and IR thermometer together so that IR thermometer will take readings automatically when someone crosses it.

The system is quite simple and very effective, especially at places like Industries, Offices and Colleges. This will be really helpful to the society.

REFERENCES

JOURNAL

1. Chamola, Vinay, Vikas Hassija, Vatsal Gupta, and Mohsen Guizani. (2020). A Comprehensive Review of the COVID-19 Pandemic and the Role of IoT, Drones, AI, Blockchain, and 5G in Managing Its Impact. *IEEE Access*, 8: 90225–90265.
2. Abdulrazaq, Mohammed N., Halim Syamsudin, Salah Al-Zubaidi, Ramli R. AKS, and Eddy Yusuf. (2020). Novel Covid-19 Detection And Diagnosis System Using IOT Based Smart Helmet. *International Journal of Psychosocial Rehabilitation*, 24: 2296–2303. doi: 10.37200/IJPR/V24I7/PR270221.
3. Rahman Md Siddikur, Noah C. Peeri, Nistha Shrestha, Rafdzah Zaki, Ubydul Haque, and Siti Hafizah A Hamid. (2020 June). Defending against the Novel Coronavirus (COVID-19) Outbreak: How Can the Internet of Things (IoT) Help to Save the World? *Health Policy and Technology*, 9(2): 136–138.
4. Singh, Ravi Pratap, Mohd Javaid, Abid Haleem, and Rajiv Suman. (2020). Internet of Things (IoT) Applications to Figureht against COVID-19 Pandemic. *Diabetes & Metabolic Syndrome*, 14(4): 521–524.
5. Singh, Ravi Pratap, Mohd Javaid, Abid Haleem, Raju Vaishya, and Shokat Al. (2020 May 15). Internet of Medical Things (IoMT) for Orthopaedic in COVID-19 Pandemic: Roles, Challenges, and Applications. *Journal of Clinical Orthopaedic and Trauma*, 11(4): 713–717.
6. Swayamsiddha, Swati and Chandana Mohanty (2020 June). Application of Cognitive Internet of Medical Things for COVID-19 Pandemic. *Journal of Diabetes and Metabolic Syndrome* 11,14(5): 911–915.

Websites

7. https://www.who.int/.
8. There Is a Current Outbreak of Coronavirus (COVID-19) Disease. https://www.who.int/health-topics/coronavirus#tab=tab_1
9. Technical Specification Infrared Thermometer PCE-JR 911. https://www.pce-instruments.com/english/slot/2/download/56251/datasheet-infrared-thermometer-pce-jr911.pdf
10. There Is a Current Outbreak of Coronavirus (COVID-19) Disease. https://www.who.int/health-topics/coronavirus#tab=tab_3
11. WHO Coronavirus Disease (COVID-19) Dashboard. https://covid19.who.int/?gclid=EAIaIQobChMIiZjat9Cc6gIVmiQrCh1qkw9GEAAYASAAEgL1k_D_BwE
12. 62 MAX Mini Infrared Thermometer. https://www.fluke.com/en-us/product/temperature-measurement/ir-thermometers/fluke-62-max

17 Rapid Detection of COVID-19 Using FET and MOSFET Biosensors

Rahul Koshti and Pravin Wararkar
SVKMs NMIMS, MPSTME, MPTP

CONTENTS

17.1 INTRODUCTION

Since the time the chief proceedings of the disease raised due to the spread of corona virus 2019 (COVID-19) in Wuhan of China in December 2019, it has influenced more than 200 nations and domains across the globe with 76,90,708 cases and in excess of 4,27,630 passing as on 14 June 2020. With this developing emergency, organizations and scientists all over the world are searching for the approaches to curb the difficulties of this infection, to moderate the outbreak and build up a solution for this malady. In the perplexing fight with corona virus, technology and science is assuming an essential job. For instance, from the get-go in the flare-up when China started its reaction to infection it concentrated on artificial intelligence by depending on face acknowledgment cameras to follow the tainted sick person with movement history, robotics to convey food and drugs, automatons to sanitize open spots, to watch and communicate sound intelligence to open urging them to remain at home [1]. Artificial intelligence has been utilized broadly to find recent particles while in transit to discover help for COVID-19 infection. Numerous analysts are utilizing artificial intelligence to discover new medications and drugs for the fix, alongside some software engineering specialists concentrating on identifying the irresistible sick person by the use of image processing example X-ray beams and computed tomography scan filters [2]. Artificial intelligence is in any event creating following programming resembles observing arm bands that help in order for people groups breaking the isolate rule. Advanced cell phones and artificial intelligence have improved warm cameras utilized for distinguishing fever and contaminated individuals [3]. Taiwan imbued its national clinical protection data with contributions from the movement and customs database, thus going up against the corona virus patients based on their movement history and indications [4,5]. Similarly machine learning, deep learning, robotics, Internet of Things-installed framework, field-effect transistor (FET) and metal oxide semiconductor field effect transistor (MOSFET)-based biosensors are significant advancements which are being utilized as a weapon to battle with COVID-19 pathogens.

For rising pathogens, RT-PCR (real-time reverse transcription–polymerase chain reaction) is the essential method of determination. As of now, real-time RT-PCR is utilized for the recognition of critical respiratory syndrome, SARS-CoV-2, dependent on the latest distributed research center conventions [6]. SARS-CoV-2 is profoundly contagious and is as of now expanding quickly over the globe. The pace of transmission of COVID-19 is a lot quicker than those of Middle East respiratory Syndrome (MERS) and severe acute respiratory syndrome (SARS). Additionally, the transmission of asymptomatic COVID-19 has been accounted for [7,8]. Molecular findings utilizing ongoing RT-PCR take in any event 3 hours, including planning of the viral RNA. What is more, the step of RNA planning can influence analytic precision, henceforth, profoundly delicate immunological indicative strategies that straightforwardly distinguish viral antigens in the clinical samples. Test planning initiatives are fundamental for quick and exact analysis of corona virus. Among the numerous analytic techniques as of now accessible, biosensing gadgets which are based on FETs offer numerous benefits. In addition to the capacity to build profoundly responsive and prompt estimations utilizing limited quantities of analytes, biosensors based on FETs are found to be potentially useful in onsite detection and clinical diagnosis [9,10]. Graphene is the two-dimensional sheet in which atoms of carbon are arranged hexagonally, and all are uncovered on its surface [11]. It has demonstrated to be a helpful material for different detecting stages because of its unprecedented features that include high conductivity of electrons, greater mobility of carriers and huge area [12]. FET biosensors made up of graphene can recognize encompassing changes on its surface and give an ideal detecting condition to low-noise and ultrasensitive detection. Considering this, biosensing technology based on FET based with graphene is exceptionally alluring for uses identified with sensitive or responsive immunological findings [13,14].

17.2 FET BIOSENSORS: TYPES AND FABRICATION PROCESS AT A GLANCE

Among different potentiometric methods, detection dependent on FETs has pulled in significant consideration in light of the fact that of its potential for scaling down,

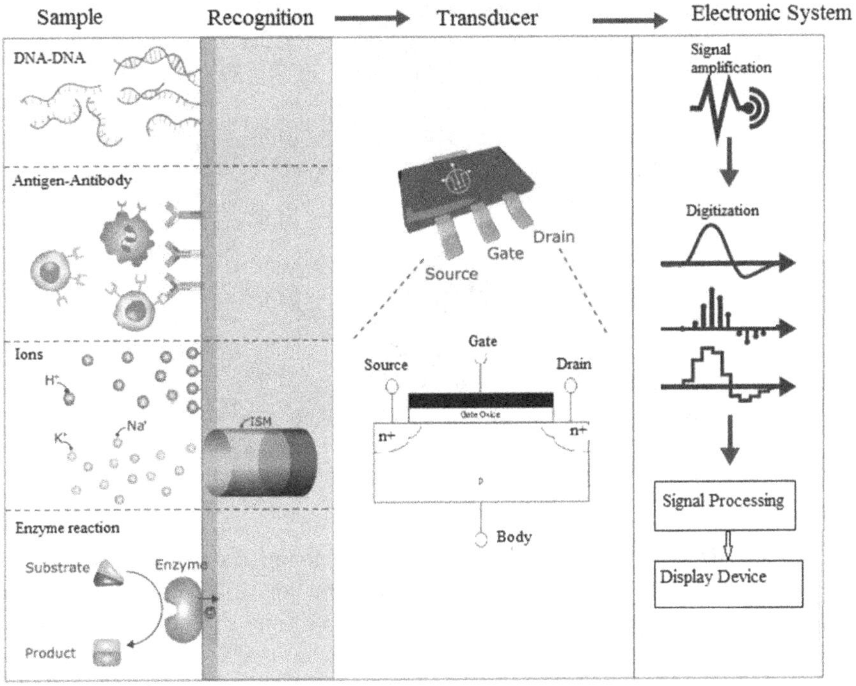

FIGURE 17.1 Illustration of FET biosensor.

equal detecting, quick response time, what's more, consistent integration with electronic assembling forms, for example, complimentary metal-oxide semiconductors. The idea of an ISFET (ion sensitive field effect transistor) was presented in the mid of 1970s and obtained from MOSFET. It was understood that a MOSFET in which removal of the underlying gate and metal gate oxide embedded in a watery arrangement alongside a reference anode could be utilized to recognize particles. There are a wide range of FET sensor structures and detecting materials. Alongside various objective analytes, it brings about a horde of various sensor framework mixes. These frameworks share the equivalent in general development that is depicted in Figure 17.1.

The data acquired from the example or sample relate to both the movement and concentration of the objective analyzed and/or the existence/amount of the biomolecule, which is converted to an electrical sign by means of the field effect. At that point, the signal can be amplified and processed, and showed or sent to the cloud contingent upon the applications [15].

17.2.1 Materials Used for FET Sensor

17.2.1.1 Silicon Nanowire Biosensors

Silicon nanowire (SiNw) is the first ultrasensitive solid-state biosensor presented by Cui et al. [16]. SiNw is accounted for showing the capacity of identifying nucleic acid [17–21], protein [22–25] and virus [26] with preeminent restriction of recognition

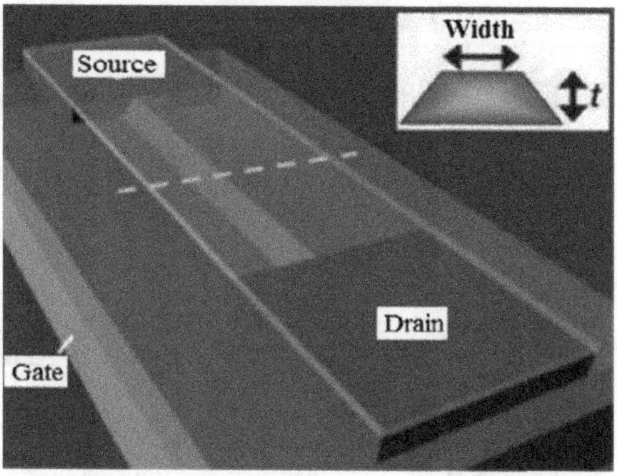

FIGURE 17.2 SiNw FET: A wired channel connects source and drain.

(e.g., limit of detection (LoD)). Biosensor of SiNW comprises a structure similar to wire as a channel with three dimensions dependent on polysilicon or crystalline. The source terminal and channel cathode interface of two sides of this wire [27] which is delineated in Figure 17.2.SiNW sensors are of two types: the transistor type and the resistor type. The transistor type has an expansion with gate at the bottom for the sensor to execute transistor functionality [28].

17.2.1.2 Organic FET Biosensors

The organic field-effect transistor (OFET) is an undifferentiated form of TFT (thin film transistor) in which an active material organic semiconductor (OSC) is used. It may be of P-type or N-type transistor relying upon substantial determination. Unique in relation to MOSFET, OFET has a place with the family of TFT, which works in accumulation mode [29] which implies that supplied gate voltage restricts the current in the channel in an immediate way. The mobility of OFET commonly lies in the range of 10^{-1} to 10^{-2}cm^2 V/s, which is significantly lower than the mobility of Si. This distinction begins from OSC's substantial structure; for example, noncovalent bonds and π-bonds are the principle bearer progress pathways which result in more awful mobility. Regardless of this downside, OSC is exceptionally perfect for adaptable substrate. The plausibility of acknowledging adaptable and electronic wearable gadgets attracts researchers' considerations on organic FET as of late. Like ISFET, OFET can be changed from customary transistor to biosensing transistor. By inundating the organic FET in an electrolyte domain, the structure of electric double layer (EDL) shows up right away. In this manner, it can control the gadget by including an electrode reference as absolution gate anode. An arrangement of organic FET biosensor (e.g., EGOFET or OFET with electrolyte gate) is appeared in Figure 17.3.

17.2.1.3 Graphene-Based FET Biosensors

The graphene is the one among the most alluring substrate since the mid of 2000s [30]. Because of its extraordinary substantial properties, for example, high mobility

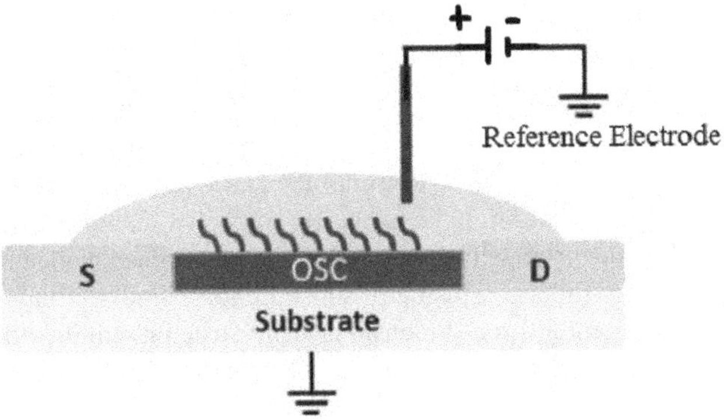

FIGURE 17.3 Illustration of organic biosensor.

of electron/hole, mechanical strength and so forth, graphene is taken into consideration as a material for the semiconductor gadgets to come. A TFT structure of graphene FET (GFET) biosensor is appeared in Figure 17.4.

From the band diagram outlook, graphene is the material with zero/minimum band gap. It effectively produces holes and electrons with +ve and −ve electric field separately (e.g., electrical doping). This wonder is referred as a bipolar characteristic, with which drain current is produced by gate voltage with positive and negative polarity. This gate voltage with the minimum drain current is known as the charge-neutral point; at this point graphene exhibits intrinsic or pure characteristics (e.g., nondoped). To work on a GFET, it is regularly structured in a single (top/base) or two-gate set-up [31]. In a regular bottom side gate, the active channel of GFET is controlled by the potential coming from the back side by means of gate with deposited dielectric. While in case of the EDL it seems to fill in dielectric with top gate as the bottom gate GFET drenched in a solvent. A small change in voltage from solution by means of top gate can create the bipolar conduct, as well. This makes a domain for specialists to watch the surface potential change because of the unsettling influence of analytes that demonstrates that GFET can fill in as a biosensing gadget.

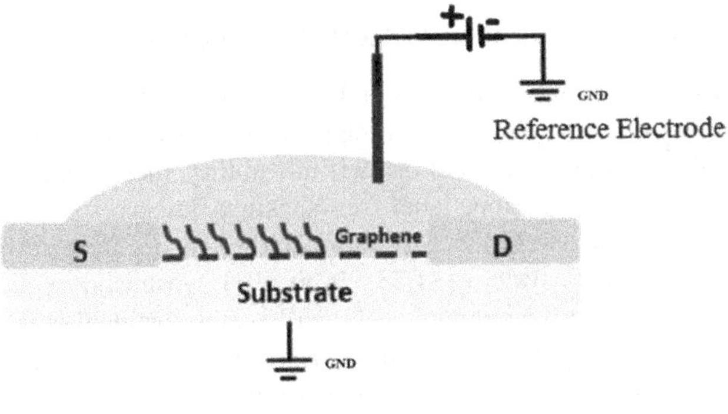

FIGURE 17.4 GFET sensor.

17.2.2 Manufacturing Process of FET Biosensors

Manufacturing processes of FET biosensors are of two types, bottom up and top-down, used to create biosensors based on SiNW. The SiNW is developed on the substrate in a chemical vapor deposition framework by means of vapor liquid solid technique. At that point, the source terminal (S) and the drain terminal (D) metals are connected. Despite what might be expected, top down manufacture method [32] structures SiNW by the process of lithography. For occurrence, E-beam lithography maybe utilized on the silicon-on-insulator substrate for nanodesigning and response ion etching. At that point a three-dimensional wire structure can be built up. In the three-dimensional wire structure, the region of detection is characterized by light doping while source and drain contacts are characterized by heavy doping by means of ion implementation technique. Of these two manufacture techniques, the top-down technique becomes mainstream as of late in light of its high suitability to standard complementary metal oxide semiconductor (CMOS) process, that can possibly be integrated with interfacing circuit and probability to be delivered on mass and cost-effectiveness [28].

17.2.3 Fabrication of Graphene-Based Sensing Devices

Conventional wet transfer method is used to fabricate graphene -based sensors in which graphene is transferred to the substrate of SiO_2/Si. Then graphene on Cu foil is spin-coated with polymethyl methacrylate (PMMA) around 500 rpm for 10 seconds and at 3000 rpm for 30 seconds. Then etching of PMMA/Graphene on Cu foil is performed. Once the copper foil is etched, the layer of PMMA/graphene is moved utilizing cleaned glass slides into the DI (deionized) water shower, and etchant of copper is removed. In this manner, the layer of PMMA/graphene is moved to a substrate of SiO_2/Si and it is dried under surrounding conditions for the time being. The layer of PMMA is removed by washing the substrate in acetone bath for 2 hour. At last, the graphene was moved onto the substrate, and then the substrate is washed by isopropyl alcohol solution and dried by nitrogen gas [33].

17.3 DETECTION MECHANISM

17.3.1 Electrical Connections of COVID-19 FET Biosensor

Electrical sensing of biomolecules dependent on its intrinsic charges, which is a proficient and the most sensitive sensing approach. In particular, biosensors using FET/MOSFETs are alluring a result of their movability, modest large-scale manufacturing, low power utilization, label-free sensing, quick responsiveness, and a voltage/potential for the on-chip incorporation of the sensors and electronics instruments and measurements system [34,35]. In an FET biosensor, explicit acceptors immobilized in the detecting channel specifically catch the ideal objective biomolecules or virus. The caught biomolecules/virus can create a gating/doping effect on the channel [36–39]. Both are changed over into an electrical wave by the FET, ordinarily as a current in drain—source or channel transconductance (Figure 17.5).

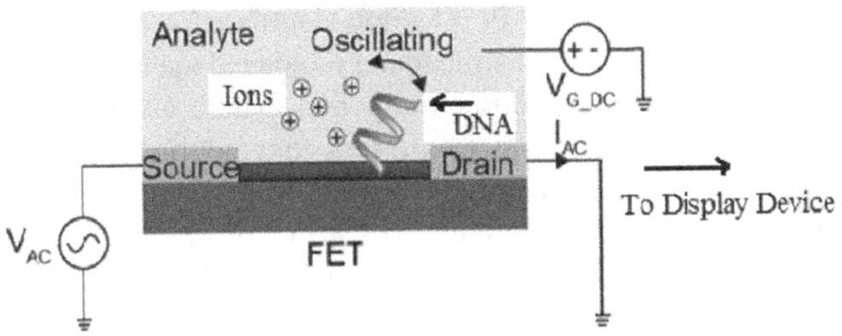

FIGURE 17.5 Electrical connections of FET biosensor.

17.3.2 Functionality of COVID-19 FET Biosensor

COVID-19 is a quickly growing sickness/infection realized by serious intense disorder of respiratory system. Since no particular medications or immunizations for infections of COVID-19 are yet accessible, its management and early detection are essential for controlling the spread of the disease. Here, a biosensing device based on FET is accounted for identifying SARS-CoV-2/COVID-19 in clinical examples. A device for sensing is created by covering sheets of graphene of the FET with a particular antibody against the virus of SARS-CoV-2 spike protein. The functioning and performance characteristics of the sensors are resolved by utilizing antigen proteins, virus and specimens of swab of nasopharyngeal, taken from corona virus COVID-19 patients. FET sensor is an exceptionally sensitive immunological analytic technique for COVID-19 which requires no naming or example premedication. The functionality of FET-based graphene biosensor used for rapid/quick detection of virus of COVID-19 is elaborated in Figure 17.6 [33].

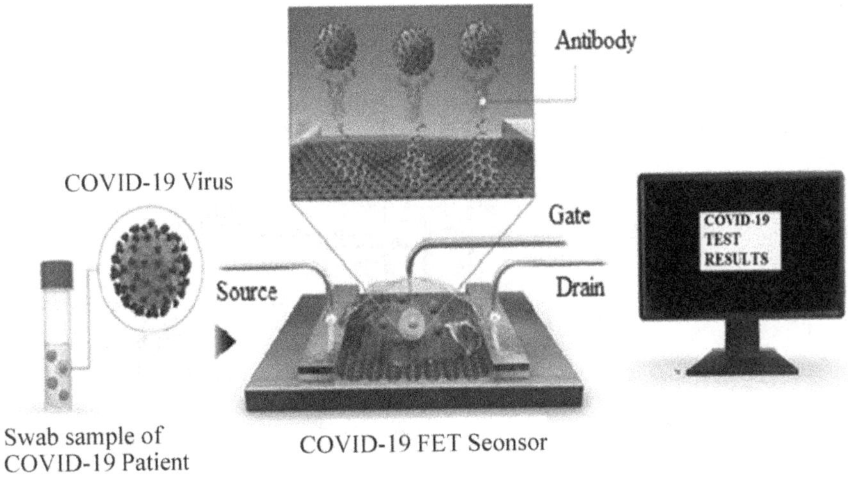

FIGURE 17.6 Functional diagram of COVID-19 FET sensor.

The innovation behind COVID-19 infection identification depends on biosensing, which means to distinguish (bio)chemical and biological agents utilizing an organically inferred/biomimetic recognition component where as either experiencing a biochemical reaction (e.g., biosensors based on enzyme) or binding of target molecules (e.g., analyte) in an exceptionally explicit approach. The biosensing inputs to the sensor are provided with respect to gate to source terminal; this input is recognized by the graphene channel and as a result of this antibodies are produced. Once the biosensing input is recognized, it is transduced into an electrical signal and this electrical signal is amplified and is converted into digital form and is sent to a display device to read, measure and record. A rapid increase in the curve of electrical signal with respect to a reference level indicates identification of COVID-19 virus. Here graphene like a detecting material is chosen, and spike antibody of SARS–CoV-2 is conjugated on the sheet of graphene. The use of graphene permits fast and accurate sensing of the virus that helps in obtaining accurate and measurable electrical signal.

Characteristics of FET Biosensor: Data obtained from X-ray photoelectronic spectroscopy and Raman spectra are availed to ensure graphene surface is functionalized with chemicals using PBASE as shown in Figure 17.7a,b [33]. Be that as it may, the D and D′ peak clearly showed up because of the pyrene group within the binding and orbital hybridization of PABSE to the surface of graphene. After the PBASE is applied to the graphene sheet, the factual information of the peak positions of 2D was moved to a high frequency (approximately 3.26 cm^{-1}), demonstrating P-type doping by using π–π interaction [40,41] among PBASE and graphene (Figure 17.7b) [40,41].

Electrical measurements can be performed for the sensing and detection of SARS-CoV-2 spike antibody on the graphene surface, according to I-V curves shown in Figure 17.8b [33]. After the functionalization and immobilized PBASE of the antibody response on the channel of graphene, the slope (dI/dV) diminished. This difference in the slope demonstrates the effective presentation of the SARS-CoV-2 virus spikes antibody.

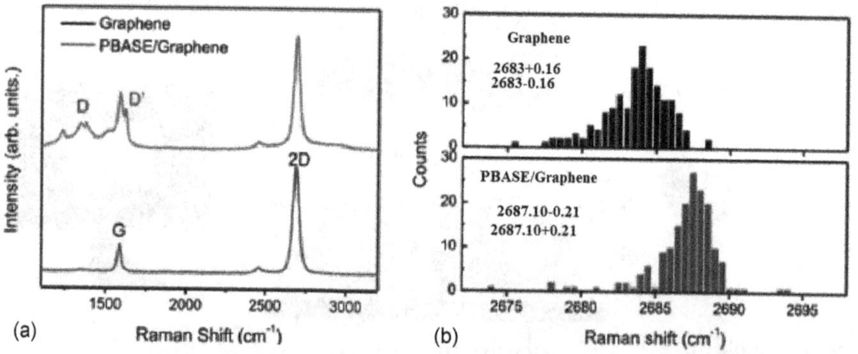

FIGURE 17.7 Surface analysis of PBASE-modified graphene using Raman spectroscopy.

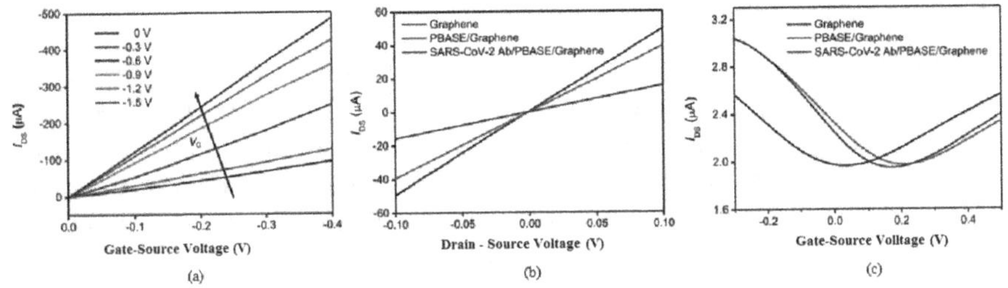

FIGURE 17.8 I–V characteristics of COVID-19 FET sensor.

Figure 17.8a shows the I-V response curve of the FET sensor for the detection of COVID-19 as a component of VG (variable gate voltage) ranging from 0 to −1.5 volts in steps of −0.3 volts. It is seen that drain-source current I_{DS} increased negatively negative increase in gate voltage [42,43]. Moreover, the straight I–V curve showed exceptionally constant ohmic contact, demonstrating that the sensor with FET for COVID-19 gave a reliable electrical sign to recognition of the SARS-CoV-2 antigen protein, refined SARS-CoV-2 infection or SARS-CoV-2 infection from clinical specimens which are target analytes. As shown in Figure 17.7c, after PBASE functionalization, a conspicuous positive move was seen because of the P-type doping of the group pyrene. Be that as it may, the transfer characteristic is moved negatively, proposing that the positive charge of the antibody applied an N-type doping on the graphene after the immobilization of antibody.

17.4 RESULT AND DISCUSSION

For the quick detection of COVID-19 virus FET-based biosensors are of significant use. Among the three types of the sensors discussed here, graphene-based FET sensor has proven results and hence can be effectively used. In the mid-1970s the concept of ISFET is developed which was obtained from MOSFETs. For the evaluation of effectiveness of the sensor various tests can be performed and based on the result of the test effectiveness can be determined. To ensure that graphene surface is chemically functionalized surface analysis is performed. The presence of SARS-CoV-2 spikes antibody, electrical measurements can be performed and it is concluded that if the slope of the di/dv decreases, SARS-CoV 2 spike antibody is presented on the graphene. The electrical measurements can be performed over various ranges of gate voltage.

Conclusion: Field-effect transistors which are introduced as naming free, specific and sensitive electronics biosensors. In this review, the working components of different kinds of devices used for biosensing, materials and functionality are discussed. Also the nature of the sensor is examined with the surface analysis and electrical measurements, and it is found that FET-based biosensing devices are suitable for the quick detection of COVID-19 and it would start analytic transformation for the next-age improvement.

BIOGRAPHY

Prof. Rahul Koshti is working as an Assistant Professor in the Department of Electronics and Telecommunication Engineering of SVKMs NMIMS, MPSTME Shirpur Campus. He is a post graduate in digital communication with experience of around 12 years and pursuing Ph.D. in wireless communication. He has participated in many national and international conferences and workshops; also he had been organizer of national conference. He is a reviewer of IEEE conferences. He is the member of various technical societies and organizations. His subjects of interest are wireless communication, antenna and wave propagation, and research area includes cognitive radio-based wireless networks. He has attended many courses based on electronics engineering and wireless communications from well-known international universities.

Dr. Pravin Wararkar is an Assistant Professor in the Department of Electronics and Telecommunication Engineering at Mukesh Patel School of Technology Management and Engineering, SVKMs Narsee Monjee Institute of Management Studies (Deemed-to-be UNIVERSITY), Mumbai, India. He is serving in the SVKMs NMIMS, Mumbai since July 1, 2015. Currently, he is associated with SVKM's NMIMS, MPSTME, Shirpur Campus. He received his Bachelor of Engineering degree in Electronics and Telecommunication Engineering, Master of Technology degree in VLSI and Doctoral Degree in Electronics Engineering from Nagpur University, India. He also completed Various PG Diplomas like PGDBM, PGDIRPM, PGDOM, PGDN and PGDCCA in Management and Technology domain from Nagpur University. With over 10 years of experience in industry and academics, his expertise lies in Electronics Engineering and Designing of HR Management Policy. He has completed various sponsored research projects from Government funding agency and Industry. He has authored Several Books in Engineering and Management domain. He has also authored over 65 papers in the journals and conferences of international repute and approx. 75 citations, h-index: 6; i10-index: 1 (Google Scholar). He is an Editorial Board Member of various journals of Engineering and Technology. He acts as a regular reviewer for reputed journals like "Elsevier, Springer, Taylor & Francis" and IEEE Conferences. He is a Session Chair for Various Technical Events, Seminar and Workshops. He is a member of Various professional Societies like ISTE, IEI, IETE, IAENG, SCIEI, MEASCE, NSPE, UACEE-IRED, ICSES, ASR, IAER, IAASSE, BR2R, INSCIAT, SDIWC, IETA, IS and E4C. He can be reached at pravinwararkar.webs.com/pravin.wararkar@nmims.edu or pwararkar@gmail.com.

REFERENCES

1. Estrada R., Arturo M. (2020) The uses of drones in case of massive epidemics contagious diseases relief humanitarian aid: Wuhan-COVID-19 crisis. University of Malaya (UM) - Social Security Research Centre (SSRC) SSRN. DOI: 10.2139/ssrn.3546547.
2. Nguyen T.T., Waurn G., Campus P. (2020) Artificial intelligence in the battle against coronavirus (COVID-19): A survey and future research directions. DOI: 10.13140/RG.2.2.36491.23846.
3. Maghdid H.S., Ghafoor K.Z., Sadiq A.S., Curran K., Rabie K. (2020) A novel AI-enabled framework to diagnose coronavirus COVID 19 using smartphone embedded sensors: Design study. 1–5, Cornell University. arXiv.org > cs > arXiv:2003.07434v2.
4. Wang C.J., Ng CY., Brook R.H. (2020) Response to COVID-19 in Taiwan big data analytics, new technology, and proactive testing. *Journal of JAMA Network*, DOI: 10.1001/jama.2020.3151
5. https://www.techuk.org/insights/news/item/17187-how-taiwan-used-tech-to-fight-covid-19.
6. WHO. Coronavirus disease (COVID-19) technical guidance: Laboratory testing for 2019-nCoV in humans. https://www.who.int/emergencies/diseases/novel-coronavirus-2019/technical-guidance/laboratory-guidance
7. Bai Y., Yao L., Wei T., Tian F., Jin, D. Y., Chen L., Wang M. (2020) Presumed asymptomatic carrier transmission of COVID-19. *JAMA*, 323, 1406–1407, DOI: 10.1001/jama.2020.2565.
8. Zou, L., Ruan F., Huang M., Liang L., Huang H., Hong Z., Yu J., Kang M., Song Y., Xia J., Guo Q., Song T., He J., Yen H.L., Peiris M., Wu J. (2020) SARS-CoV-2 viral load in upper respiratory specimens of infected patients. *New England Journal of Medicine*, 382, 1177–1179, DOI: 10.1056/NEJMc2001737.
9. Janissen R., Sahoo P.K., Santos C.A., da Silva A.M., von Zuben A.A.G., Souto D.E.P., Costa A.D.T., Celedon P., Zanchin N.I.T., Almeida D.B., Oliveira D.S., Kubota L.T., Cesar C.L., Souza A.P., Cotta M.A. (2017) InP nanowire biosensor with tailored biofunctionalization: Ultrasensitive and highly selective disease biomarker detection. *Nano Letter*, ACS Publications 17, 5938–5949, DOI: 10.1021/acs.nanolett..7b01803.

10. Liu J., Chen X., Wang Q., Xiao M., Zhong D., Sun, W., Zhang, G., Zhang Z. (2019) Ultrasensitive monolayer MoS2 field-effect transistor based DNA sensors for screening of down syndrome. *Nano Letter*, ACS Publications 1437–1444, DOI: 10.1021/acs.nanolett.8b03818.
11. Cooper D.R., D'Anjou B., Ghattamaneni N., Harack B., Hilke M., Horth A., Majlis N., Massicotte M., Vandsburger L., Whiteway E., Yu V. (2012) Experimental review of graphene. International scholarly research notice, ISRN Condens. *Matter Physics*, DOI: 10.5402/2012/501686.
12. Geim A.K., Novoselov K.S. (2007) The rise of graphene. *Nature Materials*, 6, 183–191, DOI: 10.1038/nmat1849.
13. Lei Y.M., Xiao M.M., Li Y.T., Xu L., Zhang H., Zhang Z.Y., Zhang G.J. (2017) Detection of heart failure-related biomarker in whole blood with graphene field effect transistor biosensor. *Biosensors and Bioelectronics*, Elsevier Publications 91, 1–7, DOI: 10.1016/j.bios.2016.12.018.
14. Zhou L., Mao H., Wu C., Tang L., Wu Z., Sun H., Zhang H., Zhou H., Jia C., Jin Q., Chen X., Zhao J. (2017) Graphene biosensor targeting cancer molecules based on non-covalent modification. *Biosensors and Bioelectronics*, Elsevier Publications, 87, 701–707, DOI: 10.1016/j.bios.2016.09.025
15. Kaisti M. (2017) Detection principles of biological and chemical FET sensors. *Biosensors and Bioelectronics*, Elsevier Publications 98, 437–448.
16. Cui Y., Lieber C.M. (2001) Functional nanoscale electronic devices assembled using silicon nanowire building blocks. *National Library of Medicine*, 291(5505), 851–853, DOI: 10.1126/science.291.5505.851.
17. Halm J.I., Lieber C.M., (2004) Direct ultrasensitive electrical detection of DNA and DNA sequence variations using nanowire nanosensors. *Nano Letters*, ACS Publications 4(1), 51–54, DOI: 10.1021/nl034853b.
18. Zhang G.J., Chua J.H., Chee R.E., Agarwal A., Wong S.M. (2009) Label-free direct detection of MiRNAs with silicon nanowire biosensors. *Biosensors and Bioelectronics*, 24(8), 2504–2508.
19. Agarwal A., BuddharajuI K., Lao I.K., Singh N., Balasubramanian N., Kwong D.L. (2008) Silicon nanowire sensor array using top–down CMOS technology. *Sensors and Actuators Science Direct*, Elsevier Publications 145–146, 207–213.
20. Gao Z., Agarwal A., Trigg A.D., Singh N., Fang C., Tung C.H., Fan Y., Buddharaju K.D., Kong J. (2007) Silicon nanowire arrays for label-free detection of DNA. *Analytical Chemistry*, ACS Publications 79(9), 3291–3297, DOI: 10..1021/ac061808q.
21. Gao A., Lu N., Dai P., Li T., Pei H., Gao X., Gong Y., Wang Y., Fan C. (2011) Silicon-nanowire-based CMOS-compatible field-effect transistor nanosensors for ultrasensitive electrical detection of nucleic acids. *Nano Letters*, ACS Publications 11(9), 3974–3978.
22. Zheng G., Patolsky F., Cui Y., Wang W.U., Lieber C.M. (2005) Multiplexed electrical detection of cancer markers with nanowire sensor arrays. *Nature Publishing, Biotechnology*, 23(10), 1294–1301.
23. Mishra N.N., Maki W.C., Cameron E., Nelson R., Winterrowd P., Rastogi S.K., Filanoski B., Maki G.K. (2008) Ultra-sensitive detection of bacterial toxin with silicon nanowire transistor. *Journal Lab on a Chip*, 8(6), 868–871.
24. Chua J.H., Chee R.E., Agarwal A., Wong S.M., Zhang G.J. (2009) Label-free electrical detection of cardiac biomarker with complementary metal-oxide semiconductor-compatible silicon nanowire sensor arrays. *Analytical Chemistry*, ACS Publications 81(15), 6266–6271.
25. Hakim M.M.A., Lombardini M., Sun K., Giustiniano F., Roach P.L., Davies D.E., Howarth P.H., Planque M.M.R., Morgan H., Ashburn P. (2012) Thin film polycrystalline silicon nanowire biosensors. *Nano Letters*, ACS Publications 12(4), 1868–1872. DOI: 10.1021/nl2042276.

26. Patolsky F., Zheng G., Hayden O., Lakadamyali M., Zhuang X., Lieber C.M. (2004) Electrical detection of single viruses. *Proceedings of the National Academy of Science of United State of the America (PNAS)*, 101(39) 14017–14022, DOI: 10.1073/pnas.0406159101.
27. Stern E., Klemic J.F., Routenberg D.A., Wyrembak P.N., Turner-Evans D.B., Hamilton A.D., LaVan D.A., Fahmy T.M., Reed M.A. (2007) Label-free immunodetection with CMOS-compatible semiconducting nanowires. *Nature Material Science Journal*, 445, 519–522.
28. Syu Y.C., Hsu W.E., Lin C.T. (2018) Review—field-effect transistor biosensing: Devices and clinical applications. *ECS Journal of Solid State Science and Technology*, 7(7), 3196–3207.
29. Torsi L., Magliulo M., Manoli K., Palazzo G. (2013) Organic field-effect transistor sensors: A tutorial review. *Journal of Chemical Society Reviews*, 22, 8612–8628.
30. Novoselov K.S., Geim A.K., Morozov S.V., Jiang D., Zhang Y., Dubonos S.V., Grigorieva I.V., Firsov A.A.(2004) Electric field effect in atomically thin carbon films. *Science*, 306(5696), 666–669, DOI: 10.1126/science.1102896.
31. Zhan B., Li C., Yang J., Jenkins G., Huang W., Dong X. (2014) Graphene field-effect transistor and its application for electronic sensing. *Small*, 10(20), 4042–4065, DOI: 10.1002/smll.201400463.
32. Elibol O.H., Morisette D., Akin D., Denton J.P., Bashir R. (2003) Integrated nanoscale silicon sensors using top-down fabrication. *AIP, Applied Physics Letters*, 83, 4613, DOI: 10.1063/1.1630853.
33. Seo G., Lee G., Kim M., Baek S.H., Choi M., Ku K.B., Lee C.S., Jun S., Park D., Kim H.G., Kim S.J., Lee J-O., Kim B.T., Park E.C., Kim S. (2020) Rapid detection of COVID-19 causative virus (SARS-CoV-2) in human nasopharyngeal swab specimens using field-effect transistor-based biosensor. *ACS Nano*, 14(4), 5135–5142.
34. Xu G., Abbott J., Qin L., Yeung K.Y., Song Y. (2014) Electrophoretic and field-effect graphene for all-electrical DNA array technology. *Nature Communications*, 5, 4866.
35. Kim J.E, No Y.H., Kim J.N., Shin Y.S., Kang W.T. et al. (2017) Highly sensitive graphene biosensor by monomolecular self-assembly of receptors on graphene surface. *AIP, Applied Physics Letters*, 110, 203702, DOI: 10.1063/1.4983084.
36. Cui Y., Wei Q., Park H., Lieber C.M. (2001) Nanowire nanosensors for highly sensitive and selective detection of biological and chemical species. *Science*, 293(5533), 1289–1292, DOI: 10.1126/science.1062711.
37. Wang W.U., Chen C., Lin K.H., Fang Y., Lieber C.M. (2005) Label-free detection of small-molecule–protein interactions by using nanowire nanosensors. *Proceedings of the National Academy of Sciences of the United States of America*, 102(9), 3208–3012. DOI: 10.1073/pnas..0406368102.
38. Chen R.J., Bangsaruntip S., Drouvalakis K.A., Kam N.W., Shim M. et al. (2003) Noncovalent functionalization of carbon nanotubes for highly specific electronic biosensors. *Proceedings of the National Academy of Sciences of the United States of America*, 100(9), 4984–4989. DOI: 10.1073/pnas..0837064100.
39. Xu G., Abbott J., Ham D.H. (2016) Optimization of CMOS-ISFET-based biomolecular sensing: Analysis and demonstration in DNA detection. *IEEE Transactions Electron Devices*, 63(8), 3249–3256.
40. Guangfu W., Xin T., Meyyappan, M., Lai K.W.C. (2017) Doping effects of surface functionalization on graphene with aromatic molecule and organic solvents. *Journal of Applied Surface Science*, 425, 713–721, DOI: 10.1016/j.apsusc.2017.07.048.
41. Liu Y., Yuan L., Yang M.., Zheng Y., Li L., Gao L., Nerngchamnong N., Na, C.T., Sangeeth C.S., Feng, Y.P., Nijhuis, C.A., Loh K.P. (2014) Giant enhancement in vertical conductivity of stacked CVD graphene sheets by self-assembled molecular layers. *Nature Communications*, 5, 5461, DOI: 10.1038/ncomms6461.

42. Choi Y., Kang J., Jariwala D., Kang M.S., Marks T.J., Hersam M.C., Cho J.H. (2016) Low-voltage complementary electronics from Ion-Gel-Gated vertical van der Waals heterostructures. *Advanced Materials*, 28, 3742–3748, DOI: 10.1002/adma.201506450.
43. Ten, F., Hu K., Ouyang, W., Fang, X. (2018) Photoelectric detectors based on inorganic p-type semiconductor materials. *Advanced Materials*, 30, 1706262, DOI: 10.1002/adma.201706262.

18 Progress of COVID-19 Epidemic in India

Pooja Pathak, Avinash Dubey, and Yash Srivastava
GLA University

CONTENTS

18.1 INTRODUCTION

These days, the world is facing an emerging infectious disease "COVID-19 or 2019 Coronavirus". COVID-19 belongs to the family of SARS which also emerged in China in 2003. Corona is not a single virus but it is a group of viruses that belong to the same family. COVID-19 belongs to the same family of viruses. It was spotted recently in Wuhan, Hubei Province, China, on 31 December, 2019. The rate of spreading of coronavirus is a quick process because of its replicative nature as shown in Figure 18.1. Also the virus can transmit from human to human and from human to material surfaces. Virus can live on material surfaces for a longer time period. It is an inexperienced virus and has not been fully studied hence there is no vaccine ready as a cure. All the medical authorities in the world are engaged in finding a vaccine.

In December, 2019, Wuhan, Hubei Province, China, there occurred an outbreak of a new pathogen whose infection was quite similar to that of pneumonia. This new pathogen influenced the whole world due to its dreadful attack. The authorities in China did all the examinations and analysis to characterize and limit the disease, but they were unaware of how the virus spread? The virus rapidly initiated its spread. China took some genuine precautions to save its people. They segregated the suspicious people, then they closely monitored the patients and his contacts, studied the symptoms shown by the patients prudently and then developed a diagnostic and medication strategy. The government acted seriously and issued some set of rules for the people to protect themselves.

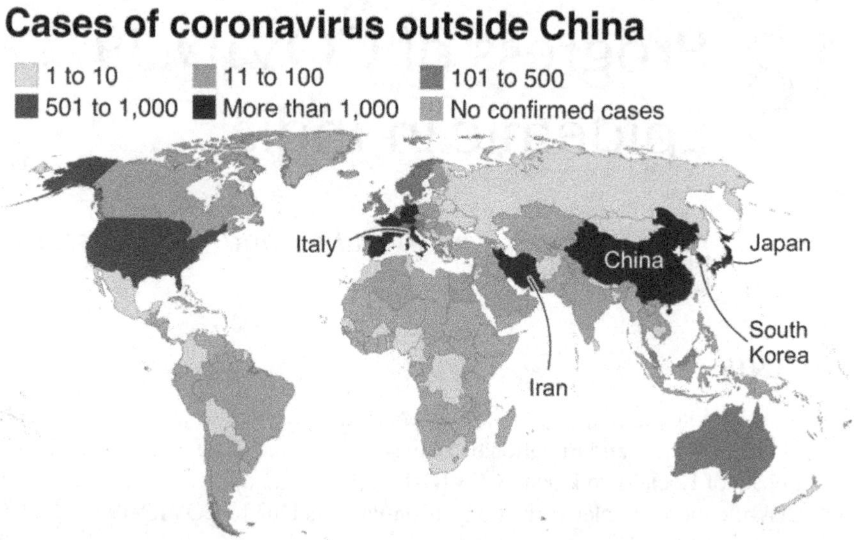

FIGURE 18.1 Corona effect on world map.

18.2 EPIDEMIOLOGICAL AND SITUATIONAL ANALYSIS

India's last update[4] at 6:00 a.m. on June 10, 2,74,780 confirmed COVID-19 virus cases were reported in India. Of the cases, 1,32,880 patients are recovered (48.8%), 1,32,880 remain active (48.3%), 8,944 remain critical (3.2%) and 7,750 died (2.8%). Total cases per million populations are 200 and deaths per million populations are 6. Total tests that are performed are 4,916,116 and tests performed per million populations are 3564 [1].

The reported cases of COVID-19 in Maharashtra increased to 90,787. Tamil Nadu reported the second highest number of cases in India as shown in Figure 18.2.

18.3 IMPACT OF LOCKDOWN DUE TO COVID-19

The lockdowns in India were started mainly from 22 March, 2020, when a slight rise in the COVID-19 cases was observed, by the government, realizing India is a country with a population of approximately 1369.56 million people and with a very high density of 382 per sq. km (in 2011). The first case of coronavirus in India was spotted on 30 January, 2020. The patient was a student who returned from Wuhan which was an epicenter of corona patients at that time. The government took all the possible safety measures as it was aware about the nature of virus. Afterward, by looking at the spread in the western countries, we have avoided all the inaccuracies that initiated its spread. The infection in India is of a primary rate so the patients are recovered faster with less casualties. The recovery rate of India is around 49% which is highly appreciable. The country medical staff, Indian police and other people who worked in this time of lockdown acted as a shield to

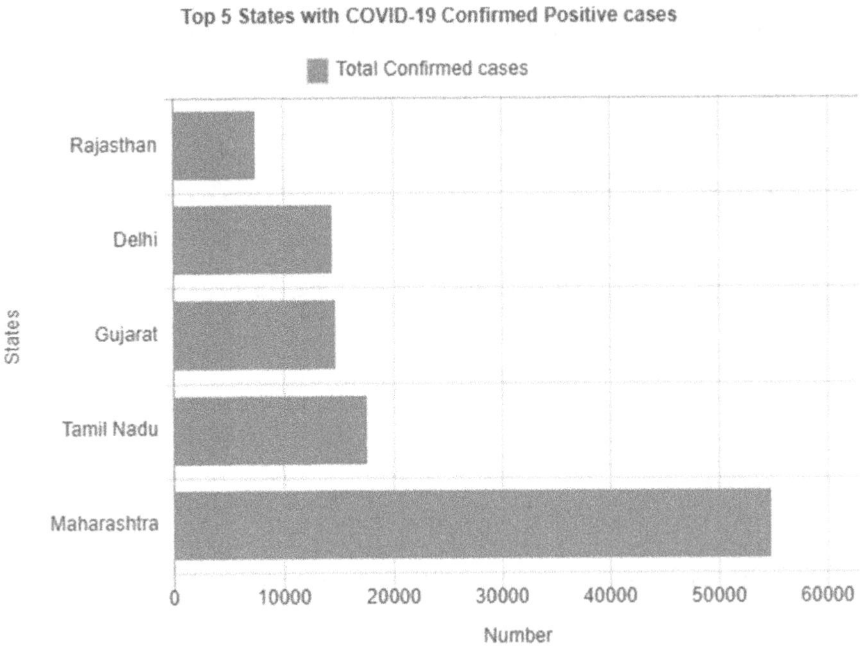

FIGURE 18.2 Top five states data of COVID-19 confirm cases.

safeguard the country by following all the guidelines that were generated by the government. Responsibilities announced for the lockdown from 24 March, 2020. First, there was a "Janta Curfew" which aimed to curtail the list of COVID-19 cases and also it aimed to limit the epidemic. In order to implement this "Janta Curfew" several crucial guidelines were generated. All the educational institutions, markets, restaurants, parks, etc., were strictly shut.

It was a successful event. Citizens rigorously followed and appreciated the guidelines. Just after 2 days of this "Janta Curfew," a 21-days lockdown was announced by the central government which entirely closed the system. The lockdowns banned the migration of the people and restricted the citizens to their homes. The residents were not allowed to step out of their apartments/residence unless they had an emergency. These guidelines were generated with the aim to flatten the curve of COVID-19 cases and to suppress the rise of patients exponentially.

The cases were growing day by day. 5,000 confirmed cases with 90% active cases and 3% deaths were reported till the morning of 7th April. Now observe Figure 18.3, which shows the rise in the number of active cases and confirmed cases. Observing the graph, we can conclude that the government took right decisions of implementing the series lockdown from 24th March. The medical authorities of India took rapid decisions by perceiving the cruciality of the situation to safeguard the country.

Now coming to the growth rate of COVID-19 cases on a regular basis, refer to Figure 18.4. It is observed that earlier the growth rate seems to form spikes as the numbers of cases were less. The deaths occurred in single digits in early stages so

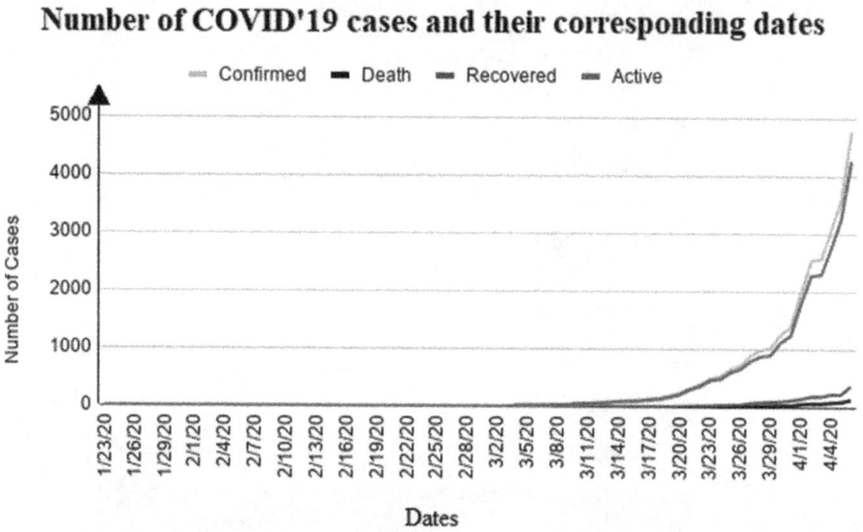

FIGURE 18.3 Date-wise combined COVID-19 cases of confirmed, death, recovered and active.

the infection rate was quite high. This stage had very few patients. The main rise in cases is observed from 22nd March, 2020. These rise in cases led the government to take such crucial steps of lockdowns. These lockdowns are major reasons for today's conditions. These lockdowns restricted the patients to less numbers with such a huge population. This provided all the governments and health authorities to supervise this situation clearly to limit the rising number of cases.

18.4 CLASSIFICATION AND NETWORK ANALYSIS ON PATIENTS DUE TO COVID-19

Patient's network and their demographic details were created on the basis of data set available at COVID-19-India dot org, nodes were patient ID, the countries of travel, and any large gathering events and edges were the connection between each patient and his event attending history or travelling history. It was noticed that Delhi, Italy, Gulf Countries, United Kingdom, Mumbai and Saudi Arabia were the major hotspots responsible for the quick spread of coronavirus in India. Classification model was created based on the database, to check that whether a patient will die or not on the basis of demographic features. This model was created with the help of Decision Tree. It was noticed that majority cases belong to 31–40 years age group in India as shown in Figure 18.5.

The patients who died are mainly above 60 years old. The infected people were either a traveler or attended spiritual ceremonies in Delhi. Age of the Patient, Gender of the patient and the state/region of the patient are three features which were found to be significant [2].

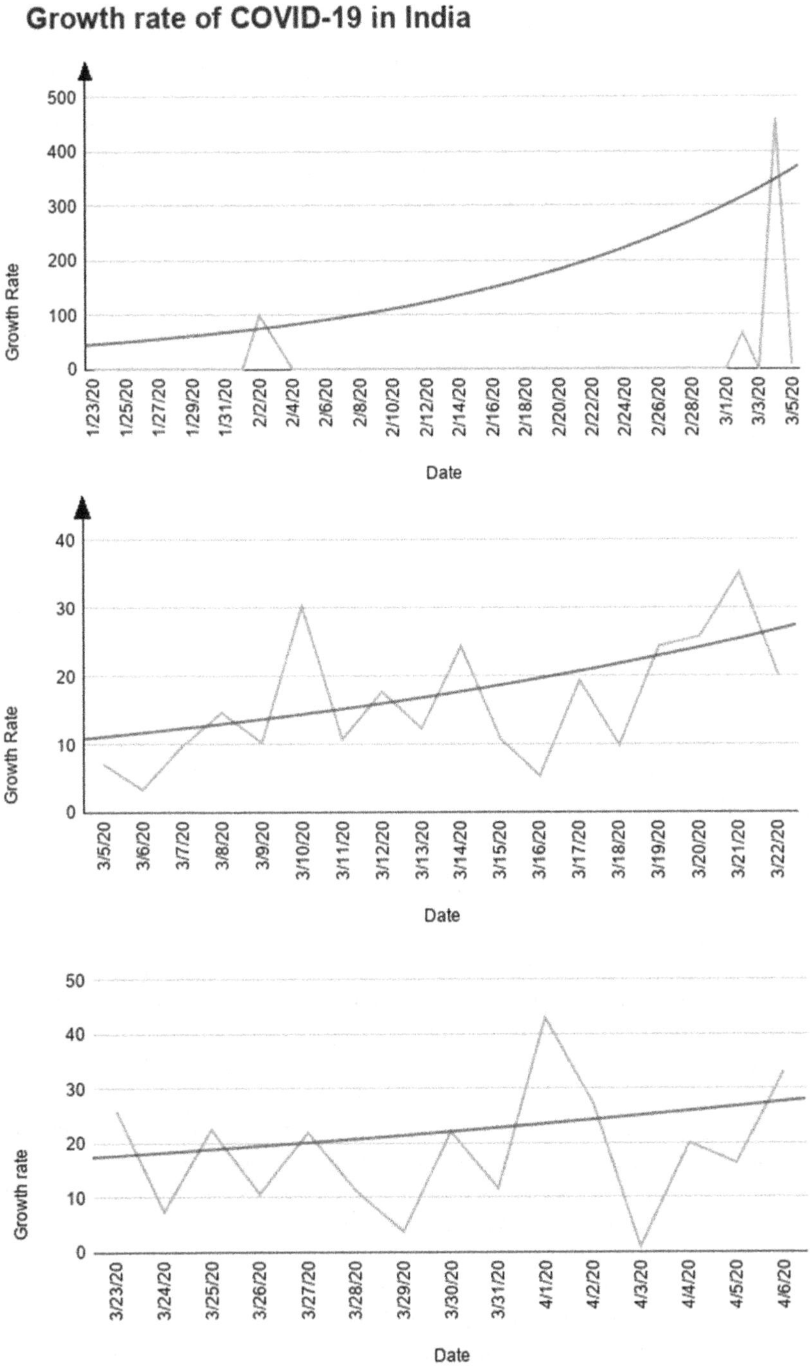

FIGURE 18.4 Date-wise combined COVID-19 cases.

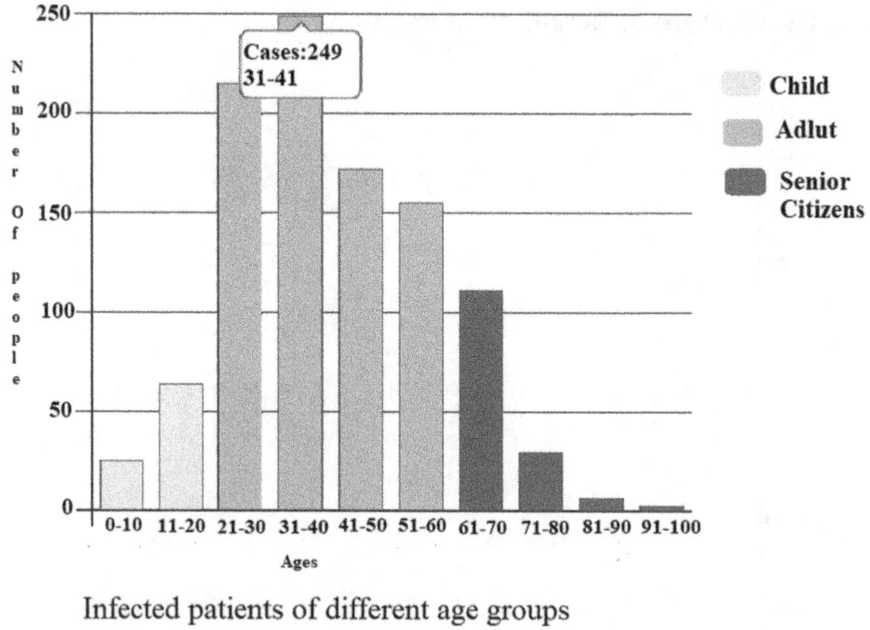

FIGURE 18.5 Age-wise infected COVID-19 cases.

Several inferences cannot be drawn from this model as data were much efficient, but as dead patient's record, a hint can be drawn, which will be very helpful to the medical and administrative authorities.

18.5 IMPACT OF MASS EVENTS

There were two mass events come into picture in India during the lockdown period. One of this event was related to the migration of the laborers to their respective states in India [3,4] and another was Tablighi Jamaat event which held at Nizamuddin Markaz Mosque, Delhi [5,6]. Approximately 25% citizens are living below poverty line, which are dependent on their regular wages. The government of India came with a plan to help these workers and laborers by releasing a package of 22 billion USD. Proper planned strategies, i.e., different infrastructural set-ups were brought in use to help needy citizens. Some agencies noticed that the country is facing heavy crisis. However, the effect of infected cases was not much by mass movement because they were not infected during their movements from workplaces to their native places (Figure 18.6).

Figure 18.6 shows at least 36% of total infected cases were related to spiritual ceremonies in Delhi [7]. Several areas were isolated from the impact of coronavirus. Therefore it is observed that spiritual events became one of the vital reasons for the spread of COVID-19. This proved to be an urgent need for the administration and medical authorities to stay highly alert for the upcoming situation that had to face.

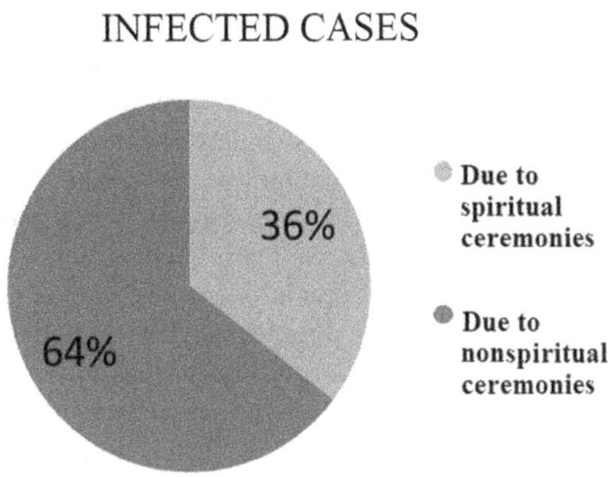

FIGURE 18.6 Effect of spiritual ceremony on COVID-19 cases.

18.6 DISCUSSION

Recently, the incidence of COVID-19 has increased due to its migration to other parts of the world. The world has never faced such enormous crises in the recent decades. Majority of the European countries and the North American countries are overblown with this virus attack. Researches are going on in the entire world to find an optimized solution to manage this spread. The governments are implementing every possible approach to invent proper medication strategies to this disease. They have to take strict actions to safeguard their citizens. The lockdowns seemed to be effective in controlling the diseases a lot. All the trains, international and national airports are closed so that people do not gather in masses. People are strictly quarantined in their houses. Many countries have imposed very high fines if they are caught roaming outside without any valid reason. In India, the main sources of coronavirus were the people who returned from foreign countries or the foreigners who visited India were found to be infected. The government took all the major steps at the right time that is why today the situation is under control in India. The lockdowns were implemented. All the colleges and universities, gyms, malls and all the entertainment zones were ordered to close. All the transport systems such as trains, airports, bus stands and highways were closed. All the offices whether private, semiprivate or governments were either working in shifts or the people had to work from home. Only the grocery and medical shops were allowed for certain hours in a day. The government kept a travel record of every citizen if they traveled to other cities in case of emergency. The lower section of the society was provided with all the medical equipment and financial support by the governments. Special trains are made to run so that people could return back to their home with proper social distancing norms. Thermal screening, sanitization, proper check-ups and social distancing are performed. The suspected people are quarantined for 14 days. Delhi is also quarantining people in their houses [4].

There are many reasons for the spreading of the virus. First, villagers who worked in cities had to return to their places. Second, due to some illegal mass gatherings the spread of coronavirus fastened. Third, some worldwide migrants who came to visit India were one of the major causes of India. There are many possibilities on how the virus was imported? Now, the country needs to take stringent measures to detect potential cases early in order to curb existing epidemic and tracking steps to prevent further spread. After the unexpected rise in coronavirus cases in India, the government of India has taken several crucial measures such as practicing of social distancing, proper lockdowns, all the malls are closed, no grouping at any place, etc., After three lockdowns, the guidelines for the fourth lockdown were different. The government opened the airport services with some serious guidelines. The government offices put in work with proper guidelines. The cases are continuously increasing but our corona warriors are working day and night. The recovery rate of India has increased to about 48%. The danger of virus importation in communities is very high and needs good precautions and strict steps to identify possible cases early and surveillance steps to prevent further virus transmission. India is still in the containment process. With increasing cases of immensely contagious COVID-19, India's economy has deteriorated. There is an environment of severe grief in India. The governments have managed the country very well and are trying to find optimized solutions to manage this spread. The government is working to save both the citizens and our economic status as both are very crucial for a country. The people can go out if they have a valid reason, else they are punished and fined. The trains and flights are started again with proper guidelines. It has been decided that rather than complete lockdowns, people should avoid mass gatherings, and partial lockdown should be a better option for improving the economy.

18.7 CONCLUSION

Coronavirus is spreading rapidly. A huge number of cases are being reported each day. In just less than 6 months it has become a worldwide issue. The mortality rate and morbidity has outstretched exceptionally, i.e., people are dying in large numbers [6]. The clinicians, scientists, researchers and other related medical practitioners are engaged to invent a proper vaccine. The cases are increasing day by day. The conditions could worsen in future. The primary possible solution that we know is self-isolation and following proper social distancing norms. The government is emphasizing on breaking the chain of spread. This means do not become a carrier or a source of spread of the virus instead stay self-isolated. This will save the entire community and will initiate the decline of spreading. Ensure proper sanitization in your area because it is an airborne disease as well as it can stay on material surfaces so do your work in a contactless manner. Follow the basic guidelines as washing up of hands if you go out. Do not allow children and aged people to go out because they have weak immunity systems. There is no need to panic. Following the above steps can ensure your safety from this debilitating virus. India has an immense potential in public health. If all the sectors of government work in cooperation to address the challenges with full citizen support then we can easily eradicate this problem from society.

18.8 RECOMMENDATIONS

COVID-19 is spreading in the whole world. To invent treatments and vaccine to prevent this infection, the clinicians are working efficiently. One step toward self-isolation must be taken by everyone, which could save us and the risk will decline immediately. This is a situation where each individual has to take steps towards minimizing the risk by staying in the house and immobilizing themselves. The airborne, contact transmission can only be disinfected if proper hand-washing protocols are followed and each individual carries out precautionary measures to safe other individuals from this debilitating virus. India has a tremendous potential in public health, and different sectors can work together to address the challenges by the engagement of society and community along with policy initiatives.

REFERENCES

1. Covid-19 coronavirus pandemic. Accessed from www.worldometers.info/coronavirus/ on 29 April, 2020.
2. Gupta, R., Pal, S. K., & Pandey, G. (2020). A comprehensive analysis of COVID-19 outbreak situation in India. medRxiv.org
3. BBC News (2020). Coronavirus: Search for hundreds of people after Delhi prayer meeting. BBC News. Accessed from https://www.bbc.com/news/world-asia-india-52104753 on 1 May, 2020.
4. Press Trust of India (2020). Coronavirus: 1,445 cases liked to Tablighi Jamaat event, total crosses 4,000. India Today News. Accessed from https://www.indiatoday.in/india/story/coronavirus-1-445-cases-liked-to-tablighi-jamaat-event-total-crosses-4-000-1663962-2020-04-06 on 5 May, 2020
5. Abid, K., Bari, Y. A., Younas, M., Tahir Javaid, S., & Imran, A. (2020). Progress of COVID-19 Epidemic in Pakistan. Asia Pacific Journal of Public Health, doi: 1010539520927259.
6. AlJazeera (2020). India: Coronavirus lockdown sees exodus from cities. Aljazzera News Channel. Accessed from https://www.aljazeera.com/programmes/newsfeed/2020/04/india-coronavirus-lockdown-sees-exodus-cities-200406104405477.html on 11 May, 2020.
7. Coronavirus: World must prepare for pandemic, says WHO. Accessed from https://www.bbc.com/news/world-51611422 on 15 May, 2020.

19 Region of Interest Detection in COVID-19 CT Images Using Neutrosophic Logic

S. N. Kumar
Amal Jyothi College of Engineering

A. Lenin Fred, L. R. Jonisha Miriam, and Ajay Kumar H.
Mar Ephraem College of Engineering and Technology

Parasuraman Padmanabhan and Balazs Gulyas
Nanyang Technological University

CONTENTS

19.1 INTRODUCTION

Image processing is stated as the usage of computerized algorithms for detailed analysis. An inclusive count of segmentation algorithms is there for ROI extraction [1,2]. The fuzzy logic-based edge detector generates proficient results, when equaled with classical Sobel edge detector. The line detection was good and the curved lines were also detected properly in [3]. A fuzzy if-then approach was proposed for edge detection and compared with traditional edge detector that requires parameter tuning [4]. Interval type-2 neuro-fuzzy interference system (IT2FIS) generates proficient results for the detection of edges, when compared with the T1FIS system [5]. The general type-2 fuzzy logic systems (GT2FLS) based on Sobel edge detection with fuzzy logic generates efficient results, when compared with T1FLS and interval IT2FLS techniques [6]. A novel edge detection technique tucker decomposition-weighted histogram of oriented gradient (TD-WHOG) was proposed for the boundary detection in images and was found to be robust for video surveillance application [7]. The deep learning-based edge detection generates proficient results and it is tested on benchmark data sets [8]. Fuzzy logic-based edge detector was proficient in the edge detection of Panels at Exterior Cladding of Buildings [9]. The neutrosophic logic analyzes imprecise data more efficiently than intuitionistic fuzzy logic [10].

The neutrosophic logic-based algorithms gain prominence in image denoising and ROI extraction [11]. In X-ray imaging for the detection of edges, the amalgamation of Gaussian filter and statistical range was compared with modified Moore-Neighbor (MN), local adaptive canny edge detection (ACED), improved Sobel and Wang and Lin (WL) operators. Results of the combined technique are better in terms of peak signal-to-noise ratio (PSNR) and computational time [12]. In [13], the pulse coupled neural network was implemented for the eradication of impulse noise. For the elimination of impulse noise, differential applied median filter (DAMF) was implemented and it was compared with four other filtering methods based on PSNR and structural similarity index measure (SSIM) measures [14]. The noise adaptive fuzzy switching median filter was found to be robust in the removal of impulse noise [15].

The chapter organization is as follows. Section 19.2 shows various filtering approaches and neutrosophic logic based edge detection. Section 19.3 demonstrates the results and discussion. Finally Section 19.4 describes the conclusion.

19.2 DENOISING AND EDGE DETECTION

Images acquired from the acquisition system are affected by noise due to various factors. This research work focuses on the filtering of impulse noise by variants of median filter and edge detection by neutrosophic domain technique. The filtering method named median filter is used for impulse noise in various applications; however, in this research work, variants of median filter like hybrid median filter and differential applied median are applied. The efficient filtering technique was estimated first and then the preprocessed image was subjected to neutrosophic edge detection.

19.2.1 Denoising of Images Using Variants of Median Filter

19.2.1.1 Classical Median Filter

The classical median filter (MF) is a rank selection filter and it is widely applied in image processing for the elimination of impulse noise. It belongs to the family of nonlinear filters and relies on the moving window principle. The widely

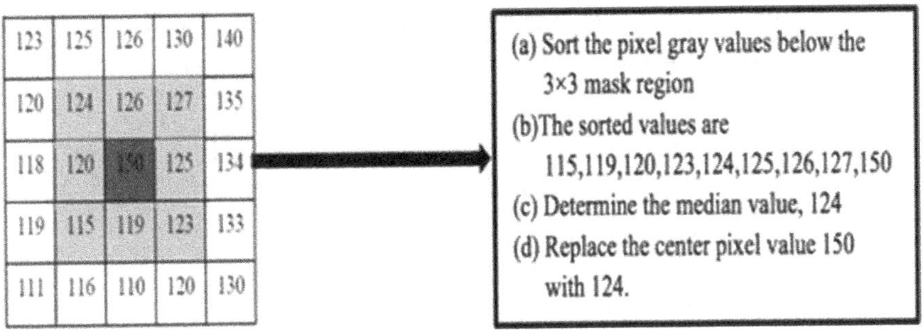

FIGURE 19.1 Median filter operation.

used mask size is 3 × 3, 5 × 5 and 7 × 7. The mask of appropriate size defined by the user is applied to the image, median of pixel gray values is estimated and the center pixel of mask region in the image is substituted with the median value (Figure 19.1).

19.2.1.2 Differential Applied Median Filter

To get rid of the impulse noise, DAMF is used and was found to be efficient, when equaled with MF [14]. Consider the matrix of an image as $\boldsymbol{X} = [\boldsymbol{x}_{mn}]_{p \times q}$ in which $\boldsymbol{x}_{mn}$ denotes the unsigned integer value and its pixel value varies between 0 and 255.

$$x_{mn} \begin{cases} \text{Noise entry of } X & \text{if } x_{mn} = 0 \text{ or } x_{mn} = 255 \\ \text{Regular entry of } X & \text{Otherwise} \end{cases}$$

Let us consider $[\boldsymbol{x}_{mn}]_{p \times q}$ and $\boldsymbol{t} \in \{1, 2, \ldots\ldots\ldots \boldsymbol{min}(\boldsymbol{p}, \boldsymbol{q})\}$. $A_X^t = [\boldsymbol{a}_{kl}]_{(p+2t)\times(q+2t)}$ is known as t-symmetric pad matrix of $\boldsymbol{X}$ and is expressed as follows.

$$\begin{bmatrix}
x_{tt} & \cdots & x_{t1} & x_{t2} & x_{t3} & \cdots & x_{tq} & x_{tq} & \cdots & x_{t(q-t+1)} \\
\vdots & \ddots & \vdots & \vdots & \vdots & \ddots & \vdots & \vdots & \ddots & \vdots \\
x_{1t} & \cdots & x_{11} & x_{11} & x_{12} & \cdots & x_{1q} & x_{1q} & \cdots & x_{1(q-t+1)} \\
x_{1t} & \cdots & x_{11} & x_{11} & x_{12} & \cdots & x_{1q} & x_{1q} & \cdots & x_{1(q-t+1)} \\
x_{3t} & \cdots & x_{21} & x_{21} & x_{22} & \cdots & x_{2q} & x_{2q} & \cdots & x_{2(q-t+1)} \\
x_{4t} & \cdots & x_{31} & x_{31} & x_{32} & \cdots & x_{3q} & x_{3q} & \cdots & x_{3(q-t+1)} \\
\vdots & \ddots & \vdots & \vdots & \vdots & \ddots & \vdots & \vdots & \ddots & \vdots \\
x_{pt} & \cdots & x_{p1} & x_{p1} & x_{p2} & \cdots & x_{pq} & x_{pq} & \cdots & x_{p(q-t+1)} \\
x_{pt} & \cdots & x_{p2} & x_{p1} & x_{p2} & \cdots & x_{pq} & x_{pq} & \cdots & x_{p(q-t+1)} \\
\vdots & \ddots & \vdots & \vdots & \vdots & \ddots & \vdots & \vdots & \ddots & \vdots \\
x_{(p-t+1)t} & \cdots & x_{(p-t+1)1} & x_{(p-t+1)1} & x_{(p-t+1)2} & \cdots & x_{(p-t+1)q} & x_{(p-t+1)q} & \cdots & x_{(p-t+1)(q-t+1)}
\end{bmatrix}$$

Consider $X = [x_{mn}]_{p \times q}$, A_X^t indicates t-symmetric pad matrix of X and $t \in [1, 2.........t]$. The r-approximate matrix of x_{mn} in A_X^t is as follows:

$$X_{mn}^r = \begin{bmatrix} a_{(m+t-r)(n+t-r)} & \cdots & a_{(m+t-r)(n+t-r)} \\ \vdots & a_{(m+t)(n+t)} & \vdots \\ a_{(m+t-r)(n+t-r)} & \cdots & a_{(m+t+r)(n+t+r)} \end{bmatrix}_{(2r+1)\times(2r+1)}$$

The row matrix of X_{mn}^r is described as

$$H_{mn}^r = [h_{1u}]_{1\times(2r+1)^2}$$

The regular row matrix of X_{mn}^r is denoted as

$$G_{mn}^r = [g_{1v}]$$

The median of G_{mn}^r is

$$\propto G_{mn}^r = \begin{cases} g_{1\left(\frac{w+1}{2}\right)}, & \dfrac{w+1}{2} \in Z \\ \dfrac{1}{2}\left[g_{1\left(\frac{w}{2}\right)} + g_{1\left(\frac{w+2}{2}\right)}\right], & \dfrac{w}{2} \in Z \end{cases}$$

19.2.1.3 Hybrid Median Filter

Hybrid median filter (HMF) belongs to the family of nonlinear filter and it performs filtering with edge preservation [16]. The objective behind this is to apply the median operation multiple times thereby changing the kernel size and the median value is determined from the calculated median values.

The operation of HMF is depicted in Figure 19.2. The HMF performance is superior with square kernel values of 3 × 3, 5 × 5 and 7 × 7. Here 3 median values are

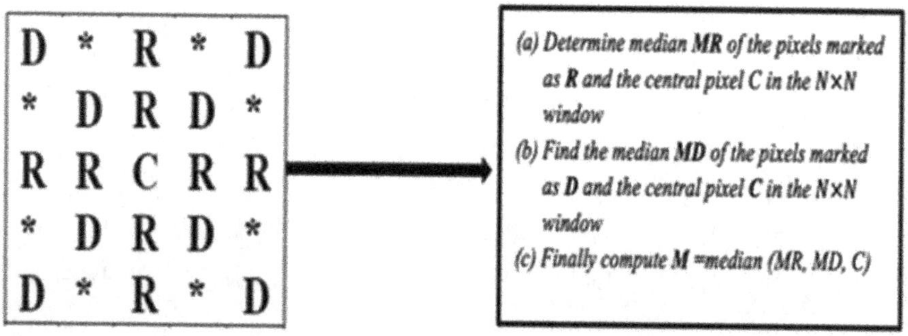

FIGURE 19.2 Operation of HMF.

determined; median of the horizontal and vertical pixel gray values (MR), median of the diagonal pixels (MD). The value obtained from filtering is the median of MR, MD and center pixel gray value.

19.2.1.4 Neutrosophic Domain-Based Edge Detection

Let $S(x, y)$ represents an image and a pixel $P(x, y)$ is represented as $P(T, I, F)$. The pixel in the image is represented by three parameters in the neutrosophic domain; true, indeterminate and false.

$$P_{NS}(a, b) = \{T(a, b), I(a, b), F(a, b)\}$$

where $\boldsymbol{T(a, b), I(a, b), and\ F(a, b)}$ represents the true, indeterminate and false pixels. The expressions for true, indeterminate and false pixels are as follows.

The expressions for $\boldsymbol{T(a, b), I(a, b), and\ F(a, b)}$ are expressed as follows:

$$T(x, y) = \frac{\bar{h}(a, b) - \bar{h}_{min}}{\bar{h}_{max} - \bar{h}_{min}}$$

where $\boldsymbol{\bar{h}(a, b)}$ is represented as follows in terms of neighborhood connectivity pixels.

$$\bar{h}(a, b) = \frac{1}{w \times w} \sum_{m=a-w/2}^{a-w/2} \sum_{n=b-w/2}^{b-w/2} h(m, n)$$

$$\boldsymbol{I(a, b) = \frac{\mu(a, b) - \mu_{min}}{\mu_{max} - \mu_{min}}}$$

where

$$\mu(a, b) = abs\left(h(a, b) - \bar{h}(a, b)\right)$$

$$\boldsymbol{F(a, b) = 1 - T(a, b)}$$

The denoised image was subjected to edge detection by neutrosophic domain approach [17]. The proposed edge detection was found to be efficient for noisy images and was found to be robust for noise free images. Various stages in the edge detection approach are as follows.

Step 1: The denoised image is converted into neutrosophic domain using neutrosophic image theory.

Step 2: Estimate the local maximum gray value for each pixel in an image based on neighborhood connectivity window (W), size of the arbitrary and hence $W = 3 \times 3$ is chosen.

Step 3: The local maximum value determined in the previous step is used for the replacement of local mean value in the neutrosophic domain.

$$\hat{T}(i, j) = \frac{\hat{h}(a, b) - \hat{h}_{min}}{\hat{h}_{max} - \hat{h}_{min}}$$

$$\hat{I}(i, j) = \frac{\hat{\delta}(a, b) - \hat{\delta}_{min}}{\hat{\delta}_{max} - \hat{\delta}_{min}}$$

$$\hat{F}(i, j) = 1 = \hat{T}(a, b) = \frac{\hat{h}_{max} - \hat{h}(a, b)}{\hat{h}_{max} - \hat{h}_{min}}$$

$$\hat{\delta}(a, b) = \left| h(a, b) - \hat{h}(a, b) \right|$$

$$\hat{\delta}_{max} = max\left\{\hat{\delta}(a, b)\right\}$$

$$\hat{\delta}_{min} = min\left\{\hat{\delta}(a, b)\right\}$$

Step 4: Estimate the gradient function of the truth image $\hat{G}(a, b)$ and is represented as $\nabla\hat{G}(a, b)$.

Step 5: The 3 × 3 neighborhood W of $\hat{\delta}(a, b)$ and optimum threshold value γ (Here we take = 2) is defined. For each neighborhood W, maximum and minimum gray values are determined.

$$W = \left\{\left\{\nabla\hat{T}(a, b)\right\}_{n=b-w/2}^{b+w/2}\right\}_{m=a-w/2}^{a+w/2}$$

$$W_{max} = maximum\{W\}$$

$$W_{mean} = mean\{W\}$$

Step 6: Define

S^{+} = Set of pixels which are greater than γ in each neighborhood

S^{-} = Set of pixels which are less than γ in each neighborhood

$N(W)$ = Number of pixels in a neighborhood W

$N(S^{+})$ = Number of pixels in the set V^{+}

$N(S^{-})$ = Number of pixels in the set V^{-}

Step 7: The edge detected images $E_{NS}(i, j)$ are represented as follows.

$$E_{NS}(i, j) = \begin{cases} 1 & \text{if } N(S^{+}) \geq N(S^{-}) \,\&\, \nabla\hat{G}(a, b) = W_{max} \; (or) \\ & N(S^{+}) \geq N(S^{-}) \,\&\, \nabla\hat{G}(a, b) = W_{mean} \\ 0 & \text{otherwise} \end{cases}$$

19.3 RESULTS AND DISCUSSION

MATLAB 2015a software was used for the development of algorithms and validated on real-time COVID-19 CT images. Prior permission was obtained from the Radiological Society of North America for the usage of real-time COVID-19 CT images in the research work. The input images are taken from the works [18,19].

19.3.1 Validation of Preprocessing Algorithms on Benchmark Images

Preprocessing approaches were tested initially on benchmark images and the results are depicted in Tables 19.1–19.3.

The input images are depicted in Figure 19.3. Prior to edge detection, preprocessing was done by variants of MF. HMF generates proficient results for the images with low noise variance and DAMF generates proficient results for images with high noise variance.

TABLE 19.1
Performance Metrics of Preprocessing Results of Median Filter on Benchmark Image

	Noise Variance Corresponding to Salt and Pepper Noise				
Performance Metrics	**0.1**	**0.2**	**0.4**	**0.6**	**0.8**
MSE	0.0323	0.0604	0.1200	0.1878	0.2665
PSNR	63.045	60.32	57.34	55.39	53.87
AD	0.0031	0.0051	0.0080	0.0073	0.0072
SC	1.0073	1.414	1.2415	1.236	1.1412
NK	0.9140	0.8498	0.7449	0.6830	0.6486
LMSE	1.0012	1.0241	1.0972	1.2224	1.3575
NAE	0.1385	0.2321	0.4194	0.5793	0.6958

TABLE 19.2
Performance Metrics of Preprocessing Results of Hybrid Median Filter on Benchmark Image

	Noise Variance Corresponding to Salt & Pepper Noise				
Performance Metrics	**0.1**	**0.2**	**0.4**	**0.6**	**0.8**
MSE	0.0155	0.03038	0.0764	0.1710	0.3343
PSNR	66.224	63.245	59.3011	55.8015	52.88
AD	−0.0250	−0.00491	−0.1111	−02087	−0.3557
SC	0.9540	0.9173	0.8327	0.8128	0.6792
NK	1	1	1	1	1
LMSE	0.4532	0.4723	0.5358	0.7460	0.9861
NAE	0.0510	0.0998	0.2254	0.4180	0.7143

TABLE 19.3
Performance Metrics of Preprocessing Results of DAMF on Benchmark Image

	Noise Variance Corresponding to Salt and Pepper Noise				
Performance Metrics	**0.1**	**0.2**	**0.4**	**0.6**	**0.8**
MSE	0.0288	0.0528	0.0911	0.1044	0.0974
PSNR	63.54	60.903	58.53	57.93	58.24
AD	0.0017	0.0028	0.0041	0.0024	0.0004
SC	1.0734	1.1344	1.2272	1.2118	1.1117
NK	0.9211	0.8635	0.7876	0.7896	0.8439
LMSE	0.8670	0.8655	0.7810	0.5288	0.4638
NAE	0.1210	0.2022	0.3229	0.3639	0.3011

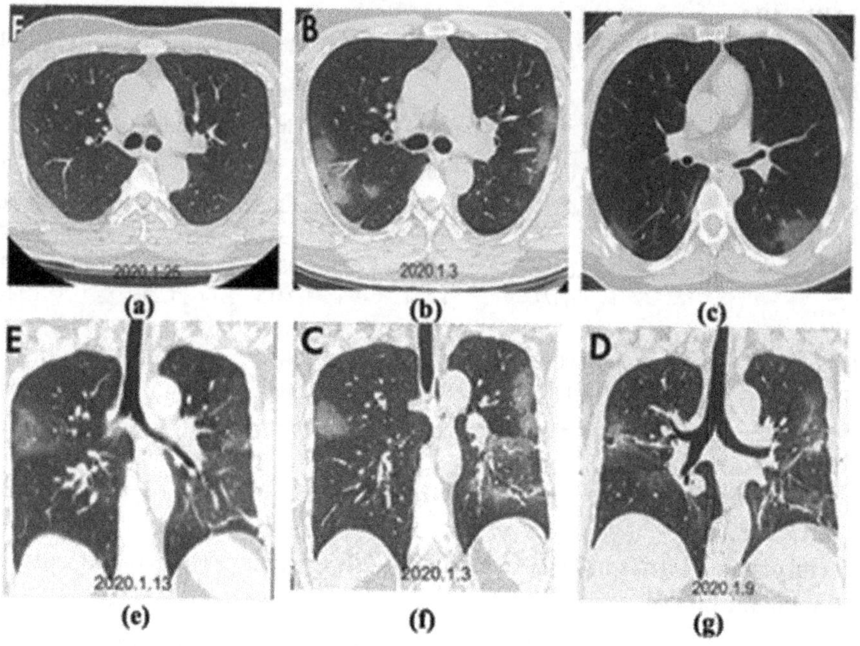

FIGURE 19.3 Input COVID CT images.

An efficient preprocessing algorithm should have high PSNR, low mean square error (MSE), Laplacian mean square error (LMSE), normalized absolute error (NAE), and average difference (AD). The SC value should be greater than 0.9 for an efficient filtering technique and should not exceed 1. The normalized cross correlation (NCC) ideal value is greater than 0.9 and all metric values rely on noise variance present in the image.

19.3.2 Test Results of Edge Detectors on Real-Time COVID-19 CT Images

Figure 19.3 indicates input COVID-19 CT images; Figures 19.4–19.6 depict the pre-processing results of median, hybrid median and DAMF. The classical edge detectors output are depicted in Figures 19.7–19.13.

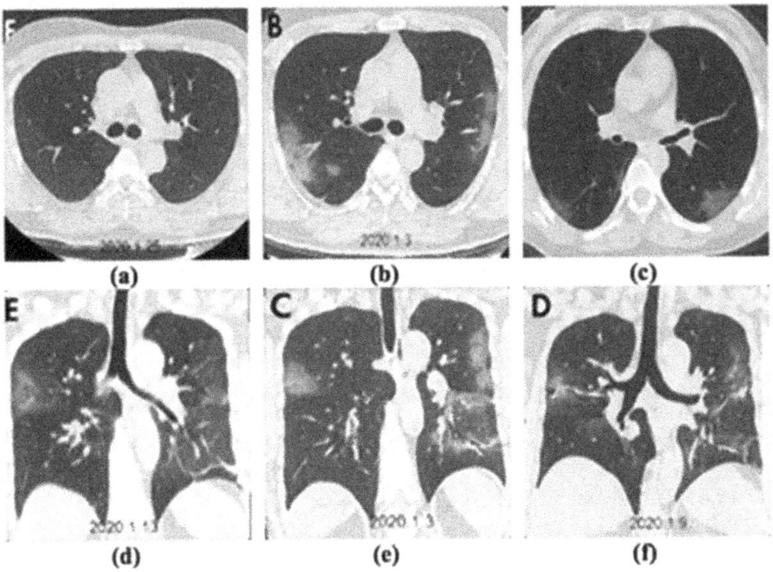

FIGURE 19.4 Median filter output for Figure 19.4.

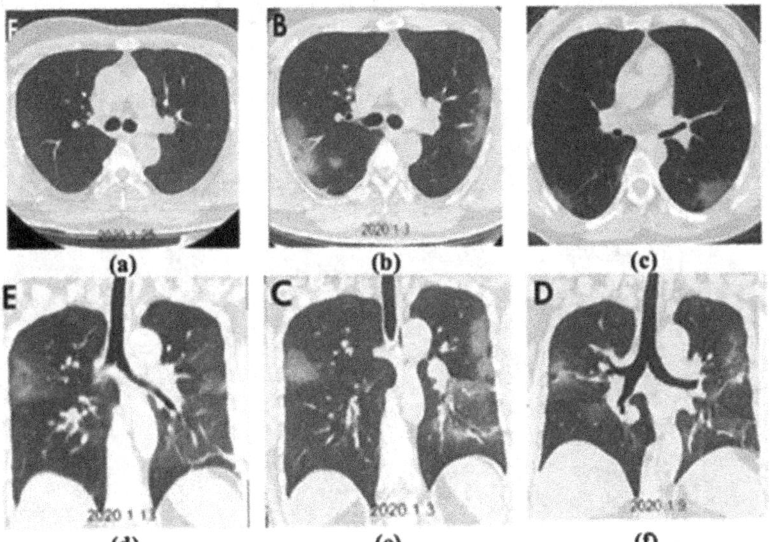

FIGURE 19.5 HMF output for Figure 19.4.

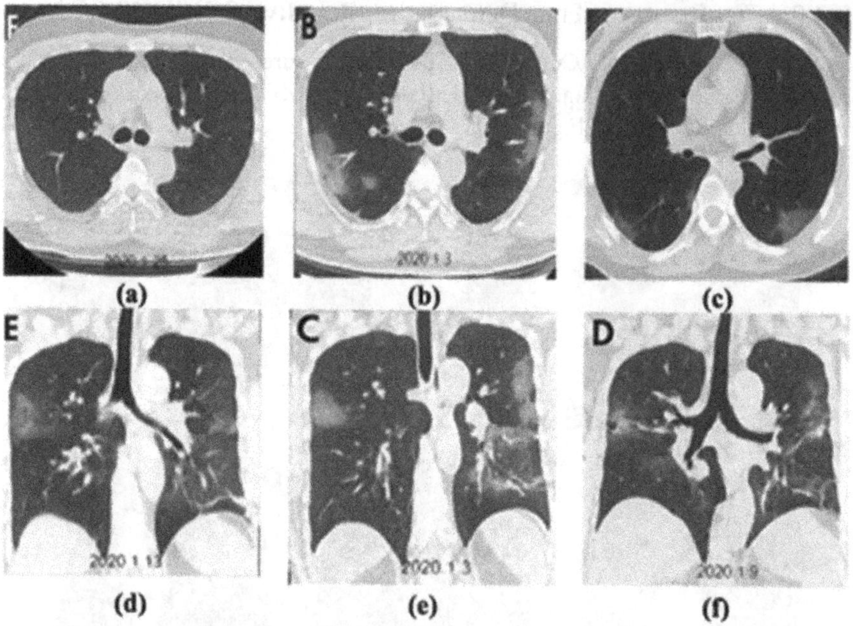

FIGURE 19.6 DAMF output for Figure 19.4.

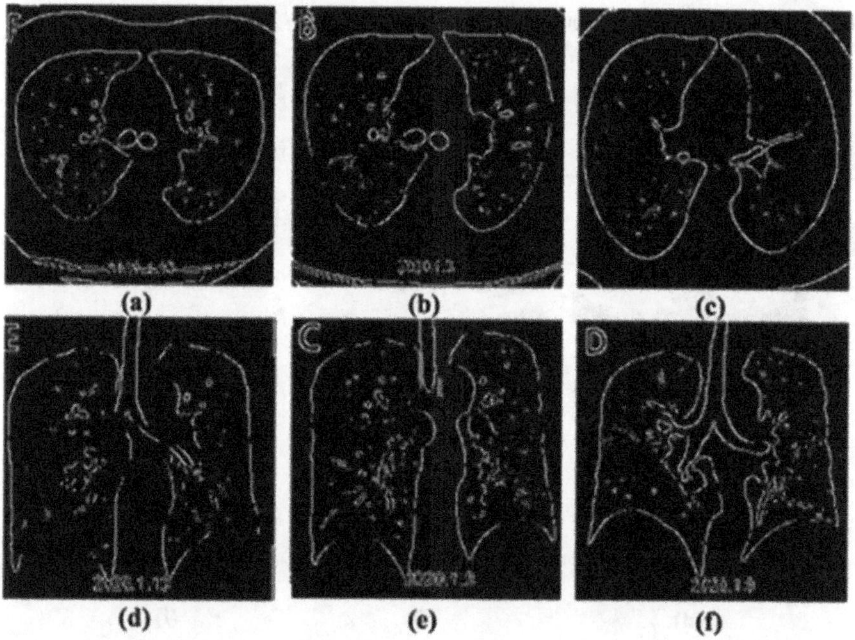

FIGURE 19.7 Prewitt algorithm edge detection results.

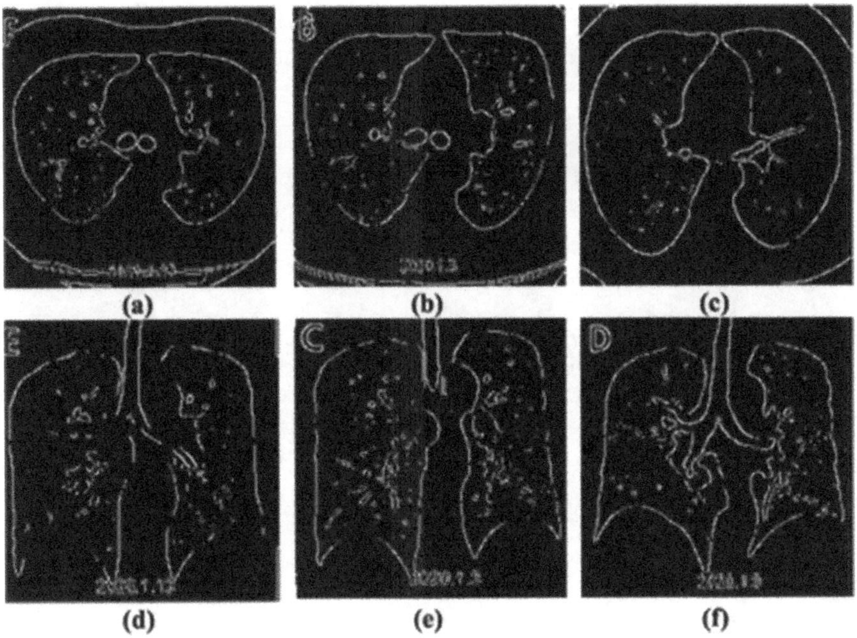

FIGURE 19.8 Roberts algorithm edge detection results.

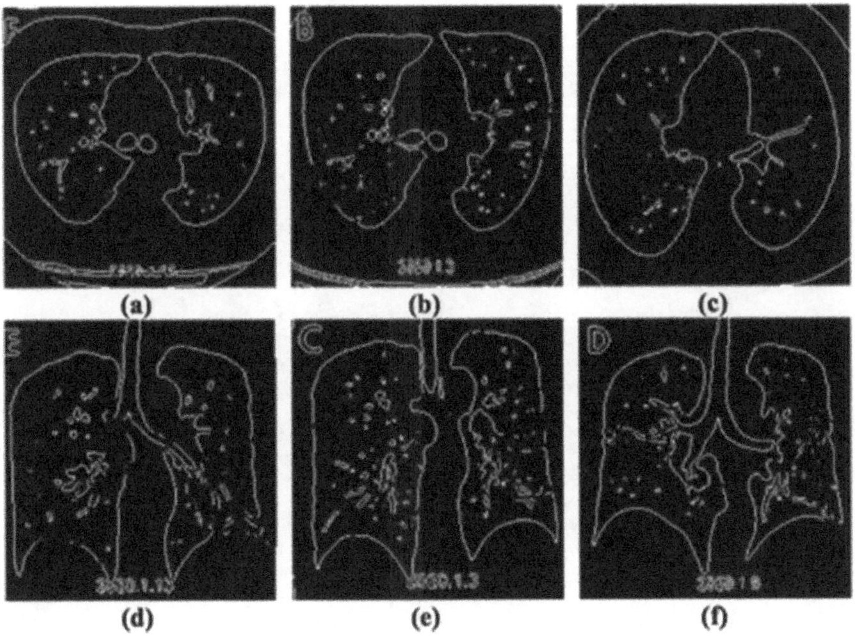

FIGURE 19.9 Sobel algorithm edge detection results.

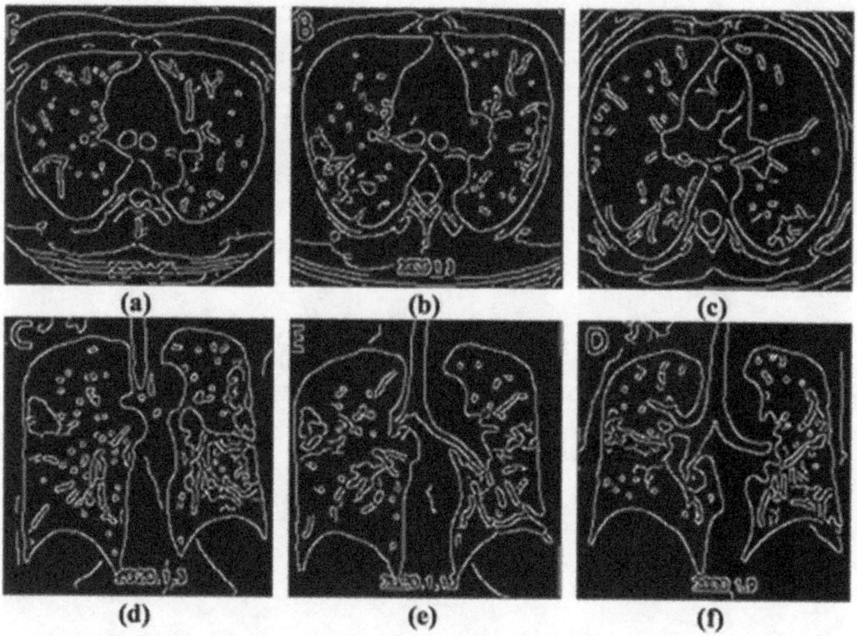

FIGURE 19.10 Canny algorithm edge detection results.

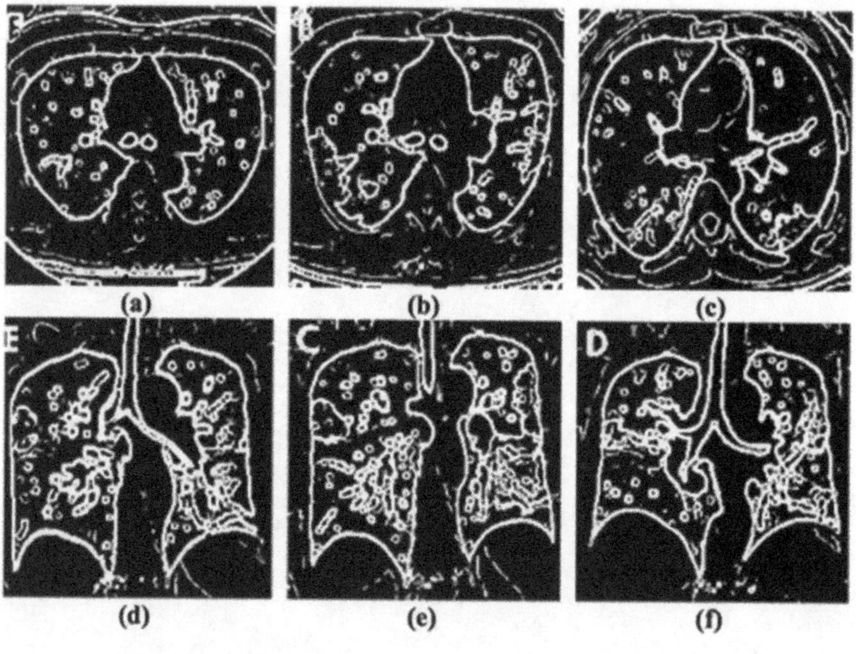

FIGURE 19.11 Neutrosophic edge detector with preprocessing by median filter.

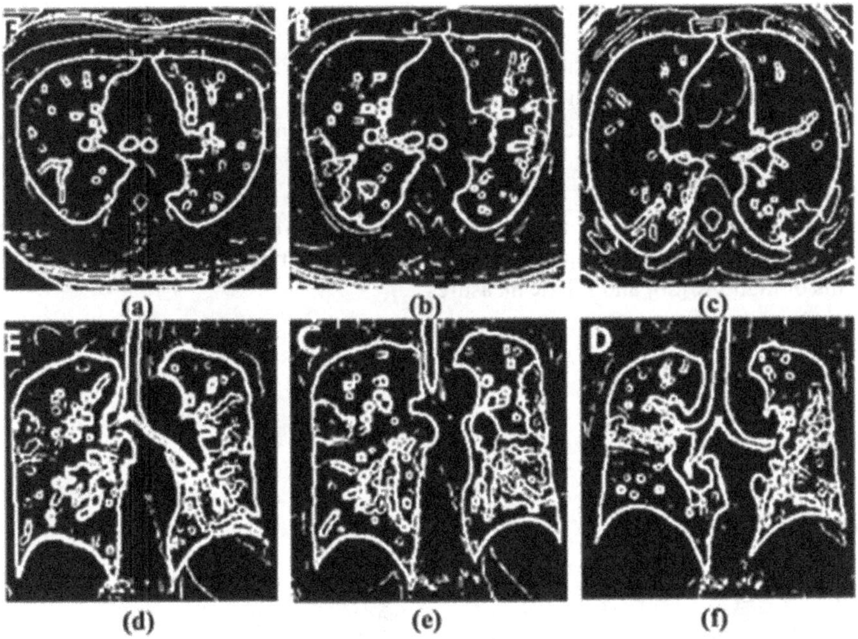

FIGURE 19.12 Neutrosophic edge detector with preprocessing by hybrid median filter.

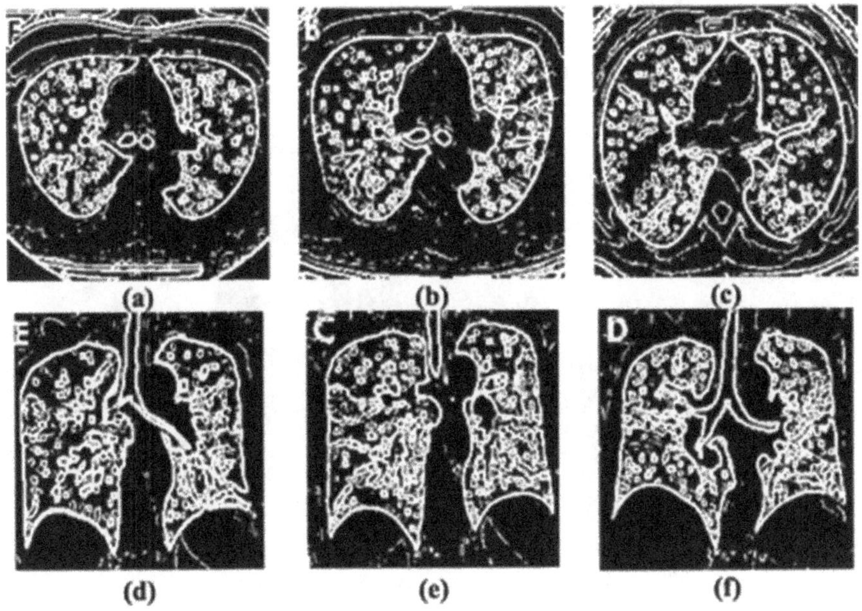

FIGURE 19.13 Neutrosophic edge detector with preprocessing by differential median filter.

19.3.3 Test Results of Edge Detectors on Benchmark Images

Algorithms were evaluated on benchmark images and PSNR; PR measures of edge detectors are depicted in Figures 19.14 and 19.15. The hybrid median filter when coupled with neutrosophic edge detector (HMF-NS) generates robust results, when equaled with median filter coupled with neutrosophic edge detector (MF-NS), decision-based median filter coupled with neutrosophic (DAMF-NS) edge detector and median filter coupled with canny edge detector (MF-Canny) techniques. The HMF-NS algorithm yields superior results, when compared with other algorithms in terms of high PR [20] and PSNR measures.

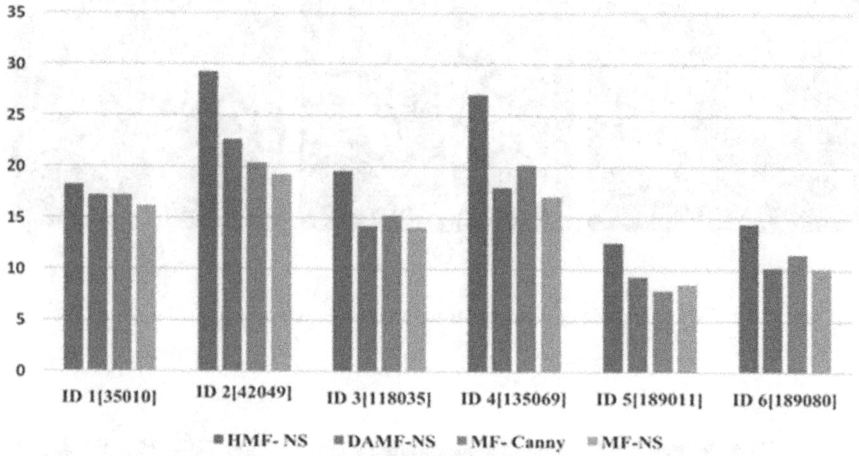

FIGURE 19.14 PR measures of edge detectors.

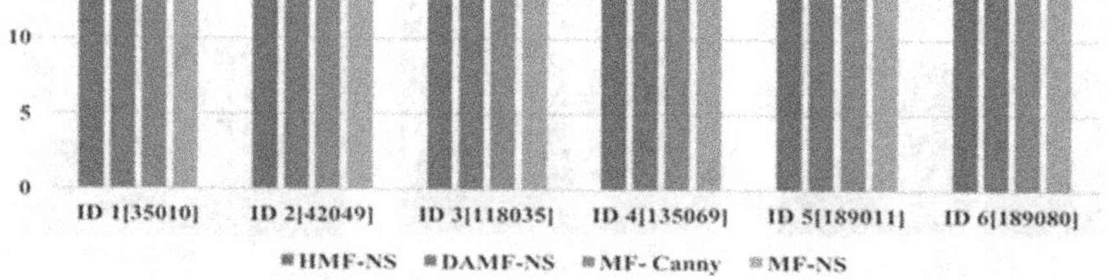

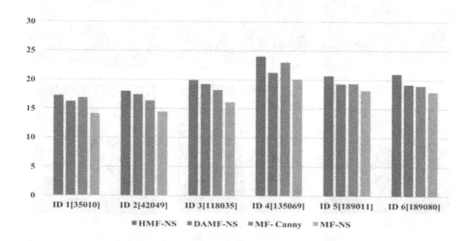

FIGURE 19.15 PSNR measures of edge detectors.

19.4 CONCLUSION

This research work proposes edge detection on COVID-19 CT images based on neutrosophic technique. Prior to edge detection, preprocessing was done by median, hybrid and decision-based median filter. The hybrid median filter when coupled with neutrosophic edge detection was superior in the tracing of boundary than the other approaches. The proposed hybrid median filter based neutrosophic edge detection yields superior results, when compared with Roberts, Prewitt, Sobel and Canny approaches. Preprocessing filters were evaluated initially on benchmark images and found that for low salt-and-pepper noise variance hybrid median filter is proficient and for high salt-and-pepper noise variance DAMF is proficient.

ACKNOWLEDGMENTS

The authors would like to acknowledge the support provided by Nanyang Technological University under NTU Ref: RCA-17/334 for providing the medical images and supporting us in the preparation of the manuscript. Parasuraman Padmanabhan and Balazs Gulyas also acknowledge the support from Lee Kong Chian School of Medicine and Data Science and AI Research (DSAIR) centre of NTU (Project Number ADH-11/2017-DSAIR and the support from the Cognitive NeuroImaging Centre (CONIC) at NTU. The author S.N. Kumar would also like to acknowledge the support provided by Schmitt Centre for Biomedical Instrumentation (SCBMI) of Amal Jyothi College of Engineering.

REFERENCES

1. Tyagi, Y., Puntambekar, T. A., Sexena, P., & Tanwani, S. (2012). A hybrid approach to edge detection using ant colony optimization and fuzzy logic. *International Journal of Hybrid Information Technology*, 5(1), 37–46.
2. Xu, Q., Zhang, Q., Hu, D., & Liu, J. (2018). Removal of salt and pepper noise in corrupted image based on multilevel weighted graphs and IGOWA operator. *Mathematical Problems in Engineering*, 2018, 1–11. Article ID 7975248.
3. Alshennawy, A. A., & Aly, A. A. (2009). Edge detection in digital images using fuzzy logic technique. *World Academy of Science, Engineering and Technology*, 51, 178–186.
4. Tao, C. W., Thompson, W. E., & Taur, J. S. (1993, March). A fuzzy if-then approach to edge detection. In [Proceedings 1993] *Second IEEE International Conference on Fuzzy Systems* (pp. 1356–1360). San Francisco, IEEE.
5. Melin, P., Mendoza, O., & Castillo, O. (2010). An improved method for edge detection based on interval type-2 fuzzy logic. *Expert Systems with Applications*, 37(12), 8527–8535.
6. Gonzalez, C. I., Melin, P., Castro, J. R., Mendoza, O., & Castillo, O. (2016). An improved sobel edge detection method based on generalized type-2 fuzzy logic. *Soft Computing*, 20(2), 773–784.
7. Epsiba, P., Suresh, G., & Kumaratharan, N. (2020). Edge detection-based depth analysis using TD-WHOG scheme. In Vijender Kumar Solanki, Manh Kha Hoang, Zhonghyu (Joan) Lu, & Prasant Kumar Pattnaik (Eds) *Intelligent Computing in Engineering* (pp. 397–407). Singapore: Springer.
8. Poma, X. S., Riba, E., & Sappa, A. (2020). Dense extreme inception network: Towards a robust cnn model for edge detection. In *The IEEE Winter Conference on Applications of Computer Vision*, (pp. 1923–1932). Snowmass Village, CO, USA.

9. Liu, C., Shirowzhan, S., Sepasgozar, S. M., & Kaboli, A. (2019). Evaluation of classical operators and fuzzy logic algorithms for edge detection of panels at exterior cladding of buildings. *Buildings*, 9(2), 40.
10. Mohan, J., Krishnaveni, V., & Guo, Y. (2013). MRI denoising using nonlocal neutrosophic set approach of Wiener filtering. *Biomedical Signal Processing and Control*, 8(6), 779–791.
11. Mohan, J., Krishnaveni, V., & Guo, Y. (2012). Performance analysis of neutrosophic set approach of median filtering for MRI denoising. *International Journal of Electronics and Communication Engineering and Technology*, 3, 148–163.
12. Bharodiya, A. K., & Gonsai, A. M. (2019). An improved edge detection algorithm for X-Ray images based on the statistical range. *Heliyon*, 5(10), e02743.
13. Deng, X., Ma, Y., & Dong, M. (2016). A new adaptive filtering method for removing salt and pepper noise based on multilayered PCNN. *Pattern Recognition Letters*, 79, 8–17.
14. Erkan, U., Gökrem, L., & Enginoğlu, S. (2018). Different applied median filter in salt and pepper noise. *Computers & Electrical Engineering*, 70, 789–798.
15. Toh, K. K. V., & Isa, N. A. M. (2009). Noise adaptive fuzzy switching median filter for salt-and-pepper noise reduction. *IEEE Signal Processing Letters*, 17(3), 281–284.
16. Rakesh, M. R., Ajeya, B., & Mohan, A. R. (2013). Hybrid median filter for impulse noise removal of an image in image restoration. *International Journal of Advanced Research in Electrical, Electronics and Instrumentation Engineering*, 2(10), 5117–5124.
17. Arulpandy, P., & Pricilla, M. T. (2020). Salt and pepper noise reduction and edge detection algorithm based on neutrosophic logic. *Computer Science*, 21(2), 179–195.
18. Lei, J., Li, J., Li, X., & Qi, X. (2020 Apr) CT imaging of the 2019 novel coronavirus (2019-nCoV) pneumonia. *Radiology*, 295(1), 18.
19. Duan, Y. N., & Qin, J. (2020). Pre-and posttreatment chest CT findings: 2019 novel coronavirus (2019-nCoV) pneumonia. *Radiology*, 295(1), 21–21.
20. Lopez-Molina, C., De Baets, B., & Bustince, H. (2013 Apr 1) Quantitative error measures for edge detection. *Pattern Recognition*, 46(4), 1125–1139.

20 Tracking the COVID-19 Suspected Cases through Web Application

S. Saiteja, Sandeep Kumar, and C. Andy Jason
Sreyas Institute of Engineering & Technology

CONTENTS

20.1 INTRODUCTION

A coronavirus is a common type of virus which usually causes diseases of the upper respiratory tract. COVID-19 is a SARS-CoV-2 infectious disease [1–3]. In Wuhan, the capital of Hubei, China, the epidemic first detected in 2019 and has subsequently spread worldwide, culminating in a coronavirus pandemic in 2019–2020. Fever, cough, and breathlessness are typical signs. Many signs may involve chest discomfort, sputum, nausea, sore throat, stomach pain, and scent and taste impairment [4,5]. Although most cases lead to minor effects, pneumonia and multiorgan failure have advanced considerably. As on 25 March 2020, the average mortality rate from confirmed patients was 4.5%, but currently, across age categories and other health conditions, varied from 0.2% to 15% [6]. The virus is often transmitted through near contact and by respiratory droplets triggered by cough or sneezing. Respiratory droplets may occur during respiration, but the virus is not thought to be airborne. It can also impact COVID-19 by contacting and then removing the polluted soil [7–9]. It is more contagious if individuals become symptomatic, even though they will propagate before symptoms form. The virus will survive up to 72 hours on surfaces. The duration ranges between 2 and 14 days, for an average of 5 days, from exposure to the start of symptoms [10]. The main diagnostic technique is through a nasopharyngeal swab reverse transcription polymerase chain reaction (rRT-PCR). A mixture of signs, risk factors as well as a chest CT scan with pneumonia features may also detect the infection. Complete hand washing and social barriers (maintaining physical distance from others, in particular of symptoms), shielding tissue and/or internal elbow while coughing and sneezing, and holding unwashed hands away from the face are advised for avoiding infection. Any regional health officials encourage the usage of masks for those who believe that they are contaminated with the virus and its carers, but not for the general population, even though those who like them may do basic clothes [11–14]. No COVID-19 vaccine or antiviral therapy is appropriate. Administration includes pain management, help, loneliness, and study steps [15].

A Written Emergency of International Concern (PHEIC) on 30 January 2020 and a pandemic of 11 March 2020 is announced by the World Health Organization (WHO) about the outbreak of the 2019–20 coronavirus [16,17]. In many countries in all six WHO regions, local transmission of the disease has been reported. It is understood that six different forms of coronaviruses infect humans. Four are normal, and at least one of them would be encountered sometime in their lives by everyone [18]. The other two groups are responsible for SARS and MERS. They are less common, but much more deadly. Before SARS, coronavirus was considered to cause serious diseases in animals not especially harmful to human beings [19–21]. This prompted scientists to believe that animals had SARS-CoV passed to human beings first. Now they suspect that a new, deadlier strain has become an animal virus.

20.2 BACKGROUND

The Latin word was "virus," meaning "poison." A virus is an infectious agent sub microscopic that only replicates inside the organism's living cells. Viruses can affect any form of life, including bacteria and archaea, from animals and plants to microorganisms. Over 6,000 of the million's types of viruses in the environment have been registered. Viruses are the most numerous forms of biological organism present in almost any Earth environment [22,23]. Virology, a subspecialty of microbiology, is also the analysis of viruses. The primary factor is that this virus spreads several awful pandemic viruses such as Dengue, Ebola, Swine flu, and other deadly diseases. Cholera, bubonic plague, smallpox, and influenza are among the most vicious victims in the history of mankind. And outbreaks of these diseases across international frontiers are properly described as a pandemic, especially smallpox, which in its 12,000-year lifetime has killed between 300–500 million people throughout history. Starting in December 2019 a new ("novel") coronavirus started to appear in humans in the area of Wuhan, China. It was called COVID-19, a shortened version of "Coronavirus Disease of 2019." Because of its newness, this new virus spreads extraordinarily rapidly among humans—no one on earth has COVID-19 immunity, and nobody had COVID-19 until 2019 [24]. Though initially seen as an outbreak in China, the virus spread worldwide in the span of months. COVID-19 was declared a pandemic by the WHO in March and by the end of the month, the world saw more than half a million people infected and almost 30,000 deaths. In the US and other nations, the infection rate was already spiking [25]. In Wuhan, China, Li-Wanlieng reported the first occurrence of the coronavirus. He is one of the

8 doctors who warned the coronavirus epidemic in December 2019 for the first time. It has affected the entire planet from there. A total of 4,101,641 confirmatory cases of COVID-19 coronavirus from Wuhan, China, and a death toll of 280,435 have been recorded by 212 countries and territories worldwide [26,27]. USA, Spain, Italy, UK, Russia, France, Germany, Brazil, Turkey, and Iran are the most affected countries by COVID-19 pandemic. One of the main means that this fatal virus spreads through physical contact is from one person to another. Now let us see how and what the symptoms of this virus are.

20.2.1 Steps of Corona Virus Travel Inside the Body

- **Entering as an immortal cell**: The virus reaches the organism through the ears, mouth, and eyes and then binds to the cell in the airway releasing the protein ACE2.
- **Viral RNA release**: The virus infects the cell by fusing the oily membrane to the cell membrane. A sequence of genetic material known as RNA is produced from inside this coronavirus.
- **Call**: the contaminated cell reads the RNA and begins to generate proteins that hold the immune system at bay.
- **Affecting the body**: Following this, the cell begins to spin fresh loops, creating further copies of the virus.
- **Fresh copies assembling**: Mount and exercise new copies of the virus on the outer edges of cell
- **Spread of infection**: A million copies of the virus may be released by each cell before an infected cell dies at last. The viruses can infect cells in the vicinity, or end up in lung escaped droplets.
- **Response of immune system**: The immune system can overact and attack lung cells in severe cases. Fluid and dying cells block the lungs which find it impossible to breathe. A limited percentage of infections may contribute to acute respiratory failure and death syndrome.

20.2.2 Testing of Corona

The COVID-19 can be tested in two ways. USA uses a special type of test called PCR to carry out corona testing. On the other hand China does antibody tests using blood samples.

- **PCR test**:

  Throat swab and saliva are taken and nucleic acid is extracted. If the genome of the nucleic acid matches with the genome of SARS-CoV-2 virus then it is positive.
- **Antibody testing**:

  The immune system produces unique proteins to combat the infection once the corona reaches the body. If these antibodies are found in the blood samples, it is positive. It takes a day for these experiments to be carried out and to collect findings. There are insufficient facilities for further testing. That is when we face the greatest challenge.

20.3 PROPOSED METHODOLOGY

First of all, with the help of the proposed work, we will open the proposed designed app and the entire process of proposed work is shown in Figure 20.1. After that in the user registration process everyone will enroll in the designed app. Once everyone enrolled, this proposed work collects the information to the users as shown in Figures 20.2–20.6. Now we have the entire data regarding user on whether they have COVID-19 symptoms or not. Based on the answers provided, we would calculate

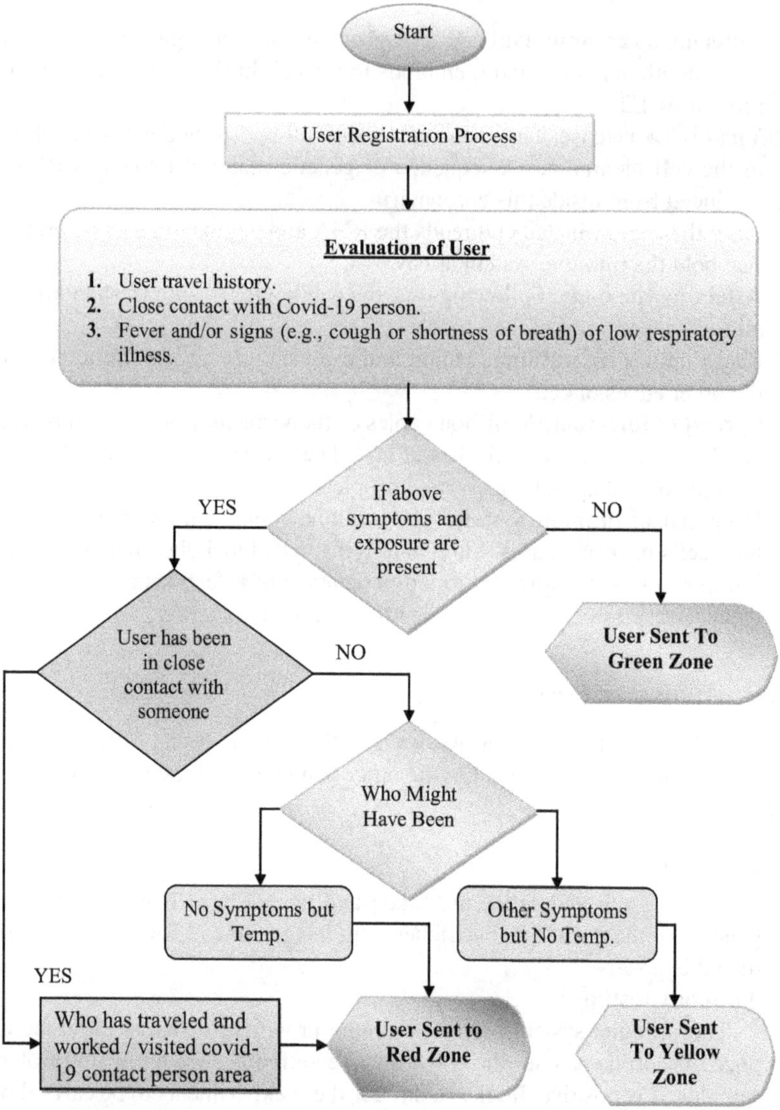

FIGURE 20.1 Flow chart of the proposed work.

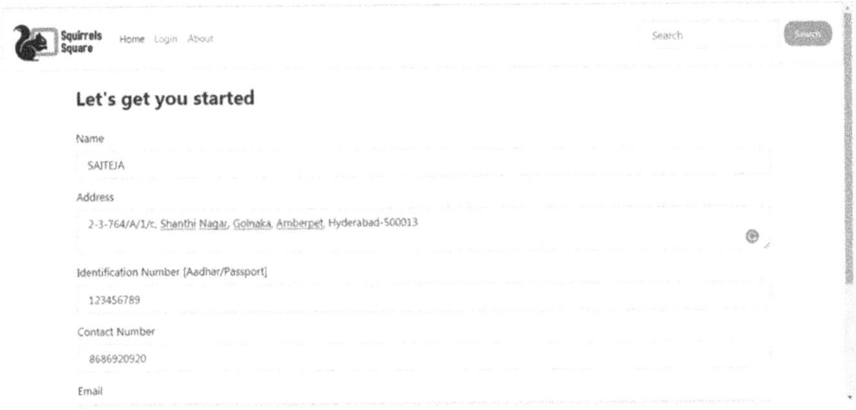

FIGURE 20.2 Filling up our details.

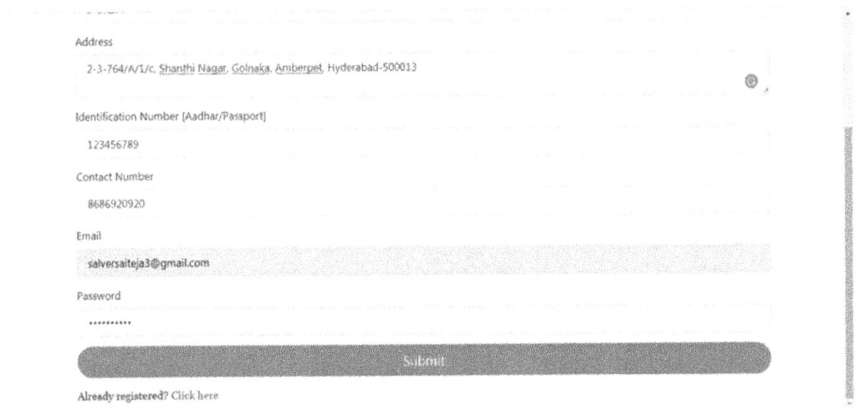

FIGURE 20.3 Submit registration of our details.

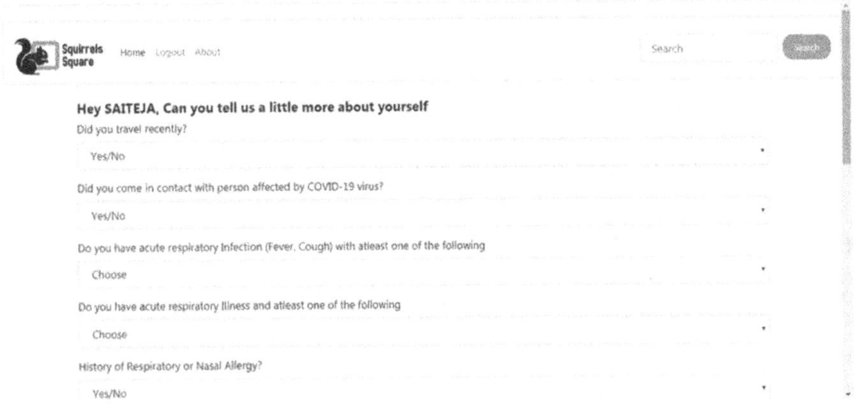

FIGURE 20.4 Little bit more information about you.

FIGURE 20.5 Little bit more information about you.

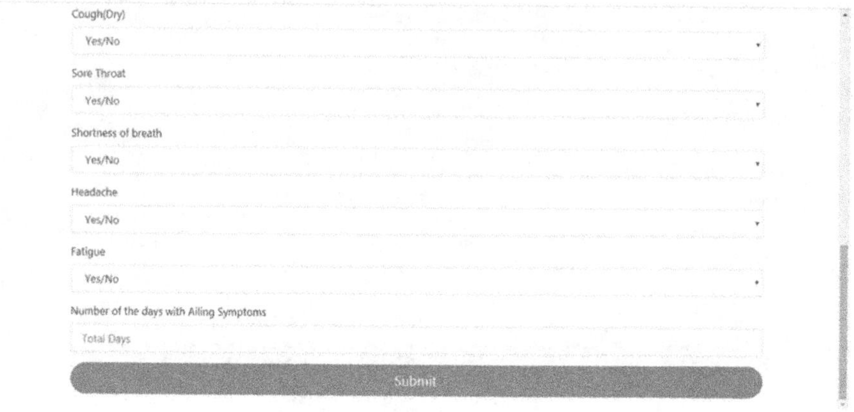

FIGURE 20.6 Submit the questionnaire details.

the threshold percentage of the coronavirus symptoms and alert the user for three categories based on the threshold percentage mentioned below:

3.1 Critical (whoever surpasses threshold %)
3.2 Medium (close to threshold %)
3.3 Low (well below threshold %)

20.3.1 Critical (Red Indication)

For people above threshold %, we would share immediate first aid and nearby corona testing center address and share details of the user with nearby testing facility and government authorities. The person who is indicated with red and traveled from abroad might have been infected with COVID-19. In addition, having temperature more than >99°F and some other symptoms as given above are shown in Figure 20.7. This user's data are been tracked by COVID-19 HOSPITAL AUTHORITIES

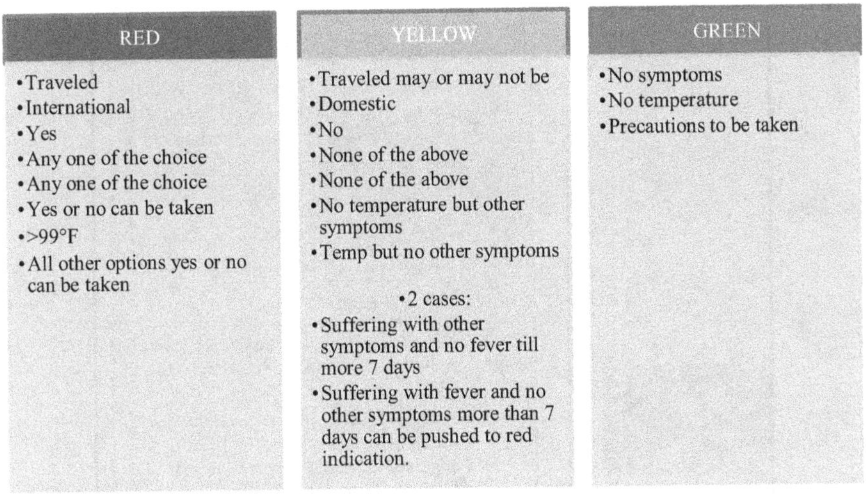

FIGURE 20.7 Represents the 3-zone.

as well government agencies and user can have addresses of nearby COVID-19 healthcare centers.

20.3.2 Medium (Yellow Indication)

For people in medium category, we would advise them to self-quarantine and share the details with government authorities and send periodical alerts to the person to submit the answers based on current health status. We would capture the person's home location and keep a track of the person's movements. The person who is indicated with yellow might have traveled domestically or else might have contact with COVID-19 affected person, or they might have visited COVID-19 healthcare centers. After person submitting his/her symptoms we get him/her by two cases.

CASE 1: No temperature but other symptoms are there then he/she must be in quarantine for 14 days and in between of 14 days after 7 days he must submit his symptom again if symptoms are meant to be normal then he/she can be at home and he will be getting remainders for submissions of symptoms and his data will be traced.

CASE 2: No other symptoms but temperature is there then he/she will be indicated in red color and the process will follow as per the procedure in red color indication.

20.3.3 Low (Green Indication)

For people on Low Risk Category, we would ask them to submit answers after 14 days and assess the situation. No symptoms and no temperature but details should be submitted for every 14 days as shown in Figure 20.7.

Finally, red zone further divided based on patients' priorities affected by COVID-19 as shown in Figure 20.8.

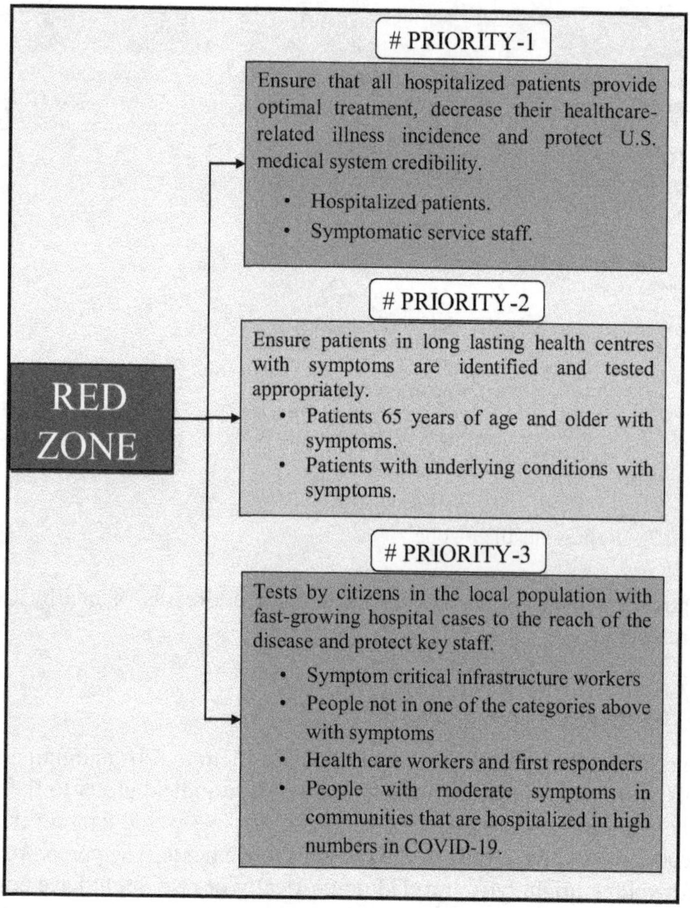

FIGURE 20.8 Further segmentation of red zone.

TESTING: A typical examination space with a door shut can be used to extract diagnostic respiratory specimen (e.g., nasopharyngeal swab). The healthcare company must select, treat, and send specimens.

- Processing of specimens of priority 1 and priority 2 in your health department, if appropriate; processing in a private laboratory.
- Processing of specimens of priority 3, where appropriate, inside your healthcare facility; processing in a company laboratory. For any additional guidance, healthcare providers may be consulted.

20.4 SUGGESTION

To control the impact of COVID-19 spread, we can make government and private and other updated testing laboratories to test few areas which cover the radius around them. The testing should be carried out through door to door and for each person so that we can maintain a data set of all positive cases at a time of that area. This can beneficial if done based on area, locality as well district, state, and the whole nation for making all positive cases data. By this we can stop the spreading of infection as shown in Figure 20.8. For example, we take Hyderabad city as an example here and we show some tabular form on how it can be workable.

20.5 CONCLUSION

The objective of our TRACK INFORMER is to reduce the count of patients and to develop hospital and clinical care units in agile ways in times of crises. TRACK INFORMER uses the user's symptoms data to know whether they are affected with COVID-19 or not. Suppose if the user confirms that he/she has similar symptoms of COVID-19 then the user will get tracked by government officials as well as they are sent to COVID-19 testing centers, before that the user will be directed to a nearby clinic/hospital of his area to submit his samples for COVID-19 check-up as well as to get confirmation on whether they are affected with COVID-19 or not. Based on the clinic/hospital report the user will be sent to red zone or yellow zone.

This method includes critical metrics, including the number of experiments, reviews and supervision, bed size and availability, and current or evolving details on therapies or care approaches. Finally using subset of hospitals and clinics testing centers where a division of positive and negative cases can be determined. By this approach we can slow down the mass spread of patients to flatten the curve of transmission at main COVID-19 testing centers and to reduce the stress of bed management and hospital equipment, as well as there will be no possibility of other people getting affected.

Figure 20.8: Represents the priority-based zone in Hyderabad (Table 20.1).

TABLE 20.1
Comparison of Our Proposed Methodology with the Existing Technology

Application Features	Arogya Sethu	Cova Punjab	T-COVID-19	Odisha COVID Dash Board	Our Track Informer
Tracking user	✓	✓	✓	✓	✓
GPS	×	×	✓	×	✓
Bluetooth	✓	×	×	×	✓
Self-assessment	✓	✓	✓	✓	✓
COVID-19 infected alerts	✓	✓	✓	✓	✓
Track infected people	✓	✓	✓	×	✓
Identify hot spot areas	✓	✓	✓	✓	✓
Self-quarantine guidelines	✓	✓	✓	✓	✓
Healthcare details	✓	✓	✓	✓	✓
Health and emergency info	✓	✓	✓	✓	✓
COVID-19 updates	✓	✓	✓	✓	✓
Government announcements	✓	✓	✓	✓	✓
Helpline numbers	✓	✓	✓	✓	✓

REFERENCES

1. Fang, Yicheng, Huangqi Zhang, Jicheng Xie, Minjie Lin, Lingjun Ying, Peipei Pang, and Wenbin Ji. (2020). "Sensitivity of chest CT for COVID-19: Comparison to RT-PCR." *Radiology*, 19, 200432.
2. World Health Organization. (2020). "Corona virus disease 2019 (COVID-19): Situation report, 72."
3. World Health Organization. (2020). "Corona virus disease 2019 (COVID-19): Situation report, 74."
4. Pan, Feng, Tianhe Ye, Peng Sun, Shan Gui, Bo Liang, Lingli Li, Dandan Zheng et al. (2020). "Time course of lung changes on chest CT during recovery from 2019 novel coronavirus (COVID-19) pneumonia." *Radiology*, 295, 200370.
5. Zu, Zi Yue, Meng Di Jiang, Peng Peng Xu, Wen Chen, Qian Qian Ni, Guang Ming Lu, and Long Jiang Zhang. (2020). "Coronavirus disease 2019 (COVID-19): A perspective from China." *Radiology*, 296, 200490.
6. Lipsitch, Marc, David L. Swerdlow, and Lyn Finelli. (2020). "Defining the epidemiology of Covid-19—studies needed." *New England Journal of Medicine*, 382(13), 1194–1196.
7. Liu, Ying, Albert A. Gayle, Annelies Wilder-Smith, and Joacim Rocklöv. (2020). "The reproductive number of COVID-19 is higher compared to SARS coronavirus." *Journal of Travel Medicine* 2(27), 1–4.
8. Chen, Huijun, Juanjuan Guo, Chen Wang, Fan Luo, Xuechen Yu, Wei Zhang, Jiafu Li et al. (2020). "Clinical characteristics and intrauterine vertical transmission potential of COVID-19 infection in nine pregnant women: A retrospective review of medical records." *The Lancet*, 395(9), 809–815.
9. Bai, Yan, Lingsheng Yao, Tao Wei, Fei Tian, Dong-Yan Jin, Lijuan Chen, and Meiyun Wang. (2020). "Presumed asymptomatic carrier transmission of COVID-19." *JAMA*, 323(14), 1406–1407.
10. Remuzzi, Andrea, and Giuseppe Remuzzi. (2020) "COVID-19 and Italy: What next?" *The Lancet*, 395(10231), 1225–1228.
11. Gao, Jianjun, Zhenxue Tian, and Xu Yang. (2020). "Breakthrough: Chloroquine phosphate has shown apparent efficacy in treatment of COVID-19 associated pneumonia in clinical studies." *Bioscience Trends*, 14(1), 72–73.
12. Mehta, Puja, Daniel F. McAuley, Michael Brown, Emilie Sanchez, Rachel S. Tattersall, and Jessica J. Manson. (2020). "COVID-19: Consider cytokine storm syndromes and immunosuppression." *The Lancet*, 395(12), 1033–1034.
13. Surveillances, Vital. (2020). "The epidemiological characteristics of an outbreak of 2019 novel coronavirus diseases (COVID-19)—China, (2020)." *China CDC Weekly*, 2(8), 113–122.
14. Zheng, Ying-Ying, Yi-Tong Ma, Jin-Ying Zhang, and Xiang Xie. (2020). "COVID-19 and the cardiovascular system." *Nature Reviews Cardiology*, 17(5), 1–2.
15. Chen, Huijun, Juanjuan Guo, Chen Wang, Fan Luo, Xuechen Yu, Wei Zhang, Jiafu Li et al. (2020). "Clinical characteristics and intrauterine vertical transmission potential of COVID-19 infection in nine pregnant women: A retrospective review of medical records." *The Lancet*, 395(11), 809–815.
16. Fang, Lei, George Karakiulakis, and Michael Roth. (2020). "Are patients with hypertension and diabetes mellitus at increased risk for COVID-19 infection?" *The Lancet. Respiratory Medicine*, 8(4), e21.
17. Anderson, Roy M., Hans Heesterbeek, Don Klinkenberg, and T. Déirdre Hollingsworth. (2020). "How will country-based mitigation measures influence the course of the COVID-19 epidemic?" *The Lancet*, 395(10), 931–934.

18. Cao, Bin, Yeming Wang, Danning Wen, Wen Liu, Jingli Wang, Guohui Fan, Lianguo Ruan et al. (2020). "A trial of lopinavir–ritonavir in adults hospitalized with severe Covid-19." *New England Journal of Medicine*, 382(19), 1787–1799.
19. Fang, Yicheng, Huangqi Zhang, Jicheng Xie, Minjie Lin, Lingjun Ying, Peipei Pang, and Wenbin Ji. (2020). "Sensitivity of chest CT for COVID-19: Comparison to RT-PCR." *Radiology*, 296(2), 200432.
20. Gautret, Philippe, Jean-Christophe Lagier, Philippe Parola, Line Meddeb, Morgane Mailhe, Barbara Doudier, Johan Courjon et al. (2020). "Hydroxychloroquine and azithromycin as a treatment of COVID-19: Results of an open-label non-randomized clinical trial." *International Journal of Antimicrobial Agents*, 56(1), 105949.
21. Rothan, Hussin A., and Siddappa N. Byrareddy. (2020). "The epidemiology and pathogenesis of coronavirus disease (COVID-19) outbreak." *Journal of Autoimmunity*, 109, 102433.
22. Bernheim, Adam, Xueyan Mei, Mingqian Huang, Yang Yang, Zahi A. Fayad, Ning Zhang, Kaiyue Diao et al. (2020). "Chest CT findings in coronavirus disease-19 (COVID-19): Relationship to duration of infection." *Radiology*, 295, 200463.
23. COVID, CDC, and Response Team. (2020). "Severe outcomes among patients with coronavirus disease 2019 (COVID-19)—United States, February 12–March 16, 2020." *MMWR. Morbidity and Mortality Weekly Report*, 69(12), 343–346.
24. Shi, Heshui, Xiaoyu Han, Nanchuan Jiang, Yukun Cao, Osamah Alwalid, Jin Gu, Yanqing Fan, and Chuansheng Zheng. (2020). "Radiological findings from 81 patients with COVID-19 pneumonia in Wuhan, China: A descriptive study." *The Lancet Infectious Diseases*, 20(4), 425–434.
25. Cascella, Marco, Michael Rajnik, Arturo Cuomo, Scott C. Dulebohn, and Raffaela Di Napoli. (2020). "Features, evaluation and treatment coronavirus (COVID-19)." In StatPearls [Internet]. StatPearls Publishing.
26. Ai, Tao, Zhenlu Yang, Hongyan Hou, Chenao Zhan, Chong Chen, Wenzhi Lv, Qian Tao, Ziyong Sun, and Liming Xia. (2020). "Correlation of chest CT and RT-PCR testing in coronavirus disease 2019 (COVID-19) in China: A report of 1014 cases." *Radiology*, 296, 200642.
27. Hellewell, Joel, Sam Abbott, Amy Gimma, Nikos I. Bosse, Christopher I. Jarvis, Timothy W. Russell, James D. Munday et al. (2020). "Feasibility of controlling COVID-19 outbreaks by isolation of cases and contacts." *The Lancet Global Health*, 8(4), 488–496.

18. Cao, Bin, Yeming Wang, Danning Wen, Wen Liu, Jingli Wang, Guohui Fan, Lianguo Ruan, et al. (2020). "A trial of lopinavir–ritonavir in adults hospitalized with severe Covid-19." *New England Journal of Medicine* 382(19): 1787–1799.
19. Fang, Yicheng, Huangqi Zhang, Jicheng Xie, Minjie Lin, Lingjun Ying, Peipei Pang, and Wenbin Ji (2020). "Sensitivity of chest CT for COVID-19: Comparison to RT-PCR." *Radiology*: 200432.
20. Gautret, Philippe, Jean-Christophe Lagier, Philippe Parola, Line Meddeb, Morgane Mailhe, Barbara Doudier, Johan Courjon, et al. (2020). "Hydroxychloroquine and azithromycin as a treatment of COVID-19: Results of an open-label non-randomized clinical trial." *International Journal of Antimicrobial Agents*: 105949.
21. Rothan, Hussin A., and Siddappa N. Byrareddy (2020). "The epidemiology and pathogenesis of coronavirus disease (COVID-19) outbreak." *Journal of Autoimmunity*: 102433.
22. Bernheim, Adam, Xueyan Mei, Mingqian Huang, Yang Yang, Zahi A. Fayad, Ning Zhang, Kaiyue Diao, et al. (2020). "Chest CT findings in coronavirus disease-19 (COVID-19): Relationship to duration of infection." *Radiology*: 200463.
23. CDC COVID-19 Response Team (2020). "Severe outcomes among patients with coronavirus disease 2019 (COVID-19)—United States, February 12–March 16, 2020." *MMWR Morbidity and Mortality Weekly Report* 69(12): 343–346.
24. Shi, Heshui, Xiaoyu Han, Nanchuan Jiang, Yukun Cao, Osamah Alwalid, Jin Gu, Yanqing Fan, and Chuansheng Zheng (2020). "Radiological findings from 81 patients with COVID-19 pneumonia in Wuhan, China: A descriptive study." *The Lancet Infectious Diseases* 20(4): 425–434.
25. Cascella, Marco, Michael Rajnik, Arturo Cuomo, Scott C. Dulebohn, and Raffaela Di Napoli (2020). "Features, evaluation and treatment coronavirus (COVID-19)." In *StatPearls [Internet]*. StatPearls Publishing.
26. Ai, Tao, Zhenlu Yang, Hongyan Hou, Chenao Zhan, Chong Chen, Wenzhi Lv, Qian Tao, Ziyong Sun, and Liming Xia (2020). "Correlation of chest CT and RT-PCR testing in coronavirus disease 2019 (COVID-19) in China: A report of 1014 cases." *Radiology*: 200642.
27. Hellewell, Joel, Sam Abbott, Amy Gimma, Nikos I. Bosse, Christopher I. Jarvis, Timothy W. Russell, James D. Munday, et al. (2020). "Feasibility of controlling COVID-19 outbreaks by isolation of cases and contacts." *The Lancet Global Health*.

21 Biomedical Imaging and Sensor Fusion

Satyendra Pratap Singh and Shalini Soman
Amity University

CONTENTS

21.1 INTRODUCTION

Picture and video preparation or refinement discovers its wide utility in scientific technology. Medical imaging methods possess their distinguished part each in ailment and remedial field. These strategies had substantially assisted to improvise the medical management of outpatients [1]. Therapies using picture guidance has substantially decreased the excessive threat of human mistakes with enhanced precision in disorder observation and resection techniques. The records of scientific imaging began in 1890s, step by step uplifted in 1980s and significantly investigated in the last few years with technical improvements. The most important aim of this chapter is to provide a fundamental understanding to the learner of various biomedical imaging methods and its improvements in the present world; secondly, this chapter discusses about sensor fusion technology [1] and lastly combining both biomedical imaging and sensor fusion for more advanced technologies.

21.2 BACKGROUND DATA OF BIOMEDICAL IMAGING

X-ray was discovered by Wilhelm Conrad Rontgen which became the start epoch for clinical imaging. In 1895, Wilhelm Conrad Rontgen determined the phenomenon of electromagnetic radiation in the wavelength limit while operating an experiment with Hittorf–Crookes tube [1]. He labelled the newly found beam as X-rays, along which he took the very prior photos of his spouse Anna Bertha's hand [1]. After this the X-rays were evolved by William Coolidge with his discovery of Coolidge tube which permits extra effective image of fundamental analysis and tumors [1]. Usage of Coolidge duct with tungsten filament by Coolidge was one of the largest utilization of X-rays within the discipline of radioscopy. In 1946, Nuclear Magnetic Resonance (NMR) in a liquefied form was located by Felix Bloch and Edward Purcell [1]. This becomes the prior stride toward the invention of Magnetic Resonance Imaging (MRI). Along with the occurrence of nuclear magnetic resonance (NMR), Raymond Vahan Damadian presented the primary magnetic resonance (MR) anatomy sensor in 1969 [1]. He observed that tumors capable to be differentiated from usual cells utilizing NMR. Godfrey Hounsfield came forward with the proposal to design an item in portions with the use of X-rays at different tilts on all sides of the target [1]. He constructed the foremost model, i.e., computed tomographic (CT) scanner and received prior CT picture of a conserved person's mind. The initial MRI body scanner on human turned up into image in 1977. With the aid of mathematics, pc program and with technical growth in numerical and communications system, recent imaging strategies have been coming into existence. This constructed medical imaging to be an integrative area which includes physicist, biologist, mathematician, pharmacologist and a statistical biologist.

21.3 DIFFERENT IMAGING METHODS

The primary concept of every visualization strategies is identical. A ray of wave passes via the frame/location which is under diagnosis, transmits or reflects back the radiation and this emission can be apprehended by using a sensor and processed to acquire a photo sample. This sort of wave varies for distinct methods. CT includes the usage of X-rays, while broadcasting frequency waves and gamma radiation are used for MRI and single-photon emission computed tomography, respectively [1].

21.4 PLAIN-FILM RADIOGRAPHIC

Conventional radiography includes the usage of X-rays; the term "ordinary X-rays" is every now and then utilized for differentiating X-rays utilized solo over X-rays mixed with additional processes (e.g., CT). For traditional radiography, an X-ray beam is prompted and exceeded via an affected person to a chunk of film or a emission sensor, producing a photograph. Different gentle body flesh constricts X-ray photons in a different way, relying on flesh thickness; the thicker the tissue, the whiter (extra radio colorless) the photograph. The limit of thickness, from higher to lesser dense, is depicted by means of alloy (white or radio colorless), bone cortex (less white), muscle and fluid (gray), fat (darker gray), and air or gasoline (dark or radio translucent).

21.4.1 Computed Radiography

Conventional radiography was replaced by Computed Radiography (CR) or Film Replacement Technology. It makes use of a stretchable phosphor Imaging Plate (IP) to seize virtual photographs in place of traditional photographic film. Here when the IP is revealed to X-ray or gamma propagation, then the image is stored by the phosphor layer. As the IP is examined by means of the scanner, a concentrated laser lets out the saved picture inside the shape of seen light photons. These photons are poised and further moved by the scanner, transformed to a digital signal after which conveyed to a personal computer (PC) for refinement and then disposed.

With the advent of CR, Digital Radiography (DR) was emerged, that is, Direct Digital Radiography. It is the method of direct conversion of the X-rays. We can say CR is the indirect method and DR is the direct method. It is named as direct conversion because in this approach, without delay, it converts the grasped X-ray into a proportionally sized electric charge without an intervening scintillating pace. Advantages of film radiography provide fewer chances to differ the method, aside from selecting a diverse film kind. CR additionally offers special IPs but also there are numerous scanners manage parameters that may be modified to the needed exploration assignment, optimizing image standard and maximizing productiveness.

21.4.2 Computerized Tomography

Computerized or CT imaging is used to find out the cause of symptoms, disease or some sort of injury in a patient. It produces a depiction of inner organs, bones, smooth tissue and blood vessels. The cross-sectional depictions or images generated by a CT test may be transformed in more than one plane and can even generate three-dimensional photographs which we can see on computer monitor, published on film or transferred to electronic media. CT scanning is frequently the best technique for determining numerous cancers because the pictures permit your physician to confirm the presence of a tumor and decide its size and place. CT is speedy, painless, noninterfering and precise. During extremity, it is capable of revealing inner wounds and bleeding quickly sufficient to assist rescue lives.

21.4.3 Magnetic Resonance Imaging

MRI is a kind of study that utilizes powerful magnetic fields and radio waves to obtain precise pictures of inner parts of the anatomy.

An MRI scanner is a big duct that includes strong magnets. While scanning you have to lie within the duct. An MRI test may be utilized to look at nearly whichever segment of the frame, which includes the brain and spinal cord, bones and joints, breasts, heart and blood vessels, internal organs, along with the liver, womb or prostate gland. The outcomes of an MRI test may be utilized to assist detection states, plan remedies and verify how powerful old remedy has been.

21.4.4 Ultrasound

Ultrasonography utilizes excessive frequency sound (ultrasound) motions to provide photos of inner body parts and other body tissues. A gadget known as a transducer transforms electric contemporary to acoustic waves that are passed through the body's flesh. Sound waves deflect off systems inside the human body and are sent back again to the transducer, which changes the waves into electrical indicators. A central processing unit (CPU) transforms the sample of electrical alerts to a depiction, which is disposed on a display screen and telerecorded as a virtual CPU picture. X-rays are not used, so there is no radiation disclosure during this process.

Ultrasonography is pain-free, quite economical and taken into consideration as very secure, even in the course of pregnancy.

21.4.5 Nuclear Medicine

This remedy is an imaging procedure that entails injection, inhalation or injection of radioactive trackers to imagine numerous organs. The trackers or radiopharmaceutical is obtained via inclusion of a radioactive isotope to a medicine specified to the organ being depicted. The radioactive tracers emit gamma emission, which is then depicted utilizing a gamma Polaroid.

The gamma Polaroid or we can say gamma camera includes an emission sensitized crystal which senses the dissemination of the tracker within the patient's frame. The statistics is transformed to a digital form to provide a two- or three-dimensional photograph on a display. The present-day generations of gamma Polaroid are combination machineries such as a CT to permit the fusion of Nuclear Medicine and CT pictures. Most widely used applications of this technique are for bone scan to know about the activities of the bones, myocardial perfusion scan, renal scan, lung scan and thyroid scan.

Nuclear medicines are utilized for quick sickness identification and in prior identification of various levels of most cancers or the correct region of cancer and different deformities. The major cons of nuclear remedy are time intake [1]. It requires numerous hours to days for the radio tracker to acquire accrued in the particular area undergoing investigation. Additionally, the quality of the depiction of pictures is not as significant as in MRI or CT [1].

21.4.6 Some Other Imaging Methods

A range of recent radioscopy strategies are nowadays accessible and are nevertheless getting advanced. Some of the strategies are given below. These methods are less usual inside the area of scientific depiction due to the absence of several key elements wanted for systematic radiology or they are in the developing procedure [1]. But, due to their distinctive resources, individually all of these imaging strategies are acquiring enormous significance gradually.

21.4.6.1 Electric Impedance Tomography

Electric impedance tomography (EIT) is an approach that utilizes ground terminal with lower frequency powered by electricity stream that discovers the difference in organic tissues in the course of occurrence of electrical conductivity. Usually, muscle tissues and frame fluids which include blood are observed here. EIT outcomes in 2D pictures which may be received from the conductivity of minute AC moved/diffused via organic specimens [1]. At high frequencies, overdue to modifications in dielectric constant, the image outcomes are produced as portions of 2D photos. Consequently, this approach is in standard called Impedance Tomography [1].

This imaging method is secure and lower value while in comparison to MRI. EIT is presently hired in screening of breast cancers, simultaneous utilization in lung air flow inside Intensive Care Unit, performance of cardiac, brain, cervix and many others [1].

21.4.6.2 Optical Coherence Tomography

Optical coherence tomography (OCT) is an imaging method that enroll backward reflecting and dispersing of light waves of prolonged wavelength which can get intense identifying vicinity, disperses the illuminating rays and builds a standard tomogram of higher resolution [1]. The photographs are acquired in micrometer resolution with the help of lower coherence interferometry. It is normally utilized in detecting prior diagnosis of the retina in ophthalmology. Furthermore, OCT is utilized to detect plaques in coronary bloodstream, in dermatology and in dentology for root canal surgery [1].

21.4.6.3 Photoacoustic and Thermoacoustic Diagnostics

Photoacoustic diagnosis is totally built on the concept of photoacoustic impact. Photoacoustic imaging makes use of laser pulse that travels in and out of the organic specimens, conducts the power and transforms it to heat, ensuing in making up of ultrasound waves [1]. Such waves are encapsulated with the aid of actuator and are transformed to pictures.

Thermoacoustic diagnosis has similar concept as that of photoacoustic image diagnosis, but this imaging utilizes radiofrequency waves. Photoacoustic diagnosis is extensively implemented in sensing of brain lesions, dimension of congregation of hemoglobin, detection of breast cancer and many others [1]. Merging of both acoustic image diagnoses are significantly utilized in the detection of breast cancer. In this the regular cells in our body are not affected as it does not use radiation. There is a growing need to apply this method in sensing of cancers.

21.4.6.4 Microwave Image Diagnosis

This technique is a rising imaging procedure mainly in analysis of breast cancers. In comparison to photoacoustic diagnosis and ultrasound image diagnosis, this imaging has favorable time ahead for breast cancers detection [1]. Microwave image diagnosis relies upon the liquid contented in tissue specimens. On the account of angiogenesis, most cancer units typically have higher liquid content material in comparison to normal cells and higher recognized in microwave image diagnosis. A higher electrical assessment is acquired interlinking the cancerous units and ordinary units in microwave image diagnosis, building it as extra suitable for cancer detection. Microwave image diagnosis is likewise applied in osteoporosis, where it senses the variant in mineral thickness of bone. As the image diagnosis technique is lesser in price, convenient and safer, large attention is exhibited by the experimenters to expand this approach for different cancer sensing and in prognosis of different sicknesses.

21.4.6.5 Elastography

This imaging procedure requires the usage of machine-like characteristics of fleshy tissues [1]. Common visualization strategies consisting of CT and MRI are not able to detect the prior stage of breast cancer, specifically because of the firmness in wounds that are mostly unable to be detected using these methods [1]. Cancerous flesh holds extra hardness in comparison to usual flesh. Elastic hardness is the primary element that is ruminated utilizing the elastic image diagnosis, as the breast cancer flesh usually goes through pulsation in prior detection of breast cancer. Besides these most cancers, the principle body part detected using this method is the liver [1]. The defective liver can possess extra hardness in comparison to regular liver. Therefore, elastography is the most ordinary image diagnosis approach utilized for sensing liver breakdowns. Distant body parts like the muscular tissues and tendons also are identified by the aforementioned method depending on flexible characteristics.

21.5 NEED FOR DIFFERENT IMAGING TECHNIQUES

Using a particular imaging technique, we can only diagnose a single organ or the organ which is only detectable by a particular technique. CT scan is able to detect only the existence of stones in kidney plus any peculiarities in our urinary tract [1]. But to get a view of arteries in kidney we need Magnetic Resonance Angiogram but to find the whereabouts of some tumor, infection or cyst we need to use CT scan. Although many imaging techniques are there, we use only a particular one based on the patient's history and problem.

Different techniques have its own usage and properties to detect particular problem. Also, all these techniques have its own cons too; as CT scan cannot provide with the blood gush and action of cerebrum cell, whereas MRI can [1]. Also due to emerging technologies these techniques can be combined to get more finer information about the disease or problem. Some coupled techniques that are in usage are PET-MRI or CT-PET, CT-MRI; all these provide the practical characterization of body parts and tissues along with complete information about the flesh thickness, length of the organ and more [1]. Therefore, coupling of the procedures gives much better diagnosis.

21.6 IMAGE PROCESSING SYSTEM

The images that we get from different imaging techniques are further used by the clinician to think for an operation, remedy and to determine the issue [1]. Image processing is very much important as the whole treatment is based on the image we get through image techniques. A small variation in the image can totally change the treatment which has to be done by the doctor. For exact diagnosis image quality and processing play a vital role.

21.7 ELEMENTS OF IMAGE PREPARING

The data prevailed out of in-patient's anatomy through imaging to the display of picture. All these play significant role in image processing.

- **Image sensors:** Sensors such as camera, videos or scanner are used to capture the images. And these captured images are later converted to digital form using digitizer.
- **Image processing hardware:** Its purpose is to speed up the process.
- **Computer:** As an image processing system, we can use a normal PC to highly specialized computer based on our needs.
- **Mass storage:** This is used for the storage of the image we obtain from the patient. Therefore, we require better storage system so as to store thousands to millions of images obtained from patients.
- **Hardcopy devices:** This is used for recording the images.
- **Image processing software:** Commercial software are available in the market which can be used for further processing of images.
- **Image display:** To display the output or the information we use devices like monitors or television, screens, etc.

21.8 SENSOR FUSION

As mentioned above in image processing sensors are used to capture the image that we get from the patient. Nowadays in image processing sensor fusion is the widely used technology.

Sensor fusion is the association of sensory statistics or facts resulting out of diverse origins in order for the subsequent statistics has much lesser ambiguity that ought to be viable while those assets have been used personally [2]. Sensor fusion can be defined as the combination of two or more data sources in a way that generates a better understanding of the system. Data fusion systems are actually widespread and are used in numerous areas which include sensor web works, robotics, video and photo processing, and intelligent machine layout, one or two more. We can address this with many names such as sensor data fusion, information fusion, multisensory fusion, etc..

21.9 ADVANTAGES OF SENSOR FUSION

Generally, performing data fusion has specific benefits. These benefits mostly contain improvements in information genuinity and accessibility. Illustration of the data fusion helps progressed sensing, conviction and dependability, also limiting of facts and figures obscurity, while expanding dimensional and transient assurance pertaining to the ultimate class of advantages. Information merging is capable to offer particular advantages for a few software conditions. For example, cordless sensor internet mechanism is regularly made of a great amount of sensor nodules, consequently presenting a latest extensibility challenges produced by capability collisions and spreading of unnecessary statistics. Concerning power limitations, conversation must be minimized to expand the existence period of the sensor nodules. When statistics fusion takes place at some stage in the routing method, this sensor record is merged and the most effective end outcome is conveyed, the variety of text is decreased, strikes are averted and power is conserved.

21.10 DISADVANTAGES OF DATA FUSION

The basic challenges that sensor data fusion face are as follows:

- **Data imperfection:** The facts obtained by using sensors are continually affected by using a few stages of lack of precision along with ambiguity within the specifications. Data fusion procedures require to be capable of showing such imperfections successfully, and to make use of the records excessively to minimize their results [3].
- **Outliers and spurious records:** The ambiguities are not just arising due to the lack of details and noises but also due to the obscurity and uncertain behavior of environment. Data algorithms should be able to reduce the above-mentioned consequences [3].
- **Conflicting data:** confusing facts maybe complicated mainly when everything is depended on situational perspective analysis and to avoid unreasonable results. Highly conflicting data must have been treated with special care using algorithms.
- **Data correlation:** This difficulty is mainly critical and not unusual in dispensed fusion settings, e.g., wireless sensor networks, as an instance a few sensor nodules are possibly to be uncovered to the equal outer sound biasing their readings. In such case information reliance is not considered for the fusion procedures, may additionally undergo from over/beneath command in effects.

- **Data adjustment:** Facts from different sensor need to be framed as one before fusion. This kind of alignment problem is known as data registration as it deals with calibration error due to different sensor nodes.
- **Data association:** In this data association becomes difficult as it becomes impossible to distinguish from which sensor, we actually got the information. This may occur in two ways: track-to-track and measurement-to-track.
- **Preparing structure:** Information combination can be done in a centered or decentered way. The latter is commonly most suitable in Wi-Fi sensor connections because it permits every sensor nodule to system to particularly gather facts. This is a great deal and convenient in comparison to the communicational difficulty needed by means of a centered technique, while each and every record has to be moved to a main working node for fusion.
- **Functional control:** An appropriately manufactured multiple combination method should be there so as to incorporate different timescale to handle the different timing variations in data. Or else the information may traverse and the information we finally get will be out of sequence. So, in real-time applications, this concern must be overseen with care to evade probable working decrease.
- **Standard vs changing occurrence:** The changes that occur with time must be captured and updated accordingly, because it plays an important role in the validity of fusion result.
- **Data dimensionality:** Data must be compressed to lower dimensional facts either locally or globally. This would help in the transmission of data and would save the transmission range and the capacity needed for transferring the facts. And subject to global transforming, confining of statistical loads of the dental combinational nodules needs to be done [4].

21.11 ALGORITHMS TO OVERCOME CHALLENGES

There are a number of methods and algorithms covered by sensor fusion to overcome data-related challenges which includes as follows:

- **Central limit theorem:**

  In a few conditions, while independent random variables are delivered, their properly normalized sum tends toward an everyday distribution even though the authentic variables themselves are not commonly distributed. The theorem is a main concept in possibility concept as it implies that probabilistic and statistical strategies that work for normal distributions can be relevant to many issues related to other forms of distributions.
- **Kalman filter:**

  This filter is also called as Linear Quadratic Estimation [5]. This procedure utilizes alignment of rules that make use of a series of measurements discovered as time passes, including probable noise and some other imprecision, and obtains approximate of unspecified factors that incline to be very much accurate than rest formulated on one dimension only, by approximating a collective possibility dimension upon the factors for every time frame [6]. The filter is named after Rudolf E. Kálmán, one of the prior builders of this theory.

- **Bayes network:**
  It is a type of statistical model which can represent the relationship between diseases and symptoms. After giving the symptoms, this network can be used to compute the probabilities of existence of different diseases.
- **Evidence theory:**
  This theory is also called Dempster-Shafer theory. In this theory, basically one has to combine all the evidence given and come to a conclusion or a degree of belief taking into account all the evidence which was considered.
- **ConvNet:**
  This is an artificial multiple layer between I/p and O/p layers, mostly applied for the analysis of visual imagery. This has a wide application in image recognition, medical image analysis, video recognition, etc.

These are some of the algorithms used to overcome some data fusion challenges. But the above-mentioned challenges have been identified and more investigations are ongoing to resolve it. The available algorithms are not able to solve it. Works are being done to solve these deformities.

21.12 APPLICATIONS OF SENSOR FUSION

There are various applications of sensor fusion such as in robotics, biomedical imaging, equipment monitoring, remote sensing, transportation systems, micro and smart sensors and military applications. Here mainly we will discuss about the application of sensor fusion in biomedical imaging. The field of multisensor fusion is evolving and the need of it is significant. As in future it would help us to deal with problems which can affect the whole world.

21.12.1 Robotics

In many industrial applications, robotics with sensory capabilities is required to enhance productivity and capability. In the areas such as inspection, part fabrication, material handling and assembly the application of fusion techniques is suitable. Teleportation, interconnection between robots and environment, dexterous hands, visual serving, etc., all these are the recent advances in robotics.

21.12.2 Equipment Monitoring

Complex equipment needs condition-based monitoring as in automated manufacturing systems, turbomachinery and drivetrains. All these can maximize the safety and reliability. Also, this will lead to reduction of repair and manufacturing cost.

21.12.3 Remote Sensing

This has a vast application which includes the monitoring of climate, environment, water source, discovering natural origins, agriculture and fighting for the import of illegal things which includes drugs. So, for extracting useful information from satellite we need remote sensing techniques.

21.12.4 Transportation Systems

Global positioning system (GPS) plays an important role in our day-to-day life. GPS-based transport navigation, aircraft landing, train control, highway systems, all these utilize multisensor fusion methods for better efficiency and reliability. Also, the main need of it is safety.

21.12.5 Microsensors

Bigger the size of the sensor the more efficient it will be. But in markets bigger sensors are avoided just because of its size even though it has higher reliability. And smaller the size of sensor its applicability increases by the following: lower weight, lower manufacturing cost, fewer material required, greater portability and wide range of applications. Integrated circuit (IC) technology has enabled us to produce small reliable sensors. And it is very much in demand in microelectronics applications.

21.12.6 Military Applications

For force command and control, situation analysis, intelligence analysis, etc., multisensor fusion is used. For monitoring land environment, multisensor fusion of remote sensed data was used. This would help the militants to introspect the land area.

21.12.7 Biomedical Application

Moving onto the main application which we will discuss in brief is application of sensor fusion in biomedical imaging.

The most significant usage of multisensor fusion is in medical image merging. The principle of sensor fusion is used in the merging of biomedical images so that to get better and enhanced images [6]. And this method of fusion helps in the diagnostic and working of some organs and tissues. The medical pixels received from exclusive sensors are fused together to decorate the diagnostic satisfactory of the photograph modality. Hence multimodality photofusion is accomplished to combine the attributes of diverse image sensors right into a single photo. In the process of photo fusion, the discrete level of records via every and each pixel of a photo at a particular instance of time, state, position or occasions is mapped upon respective records from every pixel of every other image of the equal item at a distinctive instance. For the numerous design and creation of each sort of optical sensors placed a few limitations on the type of facts received through them. Therefore, the technique of picture fusion develops a composite image which suffices for the statistics which is not provided by a single optical sensor.

21.13 MEDICAL IMAGE FUSION METHODS

We arrange the methodological trends in biomedical image fusion strategies into those who depend on feature-level processing and people that function at decision-degree fusion. Feature-stage fusion frequently helps the enhancing photostandard rate and extracts newer features which can be in any other case hard to locate in the actual set of features.

This feature stage fusion can lead to many problems and interimage irregularities such as mismatches in rotations, shifts, scale, pixels, resolution, image noise and contrast. This would lead to destitute fusion production and lower heftiness in the feature representations. This would eventually lead to wrong conclusions that would reduce the reliability of the medical image analysis.

21.13.1 Morphological

This operator utilizes the relatedness linking pixels, either to enhance the spatial positioning of the pixel or to misshape them to draw out applicable features from the subset of spatially localized pixel features. The morphological filtering techniques for biomedical image fusion were applied, for example, in brain diagnosis [7].

21.13.2 Knowledge

In this basically the knowledge of medical practitioner is required to have domain knowledge then his/her knowledge can be utilized to design dissections, labeling along with documentation of images. The main drawback of this algorithm is human error. Limitations inflicted by human perception in the depictions which are liable to big picture element intensity fluctuations.

21.13.3 Neural Networks

Artificial neural systems (ANN) are motivated from the possibility of natural nerve system containing the capacity to gain from contributions for handling highlights and for settling on worldwide choices. The ANN illustrations want an information preparing set to recognize the arrangement of boundaries of the system alluded to as loads [7]. A few applications for clinical picture combination, for example, taking care of the issues that highlight age, classification of information combination, picture combination, miniaturized scale calcification conclusion, bosom malignant growth recognition, clinical conclusion, malignant growth finding, regular figuring strategies and classifier combination.

21.13.4 Fuzzy Logic

The combined, separated and bargain characteristics of the fuzzy rationale have been broadly investigated in picture handling and have ended up being helpful in picture combination. The fluffy rationale is applied both as an element change administrator or a choice administrator for picture combination There are a few utilizations of fluffy rationale base picture combination, for example, cerebrum conclusion, malignant growth treatment, picture division and mix, amplification shared data, profound cerebrum incitement, mind tumor division, picture recovery, spatial weighted entropy, highlight combination, multimodal picture combination, ovarian malignancy finding, sensor combination, common processing strategies and quality articulation [7].

21.13.5 Other Methods

The highlights got from the wavelet include combination procedures that have been utilized alongside other highlight extraction techniques to improve the heartiness of the wavelet-based combination draws near. A few models are combined with help of vector machines, the utilization of wavelet-surface measure, wavelet joined with attractive reverberation angiogram magnetic resonance angiogram (MRA), the utilization of wavelet-self versatile administrator, wavelet-goals with entropy, nonlinear methodology with properties of wavelet move invariant imaging, autonomous segment investigation independent component analysis (ICA) joined with wavelet, wavelet and edge highlights, wavelet with a hereditary methodology, wavelet consolidated with contourlet change, half-breed of neuron systems with fluffy rationale and wavelets, and wavelet entropy. Also, there are support vector machines (SVMs) which can be consolidated alongside other combination calculations and strategies to improve processing speed and to work with better portrayals of low-dimensional element vectors. Support vector machines (SVM) strategies incorporate SVM joined with wavelets, SVM with versatile comparability measures, SVM-information combination, and SVM joined with ANN and GMM.

21.14 CONCLUSION

There are several biomedical imaging techniques which have significantly enhanced the wellness program of patients. And imaging techniques have minimized the risk of human error with enhanced precision in surgical procedures and sensing of diseases. Lastly, some of the biomedical imaging techniques are discussed in this chapter along with sensor fusion technique with problems faced in sensor image fusion and algorithms in use with applications of sensor fusion and some of the image fusion methods.

As discoveries and findings are all about making life better for human beings, the main conclusion of this chapter is to show that sensor fusion methods can enhance the imaging techniques which would help the clinicians to come up with better treatment for their patients. And early detection of any disease can help save lives of tons of people. Still there are some difficulties that stay open in the picture combination techniques. Clinical imaging has ended up being helpful and the faith in these procedures is on the ascent, it is expected that the development and handy progressions would keep on developing in the upcoming times.

REFERENCES

1. Ilangovan, Shanmuga Sundari, Biswanath Mahanty, and Shampa Sen (2017), Emerging Technologies in Intelligent Applications for Image and Video Processing (pp. 401–421), "Chapter 16 biomedical imaging techniques", *IGI Global*.
2. Namiot, Dmitry and Manfred Sneps-Sneppe (2016) "Chapter 2 on internet of things programming models", Springer Science and Business Media LLC.
3. Khaleghi, Bahador, Alaa Khamis, Fakhreddine O. Kartay, and Saiedeh N. Razavi (2013) "Multisensor data fusion: A review of the state-of-the-art", *Information Fusion* 14: 28–44.

4. Mahmoud, Magdi.S and Yuanqing Xia (2014) "Multi-sensor data fusion", *Networked Filtering and Fusion in Wireless Sensor Networks* (1st ed.). CRC Press. https://doi.org/10.1201/b17667.
5. Wang, Yuh-Rau, Wei-Hung Lin, and Ling Yang (2012) "A novel DSP based real time license plate detection algorithm", *International Conference on Machine Learning and Cybernetics*, Xian, pp. 1779–1783, doi: 10.1109/ICMLC.2012.6359645.
6. Kijpokin Kasemsap (2017) "Chapter 13 robotics", *IGI Global*, doi: 10.4018/978-1-5225-1941-6.ch013.
7. James, Alex Pappachen and Belur V. Dasarathy (2014) "Medical image fusion: A survey of the state of the art", *Information Fusion*, 19: 4–19, ISSN 1566-2535.

22 Application of Artificial Intelligence for Coronavirus (COVID-19) Disease Management

A Preliminary Review

Saumyadip Hazra, Abhimanyu Kumar, and Souvik Ganguli
Thapar Institute of Engineering and Technology

CONTENTS

22.1 INTRODUCTION

The novel corona virus or COVID-19 has thrown the entire world on a back foot. The pandemic stretched its wings in each and every pocket of the globe. Since its outbreak in China, and later spread over Europe, America and in India as well, researchers are struggling hard to diagnose, monitor and provide successful treatment to get the persons come out of this disease. While one wing of researchers is involved in developing medicines and vaccines to combat this pandemic, the other set of researchers is aiming for an early detection of the symptoms, monitor the conditions of the affected persons, and suggest for a possible cure with the help of several artificial intelligence (AI) techniques.

Even the classification of the chest congestion occurring due to pneumonia and COVID-19 needs a quick detection for its treatment and cure [1]. Researchers in China developed a forecasting tool to predict the outbreak of this disease [2].

Moreover, mathematicians in different universities have also worked on this to check the possible damage that it may be cause to the population of the world [3]. Few researchers have also studied on how to protect the healthcare workers who are more exposed and susceptible to the disease [4]. In addition to this, AI has been phenomenal to benefit urban health monitoring and management [5]. Using CT image analysis, deep learning also found an important role in patient monitoring [6]. Further, researchers have been instrumental to determine the possible limitations, constraints and pitfalls of AI techniques against COVID-19 [7]. Even blockchain technology along with AI-based method has been applied for self-assessment of COVID-19 [8]. Some researchers worked on the prediction of the mortality risk of the patients to assist medical decision making with the help of AI technique [9]. Few researchers have investigated on the diagnosis of corona virus with the help of smartphone-embedded sensors [10]. Even hybrid AI models were evolved to predict more accurately COVID-19 [11]. Even expert systems have been employed for the diagnosis of this infectious disease [12]. Thus, a new vision on the application of technology in medicine is quite clear after the COVID-19 pandemic [13]. A diagnosis assistance of AI tool has also been proposed for suspected COVID-19 patients with fever symptoms [14]. Though an early review on the role of AI on COVID-19 and its future research directions have been presented in [15,16], this chapter provides a consolidation of the manifold applications of AI techniques to diagnose, monitor and support medical practitioners for an early detection in order to avoid high mortality rates which the present world is suffering now. Some futuristic directions have also been suggested to carry the research forward.

The remaining chapter is prepared in the following fashion. Section 22.2 provides an extensive review on the different AI-based tools being employed to diagnose, monitoring as well as to aid medical team in treatment of this disease. Section 22.3 carries out the discussions supporting the claims made and some future prospects of research to combat this menace. Section 22.4 concludes the chapter with some advice for future developments.

22.2 LITERATURE REVIEW

The literature survey is developed based on three subsections. The subsections are constituted to show the progress of the detection, diagnosis, prediction as well as monitoring of the novel corona virus focusing on the AI and the machine learning tools. Three subsections are created to highlight upon the latest developments in this field.

22.2.1 Detection and Prediction Methodologies

Hu et al. [2] applied AI method for forecasting of COVID-19 disease in China using the modified auto-encoding on the transmission dynamics of the disease. Rao et al. [17] also used machine learning technique to improvise the process of identification of COVID-19 cases using survey based made on mobile phone with an aim to reduce the spread of cases of corona virus. Vaishya et al. [18] reviewed the use of AI as a technique for analyzing, preparing for stopping of spread of COVID-19 using data from various databases and focused on the decision-making process of AI and help recover from it. Alimadadi et al. [19] integrated the data of the COVID-19 patients with advanced algorithms of machine learning in order to determine the further spreading speed of the virus, diagnose the speed of virus and developed a therapy to cure the disease. Jiang et al. [20] presented a predictive algorithm using AI to identify the characteristics of the novel corona virus and predicted its outcomes. The paper also proposed an AI tool to differentiate between the patients as less severe and more severe. Pourhomayoun et al. [9] demonstrated a discrimination strategy among the patients regarding the one who needed urgent attention and others using AI-based methods. The model showed high accuracy when tested for fresh COVID-19 positive patients. Yan et al. [21] predicted the probability of survival of patients suffering severely from COVID-19 by machine learning based approach which had accuracy around 90% when tested on the patients admitted in hospital. Rahmatizadeh et al. [22] proposed an AI based approach for the care of COVID-19 patients and proved that AI technology could have been used for the better care and decision-making. Dong et al. [23] reviewed image characteristics and models used in computing for the management and control of the corona virus. The AI methods for image processing were also explored and the characteristics were further compared. Yan et al. [24] predicted the criticality of the patient using machine learning algorithm using a prediction model and provided results with good accuracy and helped the doctors in early intervention of the COVID-19 patients. Feng et al. [14] developed a method based on AI technology for the early detection of the novel corona virus by first feeding the clinical record of the patient to the algorithm and then interpreted the results obtained. Chowdhury et al. [25] proposed a method for automatic detection of COVID-19 patients based on deep convolutional neural network. The results were then sorted by pneumonia and COVID-19 positive patients. Yan et al. [26] presented a machine learning-based method on the prediction of survival of patient who suffered from the COVID-19. The blood samples of the patients were analyzed and the survival was predicted based on the lactate dehydrogenase (LDH) content present in blood. Imran et al. [27] analyzed the cough samples for diagnosis of patients suffering from the corona virus using an AI-enabled app. The application software stored the information in a cloud network and the results were provided with good accuracy within a short duration.

22.2.2 X-ray and CT Scan-Based Diagnosis

An automatic detection method using chest CT and performance evaluation has motivated research in this direction [1]. Salman et al. [12] proposed a novel early stage COVID-19 detection method using X-ray and deep learning. The results were encouraging as a large number of images were analyzed very quickly and proved as big step toward radiology. Apostolopoulos et al. [28] evaluated the neural networking methods proposed for analyzing the X-ray images. The accuracy and sensitivity of the results obtained were very high and concluded that deep learning was an effective tool for COVID-19 detection. Ghoshal et al. [29] estimated the uncertainty in the deep learning solution based on convolutional neural network (CNN) to improve the human-machine interaction using the X-ray images and proved that the estimation improved the accuracy of detection. Wang et al. [30] introduced COVID-net which was developed using the deep convolution based on the concept of neural networks for detection of patients based on the X-ray images and gave accurate results in detection of patients. Hemdan et al. [31] developed a novel method based on deep learning for the detection of COVID-19 positive patients using X-ray images. The results proved the validity of the method and provided high accuracy as well. Castiglioni et al. [32] proposed a method based on AI for the diagnosis of COVID-19 patients based on their chest X-ray and then was tested on the patients of the COVID-19 designated hospitals. The results were also verified from radiologists as well. Narin et al. [33] proposed a method for detection based on three convolutional neural networks employing chest X-ray radiographs. All three methods provided good results in the detection of pneumonia and COVID-19. Ozturk et al. [34] gave neural network-based model for automatic detection using chest X-ray. The results proved out promising and made the screening process for radiologists relatively easier. Li et al. [35] designed a deep convolution-based neural networking which diagnosed the chest X-ray images of a patient for the determination of virus and differentiate between the normal pneumonia cases. Wang et al. [6] analyzed the CT scan images of the various patients using deep learning algorithm and determined whether the person was infected with COVID-19 or not prior to the pathogen test. Mei et al. [36] integrated the chest CT scan images using the AI algorithm and for this the medical history, exposure history, symptoms and determined whether the patient is positive or not. Bai et al. [37] gave an AI method to distinguish COVID patients from pneumonia patients based on chest CT. Zhao et al. [38] prepared a COVID-CT data set which contained 275 CT scans for COVID-19 patients for nurturing research in the field of deep learning for the detection of COVID-19 patients and for achieving this convolutional neural networking was used. Jin et al. [39] proposed AI method for COVID diagnosis with higher accuracy by taking a data set of CT scans of various patients and then employed deep convolutional neural network. Jin et al. [40] presented AI-based system to distinguish the features of COVID from those of pneumonia based on CT scanned images. Huang et al. [41] evaluated the burden on the lungs and its change over a period of time of COVID-19 patients with the help of deep learning algorithm. Chest CT scan was used and proved significantly better than other methods. Zheng et al. [42] developed deep learning-based system for the detection of virus from the chest CT images. The images were fed into the 3D system and the prediction was done with good accuracy.

22.2.3 Other Miscellaneous Applications of AI in Corona Virus Disease Management

McCall [4] focused on the use of AI as a tool for tackling with the global pandemic of COVID-19 and had also discussed about the works that AI is capable of doing during the pandemic. The paper also included the discovery of a new drug manufactured with the help of AI. Yang et al. [43] collected the migration data of people around 23rd January and plotted epidemic curve using susceptible-exposed-infected-removed (SEIR) and AI made predictions regarding the cycle of COVID-19 in China. Naudé [7] evaluated the areas where AI contributed and concluded that AI did not proved out to be helpful and discussed the reasons behind it and predicted that AI would not prove out to be effective. Peng et al. [44] developed an highly accurate method using AI for COVID-19 diagnosis in patients. Four types of AI methods were included and recorded significantly good results. Thompson et al. [8] proposed a low-cost AI and blockchain-based testing and tracking system. The paper also described the validity of the methods proposed. Maghdid et al. [10] developed a novel method for detection of infected patients using smartphone sensors. AI technology was used to read the data from the sensors and predicted whether the person was infected or not. Du et al. [11] presented a method based on AI technology for the prediction of COVID-19 and determined the carrying capacity of virus at various stages. Based on the results, a set of control and preventive measures were proposed. Dong et al. [45] reviewed the image characteristics and models used in computing the models used for management and control of COVID-19. The AI methods for image processing were also explored and the characteristics were compared. Stdbbing et al. [46] predicted a testing method for COVID-19 based on baricitinib which is an inhibitor for the treatment using AI technology. Allam et al. [47] explored the importance of early detection of COVID-19 using AI technology and also mentioned the benefits and applications of it. Naudé [15] reviewed all the attempts made with the help of AI technology to fight against COVID-19 and their status and concluded that AI did not prove out to helpful for the fight against the virus and more data of the infected people are required to make AI work. Shi et al. [48] reviewed the image acquisition techniques using AI used for to fight against and presented various fields where it can be applied along with the benefits that it presents. Allam et al. [5] surveyed the pandemic situation of COVID-19 from the perspective of urban dwellers and focused on how smart cities should work based on AI and how the sharing of data should take place to slow down the pace of COVID-19. Bullock et al. [49] overviewed the studies conducted with the help of AI trying to solve the pandemic situation of world and included the molecular scale study and epidemiological applications. Nguyen [16] reviewed the AI techniques used in various applications and presented the available AI resources. The resources available for guiding further research on AI were also presented.

22.3 DISCUSSIONS

Though one group of researchers are working very hard to develop medicines and vaccines of the virus, yet getting satisfactory solution to a virus having symptoms similar to cold and flu is still a matter of challenge. So far, no vaccine has ever developed that can combat cold and flu. Moreover, there is no vaccine till date of HIV viruses as well. People have been working on it quite a few decades. Further, there are millions of viruses available on Earth. Several of them are also available with the domestic animals and many of them are consumed by us as well. The world population may be infected by any one of them. So, researchers and medical persons have to control each of them. Hence, the very question arises, how many vaccines will people take to be safe throughout their entire life span? The list is almost endless. So early detection and diagnosis are most essential. Increasing the immunity level of the individual is thus most crucial to fight this menace. The classification of chest congestion with normal pneumonia is also equally important. The application-based software can also provide the detection of the location of the infected person. AI-based methods can truly support the detection and diagnosis process well in advance before the condition turns critical for the medical persons. Heath workers should also be monitored regularly because these persons are more susceptible to get infected as they are much exposed to the virus on a daily basis. New techniques in AI and machine learning can further aid to better and early detection to assist medical team. Some optimized approach can also be thought for predicting and detecting accurately the symptoms of this new virus in a patient.

22.4 CONCLUSIONS

The chapter presents an investigation of the various AI techniques to aid and support disease management of COVID-19. The aspects of detection, diagnosis, prediction, classification and monitoring have been highlighted in this work. Though many developments are taking place to fight efficiently the novel corona virus, yet the review cannot be exhaustive. Since plenty of research activities are occurring in different parts of the world almost concurrently, it is not possible to reflect all the reported works comprehensively as well. Still an early effort is made to consolidate most of the salient works pertaining to the topic so that the readers get benefitted by this short survey. Diagnosis and monitoring remained two prime aspects of the research for which separate subsections are formed to deliberate upon the issue. The chapter itself is an early stage of review for the interested readers to progress on the AI-based tools to handle any disease management. Further exploration on this topic can often lead to new progress in areas of AI and machine learning. Some classification tools can also be developed to distinguish ordinary pneumonia and COVID-19 infected pneumonia symptoms. Further, progress can also take place in new areas like deep neural networks, an upcoming field in the area of AI. Metaheuristic algorithms and their advantages can also be utilized to yield some optimized way of tackling the COVID-19 situation.

REFERENCES

1. Li, L., Qin, L., Xu, Z., Yin, Y., Wang, X., Kong, B.,... & Cao, K. (2020). Artificial intelligence distinguishes COVID-19 from community acquired pneumonia on chest CT. *Radiology*, 200905. doi: 10.1148/radiol.2020200905.
2. Hu, Z., Ge, Q., Jin, L., & Xiong, M. (2020). Artificial intelligence forecasting of covid-19 in china. *arXiv preprint arXiv:2002.07112*.
3. Predicting the future of the Covid-19 pandemic with data. (2020). Retrieved 27 June 2020, from https://healthcare-in-europe.com/en/news/predicting-the-future-of-the-covid-19-pandemic-with-data.html.
4. McCall, B. (2020). COVID-19 and artificial intelligence: Protecting health-care workers and curbing the spread. *The Lancet Digital Health*, 2(4), e166–e167.
5. Allam, Z., & Jones, D. S. (2020). On the coronavirus (COVID-19) outbreak and the smart city network: Universal data sharing standards coupled with artificial intelligence (AI) to benefit urban health monitoring and management. *Healthcare*, 8(1), 46. Multidisciplinary Digital Publishing Institute.

6. Wang, S., Kang, B., Ma, J., Zeng, X., Xiao, M., Guo, J.,... & Xu, B. (2020). A deep learning algorithm using CT images to screen for Corona Virus Disease (COVID-19). *MedRxiv.*
7. Naudé, W. (2020). Artificial intelligence vs COVID-19: Limitations, constraints and pitfalls. *AI & Society*, 35(3), 761–765.
8. Mashamba-Thompson, T. P., & Crayton, E. D. (2020). Blockchain and artificial intelligence technology for novel coronavirus disease-19 self-testing. *Diagnostics (Basel)*, 10 (4): 198. doi: 10.3390/diagnostics10040198.
9. Pourhomayoun, M., & Shakibi, M. (2020). Predicting mortality risk in patients with COVID-19 using artificial intelligence to help medical decision-making. *MedRxiv.*
10. Maghdid, H. S., Ghafoor, K. Z., Sadiq, A. S., Curran, K., & Rabie, K. (2020). A novel ai-enabled framework to diagnose coronavirus covid 19 using smartphone embedded sensors: Design study. *arXiv preprint arXiv:2003.07434.*
11. Du, S., Wang, J., Zhang, H., Cui, W., Kang, Z., Yang, T.,... & Yuan, Q. (2020). Predicting COVID-19 using hybrid AI model. (3/13/2020). Available at SSRN: https://ssrn.com/abstract=3555202 or doi: 10.2139/ssrn.3555202
12. Salman, F. M., & Abu-Naser, S. S. (2020). Expert system for COVID-19 diagnosis. *International Journal of Academic Information Systems Research (IJAISR)*, 4 (3): 1–13.
13 Goh, P. S., & Sandars, J. (2020). A vision of the use of technology in medical education after the COVID-19 pandemic. *MedEdPublish*, 9. doi: 10.15694/mep.2020.000049.1
14. Feng, C., Huang, Z., Wang, L., Chen, X., Zhai, Y., Zhu, F.,... & Tian, L. (2020). A novel triage tool of artificial intelligence assisted diagnosis aid system for suspected COVID-19 pneumonia in fever clinics. (3/8/2020). Available at SSRN: https://ssrn.com/abstract=3551355 or doi: 10.2139/ssrn.3551355
15. Naudé, W. (2020). Artificial Intelligence against COVID-19: An early review. IZA Discussion Paper No. 13110, Available at SSRN: https://ssrn.com/abstract=3568314
16. Nguyen, T. T. (2020). Artificial intelligence in the battle against coronavirus (COVID-19): A survey and future research directions. Preprint. doi: 10.13140/RG.2.2.36491.23846/1
17. Rao, A. S. S., & Vazquez, J. A. (2020). Identification of COVID-19 can be quicker through artificial intelligence framework using a mobile phone-based survey when cities and towns are under quarantine. *Infection Control & Hospital Epidemiology*, 41(7), 1–5.
18. Vaishya, R., Javaid, M., Khan, I. H., & Haleem, A. (2020). Artificial intelligence (AI) applications for COVID-19 pandemic. *Diabetes & Metabolic Syndrome: Clinical Research & Reviews*, 14 (4), 337–339.
19. Alimadadi, A., Aryal, S., Manandhar, I., Munroe, P. B., Joe, B., & Cheng, X. (2020). Artificial intelligence and machine learning to fight COVID–19. *Physiol Genomics*, 52, 200–202.
20. Jiang, X., Coffee, M., Bari, A., Wang, J., Jiang, X., Shi, J.,... & He, G. (2020). Towards an artificial intelligence framework for data-driven prediction of coronavirus clinical severity. *Computers, Materials & Continua*, 63(1), 537–551.
21. Yan, L., Zhang, H. T., Xiao, Y., Wang, M., Sun, C., Liang, J.,... & Tang, X. (2020). Prediction of survival for severe Covid-19 patients with three clinical features: Development of a machine learning-based prognostic model with clinical data in Wuhan. *medRxiv.*
22. Rahmatizadeh, S., Valizadeh-Haghi, S., & Dabbagh, A. (2020). The role of Artificial Intelligence in Management of Critical COVID-19 patients. *Journal of Cellular & Molecular Anesthesia*, 5(1), 16–22.
23. Dong, D., Tang, Z., Wang, S., Hui, H., Gong, L., Lu, Y.,... & Jin, R. (2020). The role of imaging in the detection and management of COVID-19: a review. *IEEE Reviews in Biomedical Engineering*, doi: 10.1109/RBME.2020.2990959.

24. Yan, L., Zhang, H. T., Xiao, Y., Wang, M., Sun, C., Liang, J.,... & Tang, X. (2020). Prediction of criticality in patients with severe Covid-19 infection using three clinical features: A machine learning-based prognostic model with clinical data in Wuhan. *MedRxiv.*
25. Chowdhury, M. E., Rahman, T., Khandakar, A., Mazhar, R., Kadir, M. A., Mahbub, Z. B.,... & Reaz, M. B. I. (2020). Can AI help in screening viral and COVID-19 pneumonia? *arXiv preprint arXiv:2003.13145.*
26. Yan, L., Zhang, H. T., Goncalves, J., Xiao, Y., Wang, M., Guo, Y.,... & Huang, X. (2020). A machine learning-based model for survival prediction in patients with severe COVID-19 infection. MedRxiv. doi: 10.1101/2020.02.27.20028027
27. Imran, A., Posokhova, I., Qureshi, H. N., Masood, U., Riaz, S., Ali, K.,... & Nabeel, M. (2020). AI4COVID-19: AI enabled preliminary diagnosis for COVID-19 from cough samples via an app. *arXiv preprint arXiv:2004.01275.*
28. Apostolopoulos, I. D., & Mpesiana, T. A. (2020). Covid-19: automatic detection from x-ray images utilizing transfer learning with convolutional neural networks. *Physical and Engineering Sciences in Medicine*, 43(2): 635–640.
29. Ghoshal, B., & Tucker, A. (2020). Estimating uncertainty and interpretability in deep learning for coronavirus (COVID-19) detection. *arXiv preprint arXiv:2003.10769.*
30. Wang, L., & Wong, A. (2020). COVID-Net: A Tailored Deep Convolutional Neural Network Design for Detection of COVID-19 Cases from Chest X-Ray Images. *arXiv preprint arXiv:2003.09871.*
31. Hemdan, E. E. D., Shouman, M. A., & Karar, M. E. (2020). Covidx-net: A framework of deep learning classifiers to diagnose covid-19 in x-ray images. *arXiv preprint arXiv:2003.11055.*
32. Castiglioni, I., Ippolito, D., Interlenghi, M., Monti, C. B., Salvatore, C., Schiaffino, S.,... & Sardanelli, F. (2020). Artificial intelligence applied on chest X-ray can aid in the diagnosis of COVID-19 infection: a first experience from Lombardy, Italy. *MedRxiv.*
33. Narin, A., Kaya, C., & Pamuk, Z. (2020). Automatic detection of coronavirus disease (covid-19) using x-ray images and deep convolutional neural networks. *arXiv preprint arXiv:2003.10849.*
34. Ozturk, T., Talo, M., Yildirim, E. A., Baloglu, U. B., Yildirim, O., & Acharya, U. R. (2020). Automated detection of COVID-19 cases using deep neural networks with X-ray images. *Computers in Biology and Medicine*, 121, 103792.
35. Li, X., & Zhu, D. (2020). Covid-xpert: An ai powered population screening of covid-19 cases using chest radiography images. *arXiv preprint arXiv:2004.03042.*
36. Mei, X., Lee, H. C., Diao, K. Y., Huang, M., Lin, B., Liu, C.,... & Bernheim, A. (2020). Artificial intelligence–enabled rapid diagnosis of patients with COVID-19. *Nature Medicine*, 26, 1224–1228. doi: 10.1038/s41591-020-0931-3
37. Bai, H. X., Wang, R., Xiong, Z., Hsieh, B., Chang, K., Halsey, K.,... & Mei, J. (2020). AI augmentation of radiologist performance in distinguishing COVID-19 from pneumonia of other etiology on chest CT. *Radiology*, 296 (3), E156–E165.
38. Zhao, J., Zhang, Y., He, X., & Xie, P. (2020). COVID-CT-dataset: A CT scan dataset about COVID-19. *arXiv preprint arXiv:2003.13865.*
39. Jin, C., Chen, W., Cao, Y., Xu, Z., Zhang, X., Deng, L.,... & Feng, J. (2020). Development and Evaluation of an AI System for COVID-19 Diagnosis. *MedRxiv.*
40. Jin, S., Wang, B., Xu, H., Luo, C., Wei, L., Zhao, W.,... & Sun, W. (2020). AI-assisted CT imaging analysis for COVID-19 screening: Building and deploying a medical AI system in four weeks. *MedRxiv.*
41. Huang, L., Han, R., Ai, T., Yu, P., Kang, H., Tao, Q., & Xia, L. (2020). Serial quantitative chest ct assessment of covid-19: Deep-learning approach. *Radiology: Cardiothoracic Imaging*, 2(2), e200075.

42. Zheng, C., Deng, X., Fu, Q., Zhou, Q., Feng, J., Ma, H.,... & Wang, X. (2020). Deep learning-based detection for COVID-19 from chest CT using weak label. *MedRxiv.*
43. Yang, Z., Zeng, Z., Wang, K., Wong, S. S., Liang, W., Zanin, M.,... & Liang, J. (2020). Modified SEIR and AI prediction of the epidemics trend of COVID-19 in China under public health interventions. *Journal of Thoracic Disease*, 12(3), 165.
44. Peng, M., Yang, J., Shi, Q., Ying, L., Zhu, H., Zhu, G.,... & Yan, H. (2020). Artificial intelligence application in COVID-19. *Diagnosis and Prediction.* Available at SSRN: https://ssrn.com/abstract=3541119 or doi: 10.2139/ssrn.3541119
45. Dong, D., Tang, Z., Wang, S., Hui, H., Gong, L., Lu, Y.,... & Jin, R. (2020). The role of imaging in the detection and management of COVID-19: A review. *IEEE Reviews in Biomedical Engineering.* doi: 10.1109/RBME.2020.2990959
46. Stebbing, J., Krishnan, V., de Bono, S., Ottaviani, S., Casalini, G., Richardson, P. J.,... & Tan, Y. J. (2020). Mechanism of baricitinib supports artificial intelligence-predicted testing in COVID-19 patients. *EMBO Molecular Medicine.* doi: 10.15252/emmm.202012697
47. Allam, Z., Dey, G., & Jones, D. S. (2020). Artificial intelligence (AI) provided early detection of the coronavirus (COVID-19) in China and will influence future Urban health policy internationally. *AI*, 1(2), 156–165.
48. Shi, F., Wang, J., Shi, J., Wu, Z., Wang, Q., Tang, Z.,... & Shen, D. (2020). Review of artificial intelligence techniques in imaging data acquisition, segmentation and diagnosis for covid-19. *IEEE Reviews in Biomedical Engineering.* doi: 10.1109/RBME.2020.2987975
49. Bullock, J., Pham, K. H., Lam, C. S. N., & Luengo-Oroz, M. (2020). Mapping the landscape of artificial intelligence applications against COVID-19. *arXiv preprint arXiv:2003.11336.*

43. [illegible] (2020) [illegible] COVID-19 [illegible]
44. [illegible] (2020) [illegible] COVID-19 [illegible]
45. [illegible] (2020). [illegible] COVID-19 [illegible]
46. [illegible] (2020) [illegible]
47. [illegible] (2020). Artificial intelligence [illegible] COVID-19 [illegible]
48. [illegible] (2020). [illegible] artificial intelligence [illegible] COVID-19 [illegible]
49. [illegible] (2020). [illegible] artificial intelligence [illegible] COVID-19 [illegible]

23 Technology and Innovation in COVID-19 Pandemic Response in the Philippines

Miriam Caryl Carada
University of the Philippines

Ginbert Permejo Cuaton
The Hong Kong University of Science and Technology

CONTENTS

23.1 BACKGROUND

The COVID-19 pandemic negatively affected the well-being, livelihoods, education, and health of Filipinos. As of writing, the disease has already infected 180,867 individuals, of which 957 died and 3909 have recovered since 31 May 2020 (CSSE-JHU, 2020 May 31). Zhang and Shaw (2020) observed that there is significant progress in coronavirus research related to infectious diseases, epidemiology, and virology but pointed that there is a gap on topics or issues associated with public health, governance, technology, and risk communication and argued that more focus needs to be given on multi-, cross- and trans-disciplinary research.

This chapter aims to contribute empirical findings on COVID-related research literature by utilizing the experiences and providing an analysis on the COVID-19 response in the archipelagic country of the Philippines. Our principal research question is: *How did the Philippines' response to the COVID-19 outbreak develop from January to May 2020?* In addition, what practices, technologies and innovations have helped in the detection, isolation, and medication of infected patients? In this chapter, we highlighted the practices implemented by communities, and the technologies developed by individuals as well as public and private organizations in the country's fight with COVID-19. By putting forth the local innovations and practices in the Philippines, we aim to provide examples to other states on how to better respond to the COVID-19 pandemic.

Given the impact of the pandemic, conducting usual fieldwork to obtain primary data was very challenging. Hence, we utilized secondary data analysis to cope with the restrictions on mobility as well as to eliminate the risk from viral exposure. We followed the steps of conducting secondary data analysis proposed by O'Leary (2017): (1) identify research questions, (2) locate data, (3) evaluate data's relevance, (4) assess data's credibility, and (5) analyze data. After identifying our principal research question, we located our secondary data from the official press releases by the Philippine government's health department, and the official websites of the companies or groups that developed the various technologies or innovations that we will discuss in the succeeding sections. We supplemented these with secondary data from published news articles by local and national publications—as well as social media postings by various local and national media entities.

23.2 OVERVIEW OF PHILIPPINE GOVERNMENT RESPONSE TO THE COVID-19 OUTBREAK

On 30 January 2020, the Department of Health (DOH) confirmed the first positive COVID-19 case in the Philippines—a 38-year-old female Chinese patient who arrived in the country from Wuhan City in China via Hong Kong on 21 January 2020 (Gregorio, 2020 February 02). The health department was quick to claim that the early detection of the patient resulted from the strong surveillance system, close coordination with the World Health Organization (WHO) and other national agencies, and the utilization of its decision tool. The DOH worked closely with the Interagency Task Force on Emerging Infectious Diseases (IATF-EID). Around five months after recording the first case of COVID-19 in the country, the number of infected individuals ballooned to 18,086 of which 957 were dead, while 3909 individuals or 21.61% of infected patients have recovered from the disease (DOH, 2020 June 01). Out of all infected individuals, 13,220 (92.1%) were active cases, with 12,285 mild cases (92.93%) and 859 (6.5%) asymptomatic cases or showing no symptoms of the virus. Moreover, a little over 14% (2606) of those infected were healthcare workers including doctors/physicians, nurses, nursing assistants, medical and radiologic technologists, and nonmedical staff.

Throughout the months of January to May there have been various technological innovations that emerged, as well as operational development that shaped the Philippines' response to the COVID-19 outbreak. The succeeding sections of this chapter provide an overview and in-depth discussion on these tech-innovations and response evolution in the country.

23.3 COVID-19 CASE CATEGORIZATION

Contact tracing is a crucial part in responding to the pandemic. Through contact tracing, the authorities involved in the COVID-19 response team can detect and isolate cases and give advice for quarantine. It also allows to track the chain of infection including the details of transmission. As of writing, the DOH announced that the country has 38,000 contact tracers and is planning to hire an additional 95,000 more to satisfy the ratio of 1 contact tracer to 800 people (1:800) of the WHO (Merez, 27 May 2020). In tracing and identifying COVID-19 cases, it is vital to classify different cases and categorize. In the country, there have been two sets of categorizations that have been used from January to May.

The first set of categorizations was used from 12 January to 8 April. There were two case classifications: (1) Person Under Investigation (PUI), and (2) Person Under Monitoring (PUM). These categorizations were separate and different from the positive cases. The PUIs were persons with COVID-19 symptoms including cough and/or fever with more than 38°C body temperature and other respiratory symptoms. The PUMs on the other hand were those who were asymptomatic (do not display symptoms). However, both categories have the following common criteria: (1) the patient has a travel history in the past 14 days to areas with issued travel restriction and/or history of virus exposure, (2) the patient has close encounter with a confirmed or probable case of COVID-19 infection, and (3) the patient has been working in a healthcare facility where confirmed and probable COVID cases were determined.

The second set of categorizations was used on 9 April 2020 as an adoption of the 20 March 2020 guidelines issued by the WHO. The three new case categorizations are (1) suspected, (2) probable, and (3) confirmed and, as of writing, is still used by the national government. A suspected case according to the DOH can be classified into four. The first classification is a person with severe acute respiratory illness (SARI), and the second classification is a person with influenza-like illness (ILI). Both SARI and ILI share similar features. Their symptoms include a person having a fever of exactly or greater than 38°C and sore throat or cough. Unique to those classified as SARI is the difficulty of breathing, and they may also have severe pneumonia with undetermined cause prior to the COVID-19 testing. ILI on the other hand is different as they have travel history in an area with reported local virus transmission within 14 days when the other symptoms show, or they have recorded contact with, or living with a person confirmed to have the virus. The third classification is a person with ILI. These people were described to have contact with a confirmed or a probable case two days preceding the previous mentioned infection or before their test cases were confirmed to be negative on repeated testing. The fourth and last classification is people with either fever, cough, shortness of breath, or respiratory symptoms. They may also belong to the vulnerable population in this pandemic which includes those in their senior years (=/>60 y. o.) or with comorbidity, those with pre-existing illness, pregnant (especially those classified to have high-risk pregnancy), and health workers.

A probable case according to the DOH can be classified into two. The first classification refers to those who were categorized as suspected case and underwent COVID-19 testing with inconclusive results. The second classification refer to those

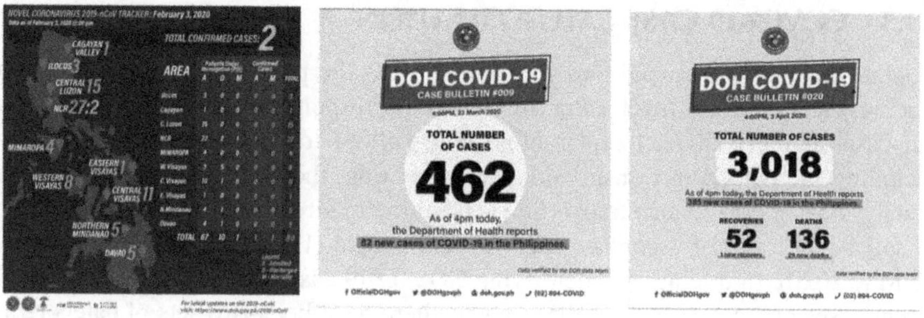

FIGURE 23.1 Evolution of COVID-19 case bulletins from January to May 2020.

who have tested positive for the test done in local laboratories but have not been tested in either a national, a subnational, or a DOH-accredited laboratory.

Confirmed cases were those with positive test results from a reference laboratory (national or subnational) or a certified DOH testing facility regardless if they were symptomatic or asymptomatic. As of 31 May 2020, the Philippines had 38 licensed laboratories capable of conducting RT-PCR or Reverse Transcription Polymerase Chain Reaction test (DOH, 1 June 2020). RT-PCR test is a nuclear-derived method for detecting the presence of specific genetic material in any pathogen, including a virus. Twenty-three (60.53%) out of these 38 COVID-19 testing laboratories were located in the National Capital Region (NCR), while the remaining fifteen 15 (39.47%) were located in the following regions: Cordillera Administrative Region, Ilocos, Central Luzon, CALABARZON, Bicol, Western Visayas, Central Visayas, Eastern Visayas, Zamboanga Peninsula, and Davao. A total of 130 subnational laboratories applied to the DOH to be licensed and registered centers capable of handling COVID-19 testing. The combined daily testing capacity of these laboratories was 8,268 tests. On the other hand, the country had 13,565 hospital beds dedicated to COVID-19 cases but only 9.7% are ICU beds, 22.3% are ward beds, and 68% are isolation beds.

In the previous category on PUIs and PUMs, only the PUIs were covered in the new classification. Reporting and monitoring based on the new category emphasized suspected cases with symptoms of SARI and ILI where the Epidemiology Bureau of DOH leads in establishing and implementing the COVID-19 Surveillance System. As of the date of writing, the case categorization of COVID-19 in the country remains evolving. These changes were influenced not only by local authorities but also by the information coming from the WHO.

23.4 HEALTH RESPONSE TECHNOLOGY AND INNOVATIONS

Following the COVID-19 response in the Philippines were various health response innovations. The immediate response of the country took form in the creation and implementation of various laws and the identification of COVID-19 cases to contain and minimize it. With these measures, public's mobility was restricted, public information about the virus were limited, so as their access to medical consultations.

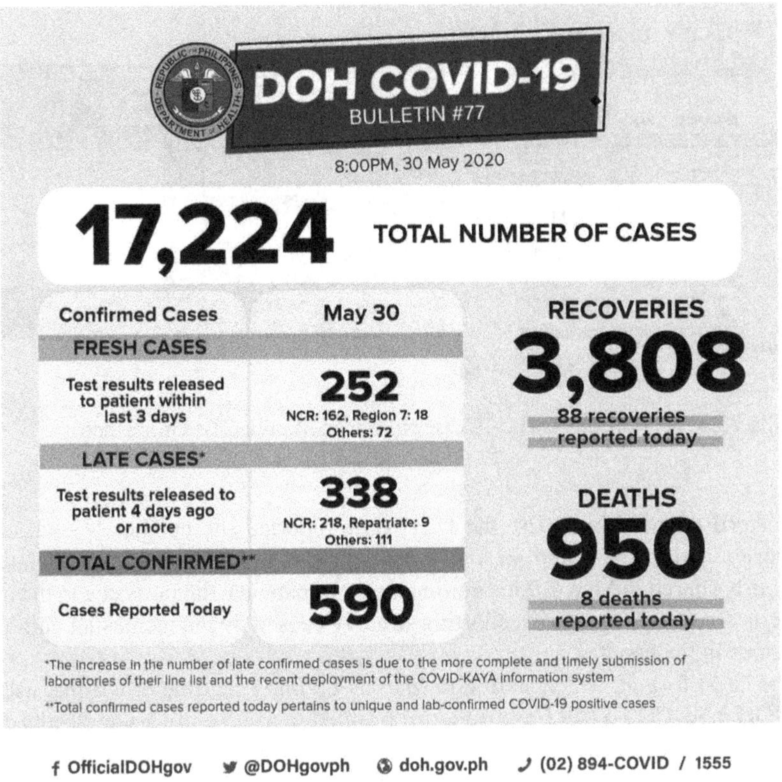

FIGURE 23.2 Current format of COVID-19 case bulletin in the Philippines.

In this section, the various health response innovations that rose in this time of pandemic were highlighted and discussed: digital reporting, digital contact tracing and alerts, and telemedicine and online health consultation.

23.4.1 From Traditional to New Media Reporting

Figure 23.1 shows that the COVID-19 reporting in the country evolved from January 2020 to May 2020. The DOH published COVID-19 situational reports since January 28, which were broadcasted between noontime to 5'o clock in the afternoon. These situational reports included the number and regional location of the PUIs along with the DOH action against the virus, the ITAF recommendation on Filipinos living in Hubei, China, and health advisories.

In March 15, the national reporting on COVID-19 changed. The DOH started its publication of the COVID-19 Case Bulletins along with its daily broadcast that shifted to 4 o'clock in the afternoon. The bulletin indicated the number of confirmed cases, the number of deaths and recoveries including relevant details of patients' age, place of residence, travel history, onset of symptoms, confirmation of virus, and the date of discharge or death.

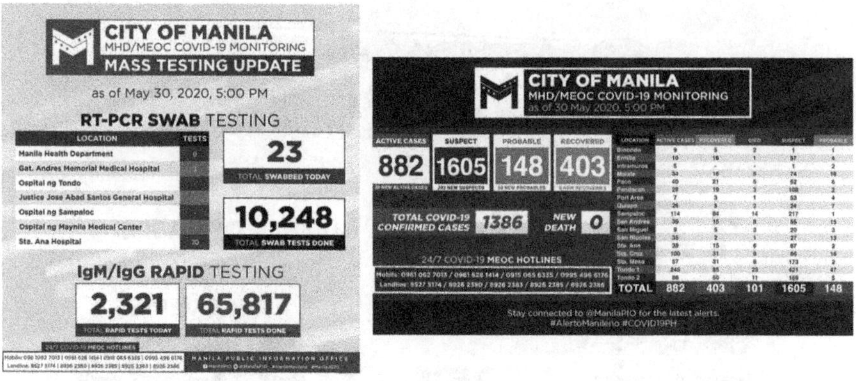

FIGURE 23.3 Format of COVID-19 case bulletin in Manila City, Philippines.

In April to mid-May 2020, the DOH only reported the number of total cases, recoveries and deaths, and its daily new data without the patients' details as previously shared in March. This non-inclusion of patients' data was due to the logistical constraint and change of reporting strategy caused by the increasing number of infections in the country.

The latest format, which started in May 29 up until the time of writing, used by the DOH in its COVID-19 Case Bulletins resulted in a surge of varying feedback, of which most were negative and critical. The agency missed its usual reporting time in May 29, and released an infographic in the evening with 1,046 newly reported cases, an increase of over 100% from the 539 confirmed COVID-19 cases reported on the previous day (May 28). Figure 23.2 shows the current reporting format that introduced the concept of *Fresh* and *Late* cases. *Fresh Cases* were test results released to patients within the last three days, while *Late Cases* were results released from tests that were conducted four days prior its results. Netizens were quick to comment that this scandal would mean that the data released by the DOH, and analyzed and utilized by the IATF-EID in developing strategic policy recommendations to the government's COVID-19 Response were faulty, wrong, and misleading, thereby, falsely portraying and assessing the virus outbreak in the country.

The Philippines has more than 73 million Facebook users as of January 2020 which accounted for 66.2% of the country's population (NapoleonCat, 2020 January). The DOH has extensively utilized this platform along with other social media platforms such as Twitter and Instagram. Apart from posting the bulletins and situationers, the DOH conducted live broadcast via Facebook. This was an innovation in the reporting system of the government particularly by the DOH and a supplement to the traditional reporting that the agency has been utilizing (television and radio platforms) to disseminate COVID-19 related news, press releases, and case reports.

Aside from the daily updates from the DOH, different local government units (LGUs) from the regional, city, and municipality levels also committed to frequently report cases in their respective areas. Various provinces along with cities and municipalities have also created their own Facebook page to reach more of their constituents.

There were also cases where *barangays* or villages created their own Facebook pages to inform their constituents on the number of COVID-19 cases in their area as well as to inform people on national strategies and policies to be observed during the pandemic. In the case of governors and mayors, they often conducted live broadcast and post updates on the number of COVID-19 cases. More often, the distribution of COVID-19 cases within the different localities was showed in their reports and broadcast. As an example, please refer to Figure 23.3 on the subnational reporting of the State capital city of Manila via twitter. The figure shows that in the city, not only the number of cases in each Barangay was reported but it also included updates on the COVID-19 testing situation. This type of reporting aids people on which locations to avoid and be more cautious during the pandemic. The new media reporting also gave way for people to directly ask their leaders about their queries and individual concerns, and at the same time, provided a chance for leaders to know their constituents more and identify their needs.

23.4.2 Websites and App-Based Contact Tracing

The Philippine government hired contact tracers as discussed previously. Though manual contact tracing was done, it was recognized to be limited, particularly costly, time-consuming, and risky. Thus, various websites and mobile applications (Apps) have been developed by private companies, individuals, and government agencies to aid in faster and low-risk contact tracing. In this section, six applications/programs for contact tracing and alerts were introduced and discussed. The applications include CovCT, Covid19tracker.ph, StaySafe.ph, WeTrace, ENDCoV, and TanodCOVID.

CovCT. CovCT is an App developed by Madison Technologies, a start-up company based in Singapore, to track COVID-19 cases. The App first became available in Malaysia in April 2020, and later was rolled out in the Philippines. Users are instructed to check-in the locations they visited by scanning the Quick Response (QR) code using their phone cameras. This way, users can help local authorities to trace and isolate cases of infection.

Covid19Tracker.ph. Covid19tracker.ph is an App that gives real-time updates regarding COVID-19 in the Philippines-confirmed cases, deaths and recoveries, and lets the users locate the nearest hospitals or facilities with available testing kits. The App has the ability to trace the route history of virus-positive individuals. Users of the App can log their symptoms, route history, and infection status. The App will automatically log other devices that virus-positive individuals have close contact to for faster contact tracing. However, the app can only be downloaded by android smart phone users.

StaySafe.ph. StaySafe.ph is an App and a website which has been developed by the Philippine Government through the support of Multisys Technologies Corporation. This is a company funded by the telecommunications giant in the country, PLDT Inc., which enables the free of charge management of the App. The App's main function is to protect oneself, the people close to you, and the community and this translates to functions on contact tracing, health condition reporting, and a system for social distancing. The website and App can generate a heat map for infections using self-reported information and information from the national government. It can

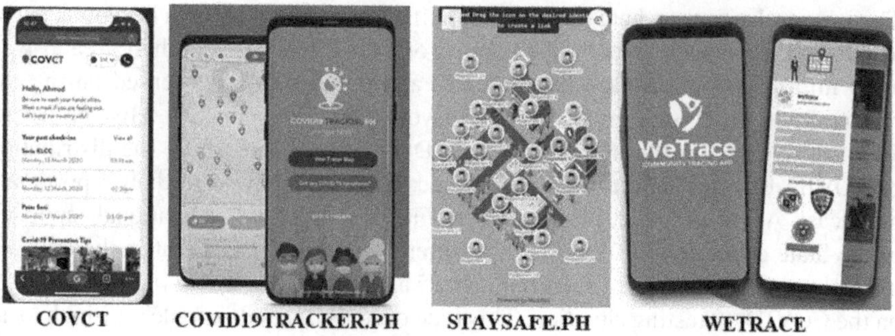

COVCT COVID19TRACKER.PH STAYSAFE.PH WETRACE

FIGURE 23.4 COVID-19 contact tracing apps in the Philippines.

also diagnose whether a user's symptoms are mild, moderate, or severe conditions. This information can help the government determine the location of the probable, suspected, and positive cases and aid in resource management and control the spread of the virus.

The usage of the App requires registration and a requirement is that you have a mobile phone number. This will be used to register to the App. Mobile phone users on the other hand are encouraged to download the App for convenience of use. Information inputs in the App are stored in a secured database which can be accessed by the government, LGUs, and frontliners. The App developers also highlight to their users about their strict compliance with the Data Privacy Act in the country.

WeTrace. WeTrace is a contact tracing App developed by a Department of Science and Technology scholar. In virtue of Executive Order P-5 of the Province of Cebu in the country, this App was required to be installed and used by all people of the province to help the LGU in faster contact tracking and reporting of people exposed with any COVID-19 positive cases. One unique feature of the App is the Report Feature which asks users to input their full name, contact information, and details of their report. Though personal information is input in the app, it generates unique individual QR codes for confidentiality purposes.

EndCoV.ph. EndCoV.ph is a web portal created by the University of the Philippines COVID-19 Pandemic Response team. The web portal has several features of which are the overview of the cases in the country (in numbers) and the location of the COVID-19 cases, hospitals, quarantine facilities, and checkpoints. Data on the epidemic curve in national, provincial, and municipal level are also available in the website. Furthermore, attached in the websites are the collection of links to donation drives, the policies and advisories relating to COVID-19 in the country, and other related resources.

Tanod COVID. The Tanod COVID is a program for contact tracing which helps LGUs to monitor and survey COVID-19 possible patients in their respective communities. *Tanod* is a Filipino word which means community-based local policy enforces, especially in barangay/village level. This program augments the crucial role of LGUs in the response through early detection and tracing of the disease carrier. This could also help the LGUs and the DOH regional offices to identify areas eligible for mass testing. In this program, the LGUs provide mobile phone numbers for people to communicate with regarding their medical conditions.

Utilizing Bluetooth and wireless connectivity via mobile programs and the Apps aforementioned can aid in faster and efficient contact tracing. These digital innovations in the reporting and contact tracing have enabled voluntary participation of the public to contain COVID-19. The first four Apps whose interface can be seen in Figure 23.4 can be downloaded for free and is used with wireless connectivity. Meanwhile, EndCoV utilizes information collected by University of the Philippines Resilience Institute (UPRI) and data are widely available to the public without having to download the App. The Tanod COVID on the other hand is more user-friendly to those who are not technology savvy as it only requires a mobile phone even without internet connection.

Nevertheless, these innovations provide people with options to register and login their information to be stored in a database creating a community within the App users. Persons classified as a confirmed case can voluntarily report through the various Apps and programs using wireless internet capability—alerting those who have the same App information about real-time interaction with a suspected/infected person. In this way, people can take measures to protect themselves and others. However, while these Apps and digital platforms are helpful, privacy concerns remain an issue. Advocates and experts on privacy issues call for strict guidelines in maintaining all the users of these platforms and Apps protected, to maintain private information secured, anonymous, and protected from potential abuse by various organizations including the government.

23.4.3 Telemedicine and Online Health Consulting

As the COVID-19 cases in the country rise, hospitals start to close their doors especially to new patients suspected of being infected with the virus reasoning that they have reached their maximum capacity. According to the news article of *Rappler* in March 28, in the duration of 3 days nearing the end of March, at least six (6) private hospitals have announced that there is no guarantee of room admission in the entirety of Manila (Tantuco, 28 March 2020). This situation became problematic as patients with non-COVID-19 related concerns have now limitations in their access to medical treatments and consultations. This situation became an opportunity for technology-related innovations in delivering healthcare services or commonly referred to as telemedicine in the country. In this section, six (6) telemedicines will be discussed, Kontra COVID Bot KIRA, Yani the ENCOVBOT, AIDE, KonsultaMD, Kitika, and MedGate.

Kontra COVID Bot KIRA. KIRA stands for *Katuwang na Impormasyon para sa Responsabling Aksyon* which in English roughly translate as *Partner in Information for Responsible Action*. KIRA functions as an anti-COVID chatbot that can be accessed through various social media platforms, in Facebook, Viber, and in the KontraCOVID.ph website and application. There are three functions of KIRA. The first function is doing Self-Check; this is to know if you are at risk of COVID-19. The second function is protection against COVID-19; KIRA shares information on what to do to fight against COVID-19. The last function is to answer queries on COVID-19 quarantine guidelines. KIRA has been operational since April 23 and was developed by the DOH in cooperation with several developmental agencies with platform support from Facebook and Viber.

AIDE KONSULTA MD MEDGATE

FIGURE 23.5 App interface of AIDE, Konsulta MD, and Medgate.

Before the development and launch of KIRA, the DOH have also opened a 24/7 telemedicine hotline where COVID-19-related consultations with medical doctors can be done along with two other hotlines for other COVID-19 concerns.

Yani the EndCoVBot. Apart from the DOH there are also various stakeholders developing and using technology to combat COVID-19 and provide telemedical support. Before the introduction of KIRA, Yani the EndCoVBot was introduced by the UPRI. Yani came from the Filipino word *Bayanihan* (Filipinos Helping Filipinos) in honor of the people fighting against COVID-19. Yani can be accessed through Facebook or the endcov.ph portal which was introduced in the previous section. Information such as the government's policies in health, education, transportation, and economic support can be answered by Yani. Psychologist and psychosocial support recommendations are also available as well as locating nearby hospitals. While KIRA uses a mix of Filipino and English, Yani gives you the option on which language to converse with. One unique feature of Yani is that it can also converse using the lesbian, gay, bisexual, and transgender or LGBT slang in the country.

AIDE. AIDE is an application developed by the Bugayong siblings in 2016. It is a booking App where users can choose healthcare professionals based on their needs. The aim of the App is to provide good quality healthcare anytime and anywhere and ultimately to become a convenient solution to everyone's medical needs. The App is free to download; however, there are certain fees to be paid depending on the services. In using the App, the user has to select a medical attention that he/she needs, fills out the details, and chooses medical professionals. The professional fees will be shown in the selection. The fees can be paid before or after availing the service.

Konsulta MD. Konsulta MD is a 24/7 health hotline service which has been developed long before the pandemic. However, it was only in the last quarter of 2017 that licensed and qualified medical professionals were included in their hotlines through partnership with AIDE. The hotline is powered by Global Telehealth, Inc., in the Philippines. The service is not free compared to the chatbot previously discussed. There is a subscription charge which ranges from Php60.00 (~USD1.20) per month for personal use to Php150.00 (~USD3.00) per month for a household or micro and small business.

Kitika. Kitika Mobile Healthcare Inc. was established in 2019 and is similar to KonsultaMD as it is also a subscription-based service that provides 24/7 primary medical care through the use of a mobile phone. The service also utilizes nurses, general practitioners, and specialist doctors. To avail the services, users need to download the App (free of charge), sign up, choose a subscription plan, and consult. The service can also be availed in two different languages, Filipino and English.

Medgate. Medgate is another company that delivers telemedicine. It differs from the previous telemedicine services as it is not developed in the Philippines. Medgate has been in-service for almost 20 years and is operating in Switzerland where it was founded, in the Middle East, and in the Philippines. In the Philippines, Medgate partners with various healthcare providers, insurance companies, and banks.

Tele Kamusta. Tele Kamusta is an innovation which is specific to the University of the Philippines—Philippine General Hospital (UP-PGH). It is an initiative of the hospital's Health Operations Director, Dr. Homer Co. The service was launched in April and is only available to COVID-19-positive patients in the hospital and their relatives for free of charge. Visitors are not allowed for COVID-19-positive patients. Thus, the UP-PGH came up with the Tele Kamusta, where visitors can do electronic visitation to their families. *Kamusta[?]* or *Kumusta[?]* is a Filipino greeting which roughly translates to *How are you[?].* This technology innovation allows patients to talk to their families for 15–20 minutes which helps alleviate loneliness while in confinement and at the same time protect their loved ones from getting infected by the virus. At a glance, this virtual meeting is part of the treatment for patients, maintaining their psycho-social well-being.

There were six telemedicines which have been discussed. In this section, the first two applications discussed are KIRA and Yani which utilize artificial intelligence and are more focused on giving advice on COVID-19-related health consultations. Most information provided by KIRA and Yani are pre-collected and are stored in their programs. This means, frequent updating is needed in their system for information to be updated. AIDE, Konsulta MD, and Kitika are Apps that need to be downloaded and can be downloaded for free in smartphones but have additional in-App purchases for doctors' consultations or subscriptions. Medgate functions similarly to AIDE, Konsulta MD, and Kitika; the main difference is that it is an application which is developed outside the country. In these Apps, patients are able to talk directly to healthcare professionals. Figure 23.5 shows the interface of AIDE, KonsultaMD, and Medgate. Among the applications, the Tele Kamusta functions differently as it is for a specific organization targeting specific persons.

23.5 THE ROLE OF TECHNOLOGY AND INNOVATION IN COVID-19 RESPONSE

The blistering development of the pandemic presents varying difficulties in assessing the severity of the outbreak, thereby impeding efforts to efficiently circulate precise–information to aid in public health planning and clinical management (Drew et al., 2020). In the Philippines, the limited mobility amid the COVID-19 pandemic has pushed human ingenuity and innovation to develop new or improve available technologies to mitigate the impacts of the disease. These technological innovations that we have discussed in three sections, (1) reporting, (2) case contact tracing, and (3) telemedicine, are in addition to intensified testing, isolation, and hygiene practices and could help in the community management of the outbreak, and ultimately save lives. New media (digital broadcasting and internet) are now tapped in the reporting of the COVID-19 situation and have overcome the use of the old media (analog), where reporting is done in government-owned platforms like television and radio channels. The government has expanded its reach to the people through innovations in their reporting system specifically in the utilization of new media such as Facebook and Twitter. Apart from the national reporting, LGUs have been creative and innovative in reaching their communities. A number of LGUs have created their own social media accounts where reporting is done in the form of live broadcast and posting infographics and news advisories. Ultimately, this innovation in the LGUS' reporting system is a big step-up as most of their old media has been mainly in the form of posters and announcements in the municipal and city halls. The utilization of new media has also enabled a two-way communication between the authorities and the public (Krishnasamy, n.d.). The authorities are able to get feedback from the public and the public can verify information from the government.

We argue that by doing so, grassroots communities have better understanding of the contextual realities they face in the fight against the disease. It also bridges the gap between the government and the electorate by providing a platform for discussion or dialog where citizens can voice out feedback, suggestions, and even discontent. The process democratizes government reporting and underscores the importance of information sharing in championing transparent and accountable governance.

Mobile phone applications or Apps are being used both in tracking and providing medical services. The data collected by these tools and Apps can then rapidly be re-deployed to inform participants of urgent health materials (Xu & Kraemer, 2020). In the case of contact tracing, usage of Apps makes it easier to locate the source of the virus transmission and determine places to which the public should avoid. In the case of telemedicine, the public is given privileges of being able to reach their doctors in the comfort of their homes. Having an innovation such as the telemedicine also opens a wider range of access to specialists for medical consultations. Patients are now able to reach doctors not only in their region but also across other regions especially doctors in the capital city. These Apps are predominantly valuable when Filipinos are advised to observe physical distancing and staying at home. These innovations in contact tracing and telemedicine minimize the risk of the public in being exposed during medical consultations and also enables faster contact tracing. The use of Apps also entails cost-cutting benefits, for instance, the use of contact tracing apps enables the data on the movement of people and the easy tracing of the persons they had contact with; by utilizing this technology, there will be less people working in the field and risking themselves for possible virus contamination.

While this aspect of virus management is encouraged, issues on data privacy and human rights should not be neglected or set aside. Even if these Apps are effective in its purpose, if the community of App users do not have the trust as well as confidence in using these Apps, efforts to integrate community participation in the COVID-19 outbreak management will not be effective and will become futile. As pointed by Leins, Culnane, and Rubinstein (2020), if adoption is insufficient, collective benefits are not guaranteed.

23.6 CONCLUSION

Having taken over by surprise, the COVID-19 pandemic created a panic across the globe, especially to various government institutions. National governments quickly crafted and put into law, various policies to protect their people especially their borders. Massive vaccine research development is also being done around the world. Along with these measures to combat the COVID-19 pandemic, technology-based innovations have been created. In the case of the Philippines, the COVID-19 pandemic has given way to technology-based innovations in the country. This particular response cuts across different sectors and involves various stakeholders. In this paper, we have discussed the health-related and technology-based innovations which served instrumental in the country's response against the virus from January to May 2020.

The innovations specifically focus on the COVID-19 reporting, contact tracing, and telemedicine. Innovations in the reporting system took form in widening the channel of reporting. In particular, the use of new media especially Facebook and Twitter could be regarded as an innovation especially to the LGUs whose reporting system before the pandemic has only taken the traditional paper and board form. Contact tracing and telemedicine both innovate in the same direction through the creation of smartphone-based applications and websites. From the human resource-heavy, high-risk, and costly traditional and manual form of contact tracing, tech-innovations not only involved individual voluntary participation and tracking of people's traffic with the use of smartphones but also democratized the response by encouraging individual participation as well as community policing. Innovations in telemedicine, on the other hand, have given the public the privilege to consult their doctors without going to the hospitals, and in a way, provided a safe avenue to access a wider pool of medical professionals.

While public health scholars widely recognize the vital role that technology plays in the control and containment of the virus, we strongly argue that contextual realities should be kept in mind. The utilization of the above-mentioned Apps has raised issues on public trust, data privacy, and human rights. We recommend that interdisciplinary research scholars explore ways to minimize and close these information gaps. This could be done through information validation of the various government agencies especially, the DOH which is mandated to lead and be proactive in safeguarding public health. Also, while these technology-based innovations promote efficiency, we argue that people under poverty and those who do not have access to these technologies should be capacitated and aided by government institutions to ensure that no one is left behind in all programs and activities in COVID-19 response in particular, and increasing public health in general.

REFERENCES

CSSE-JHU (Center for Systems Science and Engineering—Johns Hopkins University). (2020, May 31). *COVID-19 Dashboard by the Center for Systems Science and Engineering—Johns Hopkins University*. Retrieved from: https://gisanddata.maps.arcgis.com/apps/opsdashboard/index.html?fbclid=IwAR1AP8lOJmERJft_PzNtH_VqaclPNr1v7nAnewafnooMeIJgqDN_FULD1Oc#/bda7594740fd40299423467b48e9ecf6.

Drew, D., Nguyen, L., Steves, Wolf, J., Spector, T. & Chan, A. (2020). "Rapid Implementation of Mobile Technology for Real-Time Epidemiology of COVID-19". *Science*, 368(6497), 1362–1367. DOI: 10.1126/science.abc0473.

DOH (Department of Health). (2020, June 01). "Beat COVID-19 Today: A COVID-19 Philippine Situationer: Issue 35 [Press Release]". Retrieved from: https://drive.google.com/drive/folders/1Wxf8TbpSuWrGBOYitZCyFaG_NmdCooCa

DOH (Department of Health). (2020, January 28). "DOH: Still No Confirmed 2019-NCOV Case in PH [Press Release]". Retrieved from: https://www.doh.gov.ph/doh-press-release/doh-still-no-confirmed-2019-nCoV-case-in-PH

Gregorio, X. (2020, February 2). "First Novel Coronavirus Case Outside China Reported in Philippines". *CNN Philippines*. Retrieved from: https://www.cnnphilippines.com/news/2020/2/2/novel-coronavirus-cases-death-Philippines.html.

Krishnasamy, N. (n.d.). "New Media vs Traditional Media". *Asia-Pacific Institute for Broadcasting Development*. Retrieved from: https://www.aibd.org.my/node/1226.

Leins, K., Culnane, C. & Rubinstein, B. (2020). "Tracking, Tracing, Trust: Contemplating Mitigating the Impact of COVID19 through Technological Interventions". *The Medical Journal of Australia*. DOI: 10.5694/mja2.50669.

Merez, A. (2020, May 27). "Philippines Needs 95,000 More Contact Tracers for COVID-19 Response: DOH". *ABS-CBN News*. Retrieved from: https://news.abs-cbn.com/news/05/27/20/philippines-needs-95000-more-contact-tracers-for-covid-19-response-doh.

NapoleonCat. (2020, January). "Facebook Users in the Philippines". Retrieved from: https://napoleoncat.com/stats/facebook-users-in-philippines/2020/01.

O'Leary, Z. (2017). *The Essential Guide to Doing Your Research Project* (3rd ed.). Thousand Oaks, CA: SAGE.

Tantuco, V. (2020, March 28). "What Hospitals Need to Treat COVID-19 Patients". *Rappler*. Retrieved from: https://www.rappler.com/newsbreak/in-depth/256133-numbers-what-hospitals-need-treat-coronavirus-patients.

Xu, B. & Kraemer, M. U. G. (2020). Open COVID-19 Data Curation Group, Open access epidemiological data from the COVID-19 outbreak. *Lancet Infectious Disease*, 20(534). DOI: 10.1016/S1473–3099(20)30394–7.

Zhang, H., & Shaw, R. (2020). Identifying Research Trends and Gaps in the Context of COVID-19. *International Journal of Environmental Research and Public Health*, 17(10), 3370. DOI: 10.3390/ijerph17103370.

Index

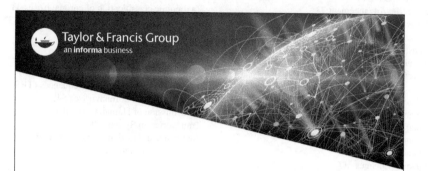